Biochemistry and Neurology

The cover design is adapted from
the *De Humani Corporis Fabrica* by Andreas Vesalius,
published in 1543.

Proceedings of a Conference organized by the Neurochemical Group of the Biochemical Society and held at Nottingham University, April 1975

Biochemistry and Neurology

Edited by

H. F. BRADFORD

Department of Biochemistry, Imperial College of Science and Technology, London

C. D. MARSDEN

University Department of Neurology, Institute of Psychiatry and King's College Hospital Medical School, London

1976

Academic Press

London . New York . San Francisco

A Subsidiary of Harcourt Brace Jovanovich, Publishers

ACADEMIC PRESS INC. (LONDON) LTD.
24/28 Oval Road,
London NW1

United States Edition published by
ACADEMIC PRESS INC.
111 Fifth Avenue
New York, New York 10003

Library of Congress Catalog Card Number: 75-37375
ISBN: 0-12-123750-8

Printed in Great Britain by Page Bros (Norwich) Ltd, Norwich,
a Member of the Harrison Group

List of Participants at Conference

G. W. Arbuthnott (Medical Research Council Brain Metabolism Unit, University Department of Pharmacology, 1 George Square, Edinburgh EH8 9J2, U.K.).

H. S. Bachelard (Department of Biochemistry, Institute of Psychiatry, De Crespigny Park, Denmark Hill, London SE5 8AF, U.K.).

E. D. Bird (Department of Neurological Surgery and Neurology, Addenbrooke's Hospital, Cambridge, and Medical Research Council Neurochemical Pharmacology Unit, Department of Pharmacology, University of Cambridge Medical School, Cambridge CB2 2QD, U.K.).

H. F. Bradford (Department of Biochemistry, Imperial College of Science and Technology, London SW7 2AZ, U.K.).

D. B. Calne (National Institute of Neurological and Communicative Disorders and Stroke, National Institutes of Health, Department of Health, Education and Welfare, Bethesda, Maryland 20014, U.S.A.).

G. Curzon (Department of Neurochemistry, Institute of Neurology, Queen Square, London WC1N 3BG, U.K.).

A. N. Davison (Department of Neurochemistry, Institute of Neurology, Queen Square, London WC1N 3BG, U.K.).

P. C. Emson (Medical Research Council Brain Metabolism Unit, University Department of Pharmacology, 1 George Square, Edinburgh EH8 9JZ, U.K.).

J. P. Fry (Agriculture Research Council Institute of Animal Physiology, Babraham, Cambridge CB2 4AT, U.K.).

P. Harris (Department of Surgical Neurology, Western General Hospital, Crewe Road, Edinburgh EH4 2XU, U.K.).

C. R. Hiley (Department of Pharmacology and Therapeutics, University of Liverpool), Liverpool L69 3BX, U.K.).

L. L. Iversen (Medical Research Council Neurochemical Pharmacology Unit, Department of Pharmacology University of Cambridge Medical School, Cambridge CB2 2QD, U.K.).

C. D. Marsden (University Department of Neurology, Institute of Psychiatry and King's College Hospital, Denmark Hill, London SE5 8AF, U.K.).

B. S. Meldrum (University Department of Neurology, Institute of Psychiatry and King's College Hospital, Denmark Hill, London SE5 8AF, U.K.).

R. J. Miller (Medical Research Council Neurochemical Pharmacology Unit, Department of Pharmacology University of Cambridge Medical School, Cambridge CB2 2QD, U.K.).

J. D. Parkes (University Department of Neurology, Institute of Psychiatry and King's College Hospital, Denmark Hill, London SE5 8AF, U.K.).

C. E. Polkey (The Neurosurgical Unit of Guy's, Maudsley and King's College Hospitals, The Maudsley Hospital, De Crespigny Park, Denmark Hill, London SE5 8AF, U.K.).

C. J. Pycock (University Department of Neurology, Institute of Psychiatry and King's College Hospital, Denmark Hill, London SE5 8AF, U.K.).
H. W. Reading (Medical Research Council Brain Metabolism Unit, University Department of Pharmacology, 1 George Square, Edinburgh EH8 9JZ, U.K.).
E. H. Reynolds (University Department of Neurology, Institute of Psychiatry and King's College Hospital, Denmark Hill, London, SE5 8AF; and Medical Research Council Clinical Research Centre, Northwick Park, Harrow, Middlesex, U.K.
C. D. Richards (The National Institute for Medical Research, Mill Hill, London NW7 1AA, U.K.).
D. F. Sharman (Agricultural Research Council Institute of Animal Physiology, Babraham, Cambridge CB2 4AT, U.K.).
D. W. Straughan (Department of Pharmacology, The School of Pharmacy, University of London, 29/39 Brunswick Square, London, WC1N IAX, U.K.).
H. R. A. Townsend (Department of Surgical Neurology, Western General Hospital, Crewe Road, Edinburgh EH4 2XO, U.K.).
J. C. Watkins (Department of Pharmacology, The Medical School, University Walk, Bristol BS8 1TD, U.K.).
C. M. Yates (Medical Research Council Brain Metabolism Unit, University Department of Pharmacology, 1 George Square, Edinburgh EH8 9JZ, U.K.).

PREFACE

The initiation of successful new therapeutic treatments in neurology, as in other branches of medicine, has in the past followed almost exclusively from chance or empirical observation, the underlying logic for their effectiveness remaining obscure. This will, no doubt, often continue to be the case in the foreseeable future since little is known about the detailed pathophysiology of most neurological disorders and also because our knowledge of the fundamental physiology and biochemistry of the nervous system is far from complete. However, the rapid rate of accumulation of information in the past decade or so is beginning to reveal the underlying plan of certain systems, such as the control of excitation and inhibition. This picture is annually becoming clearer with the burgeoning of our knowledge of the synaptic organization of the brain, the mechanisms of synapses and the nature of the transmitters with which they operate. For these reasons it is essential that frequent dialogues occur between clinicians and basic scientists in order to assess progress and highlight any opportunity of applying rational drug therapy based on a knowledge or indication of the basic lesion underlying the neurological disorder. This conference, which was held at Nottingham University on 17 and 18 April 1975, was organized with these purposes in mind; it was the second of these "Workshop" conferences to be organized by the Neurochemical Group of the Biochemical Society. The committee members and co-opted persons assisting in the organization were Dr H. S. Bachelard, Dr R. Balazs, Dr H. F. Bradford, Dr J. E. Cremer, Dr G. Curzon, Dr W. E. Davies, Professor A. N. Davison, Dr J. Lagnado (Secretary), Professor, J. N. Hawthorne (Chairman), Dr B. S. Meldrum, Professor C. D. Marsden, Dr H. W. Reading and Professor S. P. R. Rose.

We are particularly obliged to Dr John Lagnado and Professor Tim Hawthorne and various members of the Biochemistry Department at Nottingham University for the excellent organization of the mechanics of the conference, and especially to M. R. Pickard and his colleagues for recording its proceedings for publication. We are greatly indebted to the following organization for their generous financial assistance: CIBA Laboratories Ltd., ICI Ltd., Hoechst Pharmaceuticals Ltd., Reckitt and Colman (Pharmaceutical Division), Roche Products Ltd., and The Wellcome Trust.

We are also grateful to Mrs B. Gimblett and Miss P. M. Jenkyns for secretarial help in preparing the manuscripts.

HARRY BRADFORD
DAVID MARSDEN

CONTENTS

Page
List of Participants at Conference v
Preface vii

PART 1: DYSKINESIAS
Section A: General Aspects of Dyskinesias
Chapter 1: Clinical Aspects of the Dyskinesias
By C. D. Marsden 3
Chapter 2: Biochemical Aspects of Dyskinesias
By G. Curzon 13
Section B: Dopamine Receptors
Chapter 3: Clinical Pharmacology of Dopaminergic Effects in Parkinsonism
By D. B. Calne 21
Chapter 4: Supersensitivity of Dopamine Receptors
By G. W. Arbuthnott 27
Chapter 5: Biochemical Pharmacology of Dopamine Receptors in Mammalian Central Nervous System
By L. L. Iversen and R. J. Miller 35
Section C: Drug-Induced (Tardive) Dyskinesias
Chapter 6: Clinical Aspects of Tardive Dyskinesia
By J. D. Parkes 47
Chapter 7: The Effects of Tranquilizing Drugs on the Metabolism of Transmitter Substances in the Central Nervous System
By J. P. Fry and D. F. Sharman 57
Chapter 8: Long-term effects of Dyskinesia-inducing Drugs
By Celia M. Yates 73
General Discussion of Chapters 1–8 79
Section D: Huntingdon's Chorea
Chapter 9: Biochemical Studies on γ-Aminobutyric Acid Metabolism in Huntingdon's Chorea
By E. D. Bird 83
Chapter 10: Effects of Blocking Nigrostriatal γ-Aminobutyric Acid Receptors
By C. J. Pycock 93
Chapter 11: The Muscarinic Receptor for Acetylcholine in Huntingdon's Chorea
By C. R. Hiley 103
General Discussion Dyskinesias 111

PART II: EPILEPSY
Section A: General Aspects of Clinical and Experimental Epilepsy
Chapter 12: Clinical and Biochemical Aspects of Epilepsy
By P. Harris 125
Chapter 13: Experimental Epilepsy: Its Relation to Clinical Problems
By B. S. Meldrum 143
Chapter 14: Metal Implants as Models of Epilepsy
By P. C. Emson 163
Discussion Paper: Epilepsy A Clinician's View
By H. R. A. Townsend 175
Discussion Paper: Lesions in Human Epilepsy
By C. E. Polkey 179

Section B: Underlying Mechanisms of Epilepsy
Chapter 15: Observations on the Mode of Action of Antiepileptic Drugs
By C. D. Richards 185
Chapter 16: On Amino Acid Involvement in Basic Mechanisms of the Epilepsies
By H. F. Bradford 195
Chapter 17: Amino Acid Transmitters: Pharmacological and Electrophysiological Aspects
By D. W. Straughan 213
Chapter 18: Carbohydrate and Energy Metabolism in Relation to Mechanisms of Epilepsy
By H. S. Bachelard 233
Discussion Paper: Folate and Epilepsy
By E. H. Reynolds 247
Discussion Paper: A General View of Possible Causative Factors in Epilepsy and their Investigation by Means of Pharmacological Agents
By J. C. Watkins 253
General Discussion on Epilepsy 261
Author Index 267
Subject Index 281

PART I

DYSKINESIAS

Chapter 1

Clinical Aspects of the Dyskinesias

By C. DAVID MARSDEN

University Department of Neurology, Institute of Psychiatry and King's College Hospital, Denmark Hill, London SE5 8AF, U.K.

Introduction

Dyskinesias are abnormal involuntary movements which occur either spontaneously at rest or are superimposed upon and distort voluntary movements. A number of categories of dyskinesia are recognized by what the movements look like, and are described by such traditional terms as tremor, chorea, torsion dystonia or athetosis, myoclonus, etc. These descriptive terms merely imply a type of dyskinesia, not specific diseases, pathology or aetiology. Each of the categories of dyskinesia may be produced by a variety of neurological diseases, may be caused by drugs, or may occur spontaneously without evident cause. [For a review of the differential diagnosis and clinical management of the dyskinesias see Marsden & Parkes (1973).]

The dyskinesias are frequently described under 'basal ganglia diseases', as the commonest anatomical change discovered at postmortem examination in such patients is damage to one or other part of this area of the brain (which includes the corpus striatum—caudate nucleus and putamen—globus pallidus, substantia nigra, subthalmic nucleus, and red nucleus). Unlike many other functional disorders caused by brain damage, however, it has proved very difficult to relate a particular type of dyskinesia to damage in a specific part of the basal ganglia. The same dyskinesia may be produced by lesions at different sites, or may occur without any visible evidence of pathology in the brain at all. Similarly, a given lesion may cause a variety of dyskinesias in different patients, or even in the same individual. The difficulties in clinicopathological correlation of the dyskinesias are matched by a lack of understanding as to how basal ganglia damage causes abnormal movements, and ignorance as to the normal function of this part of the brain. The clinical evidence indicates that in some way the basal ganglia are concerned with the control and execution of movement, but how they do so is not known, nor is their relation to the other parts of the brain that are concerned with movement (e.g. the motor cortex and cerebellum) understood. In short, the basal ganglia are still something of a mystery to the clinician, pathologist and physiologist.

The recent biochemical and pharmacological discoveries about the basal ganglia, by approaching the problem from a different direction, have therefore been of great interest and excitement to those who look after patients with these distressing illnesses. In fact, the stimulus to such investigations arose, in part, from the clinical findings that drugs could provoke dyskinesias identical to those which occur in spontaneous disease.

Drug-induced dyskinesias, which are discussed in detail by Dr. Parkes (this volume, Chapter 6), arise most commonly in two situations, the administration of neuroleptic antipsychotic agents to psychiatric patients and during levodopa treatment of Parkinson's disease. Both illustrate one of the central themes of basal ganglia disease, namely the intimate but antagonistic relationship between the akinetic-rigid state of Parkinsonism and the hyperkinetic state of the dyskinesias. Neuroleptics provoke both Parkinsonism and dyskinesias, while levodopa, in conventional doses, produces dyskinesias in Parkinsonian, but not in normal subjects. Conversely, many of the drugs used to control dyskinesias (such as phenothiazines, butyrophenones and reserpine derivatives) do so at the expense of some degree of drug-induced Parkinsonism. Thus the akinetic-rigid Parkinsonian state and the dyskinesias appear to be opposite ends of a spectrum. This is convincingly illustrated by events during levodopa therapy, where the tremulous, stiff, bent and akinetic patient with Parkinson's disease may be converted into a mobile, supple, straight individual who now shows a wide variety of dyskinesias, such as chorea, myoclonus and torsion dystonia. If levodopa is stopped, the patient soon relapses back to his previous akinetic-rigid state. Such swings from Parkinsonism to dyskinesia and vice versa are now to be seen occurring many times a day in those who develop the disabling 'on–off' effect after many months of continuous levodopa treatment. In between the two extremes there are patients on levodopa with both signs of Parkinsonism and dyskinesias, while the same is seen in psychiatric patients treated with neuroleptics.

It is observations such as these that have lent weight to the concept that while Parkinsonism might be due to failure of activity of a part of the brain controlling movement, the dyskinesias might result from overactivity of the same region. The discovery that the death of the substantia nigra in Parkinson's disease is associated with striatal dopamine depletion (Hornykiewicz, 1973) has been complemented by the hypothesis that at least some of the dyskinesias may be due to excessive striatal dopaminergic stimulation, either by release of endogenous dopamine on to damaged 'supersensitive' receptors or by the administration of drugs affecting dopaminergic neurotransmission (Klawans, 1973). However, as discussed by Dr. Curzon (this volume, Chapter 2), while the nigrostriatal dopamine system has held pride of place in the evolution of concepts of dyskinesias, it is increasingly apparent that the many other putative neurotransmitters located within the basal ganglia are also of relevance. Current attention is focused on γ-aminobutyric acid, acetylcholine, 5-hydroxytryptamine, and other chemicals which may interact with the nigrostriatal dopamine system.

In the brief communication that follows, I will concentrate particularly on clinical aspects of those dyskinesias of current interest to biochemists and pharmacologists.

Huntington's Chorea

Chorea is a dyskinesia characterized by a continuous flow of randomly distributed and irregularly timed explosive muscle jerks that flit from one portion of the body to another. The features, trunk and limbs are continuously distorted by these brief movements, which occur at rest and interrupt normal movement. Speech, use of the hands and walking are all affected, and in severe cases the patient presents a grotesque picture of abnormal movements.

Huntington's chorea is a rare illness, affecting about 1 in 20 000 of the population, but it occurs worldwide and in all ethnic groups. The cause of Huntington's chorea is unknown. The disease is inherited as an autosomal dominant trait with full penetrance, so that 50% of children of sufferers inevitably will be affected. New mutations are rare and nearly all cases have affected relatives, although the familial nature of the disease is often concealed.

Symptoms usually appear in middle life, between the ages of 30 and 50 years, although onset in childhood and late life may occur. The disease is relentlessly progressive, leading to death on average about 14 years from onset. Either chorea or mental deterioration may be the presenting feature, although subtle changes in personality and behaviour are nearly always the earliest sign of the illness. As the disease progresses dementia becomes more profound and chorea more grotesque. Finally, it becomes necessary to commit the patient to a hospital for chronic care, where he or she becomes progressively emaciated, incapacitated, bedridden and dies, usually of intercurrent infection. No cure is available, although the chorea may be controlled by phenothiazines, butyrophenones, pimozide and tetrabenazine.

Of interest is the so-called Westphal variant of Huntington's chorea in which the disease is manifest not by dyskinesia, but by an akinetic-rigid Parkinsonian state. The disease presents in this manner rarely in adults (about 6% of cases), but is common when the illness begins in childhood. Other peculiarities of Huntington's chorea in childhood include the presence of frequent seizures and a predominant inheritance from father rather than mother.

Other abnormalities in Huntington's chorea of interest are the almost invariable weight loss and cachexia that occurs, abnormalities of eye movement, a modest impairment of the blood pressure response to tilt, and the occasional similarity of the mental changes to a schizophreniform psychosis.

At postmortem examination the brain is shrunken and atrophic. Histologically the cerebral cortex shows widespread loss of neurons, particularly in the 3rd, 5th and 6th layers, and diffuse loss of nerve fibres. Characteristically, the caudate nucleus and putamen are severely affected with loss of small neurons, particularly of the Gogli type II population. As a result, there is atrophy of the striopallidal nerve fibre bundles in the lateral segment of the globus pallidus. Less marked changes are found in other structures, but by and large, the substantia nigra and other brainstem nuclei are preserved. A reactive gliosis is apparent in affected areas, but there is no evidence of inflammatory changes.

Thus the brunt of the disease falls on the cerebral cortex and corpus striatum. Changes in the former are believed to cause the dementia, while the striatal cell loss is held to be responsible for the chorea. In advanced cases there may be near total striatal cell loss. This may explain why chorea may be replaced by an akinetic-rigid state terminally.

Whatever it is that causes Huntington's chorea damages both cerebral cortex and corpus striatum. The recent discovery of selective depletion of GABA and its synthesizing enzyme glutamic acid decarboyxlase in the basal ganglia, but not the cerebral cortex, in the brain at postmortem examination (Bird, this volume, Chapter 9) does not suggest the metabolic error responsible for the disease. Rather, these findings probably indicate the effects of massive loss of GABA-containing striatal neurons. The same is probably true for the reduction in striatal choline acetyltransferase (Bird, this volume, Chapter 9) and loss of muscarinic receptors (Hiley, this volume, Chapter 11). However, such biochemical changes obviously are of relevance to the pathogenesis of the chorea.

In summary, Huntington's chorea remains a disease whose cause is unknown, and for which no effective treatment is available. Pathological and consequent biochemical abnormalities have been identified, but their significance is not clear. With regard to the chorea of the illness, it is still unknown why, in physiological terms, damage to the striatum causes the dyskinesia. One would hope to look to animal models of the disease to help answer such questions, but, unfortunately, there is no strict animal model of Huntington's chorea and by and large animals do not develop the dyskinesias seen in man when selective lesions are made in the basal ganglia. However, biochemical manipulation of striatal function can produce dyskinesias in animals (see Pycock, this volume, Chapter 10) and may lead to greater understanding of the human illness.

[Excellent reviews of Huntington's chorea are provided by Bruyn (1968) and in the volume edited by Barbeau, Chase and Paulson (1973).]

Hormonal and Other Choreas

Chorea may be provoked by a number of drugs (other than neuroleptics), and metabolic and hormonal disorders. It is a rare complication of hypoparathyroidism, diphenylhydantoin intoxication, thyrotoxicosis, alcohol withdrawal, amphetamine abuse, and oral contraceptive administration. In such circumstances chorea may occur in otherwise normal individuals in whom no pathological change in the basal ganglia is to be expected, and the chorea disappears when the offending drug is withdrawn. There is the suspicion that under these circumstances the offending drug or illness is altering the chemical balance or sensitivity of neuronal activity in the corpus striatum.

Torsion Dystonia

The dyskinesias of torsion dystonia are sustained slow muscle spasms distorting the neck, trunk and limbs into characteristic postures. Such spasms of contraction may occur at rest or on action, and are often repetitive and even rhythmic. The neck twists to one side (torticollis) or backwards (retrocollis), the trunk is bent back (lordosis) or to one side (scoliosis); the arm usually extends and hyperpronates with flexed wrist and extended fingers; the foot plantar-flexes and turns in.

This dyskinesia may be caused by a variety of cerebral insults, such as birth injury or kernicterus ('athetoid cerebral palsy'), but may occur without any obvious cause and with an apparently normal brain at post mortem (idiopathic torsion dystonia or dystonia musculorum deformans). A similar dyskinesia can be provoked by neuroleptic drugs (acute dystonic reactions) and by levodopa in Parkinson's disease.

Dystonia musculorum deformans is a rare disease of particular interest to biochemists for no structural cause has been identified. It is often inherited, usually as an autosomal recessive trait in Ashkenazic Jews, and as an autosomal dominant trait in other ethnic groups. Symptoms usually commence in childhood or adolescence, with dystonic posturing of a leg or arm on action. Thereafter dystonic muscle spasms spread to all four limbs and axial structures over ten years or so in the majority of cases. The affected individual is often finally totally disabled by dyskinesia, although intellect and sensibility are preserved.

When the illness appears in adults it usually affects the arms or neck, and remains limited to its site of origin. Spasmodic torticollis and some forms of writer's cramp are most likely to be isolated forms of adult onset torsion dystonia; so may blepharospasm and spontaneous orofacial movements. No pathological basis for any of these dyskinesias has been identified, nor have postmortem studies been undertaken. Their cause remains a mystery. [For a review of torsion dystonia see Marsden & Harrison (1974).]

Myoclonus

The term myoclonus describes brief, shock-like muscle jerks, which usually occur repetitively in the same muscles, thus distinguishing them from the chaos of chorea. Such myoclonic jerks often occur in bed at night while falling asleep in normal people.

Myoclonus is a feature of many neurological diseases. It is sometimes associated with epilepsy, or with progressive cerebral degenerations, but may occur spontaneously. It may follow a period of cerebral anoxia of whatever cause, in which case the myoclonus distinctively occurs on movement (action mycoclonus). As the patient attempts to talk, use the arms, or walk, sudden muscle jerks distort the face or limb. Speech may be incoherent, the patient may be unable to grasp objects without throwing them and may be unable to stand or walk without being thrown to the ground. Postanoxic action myoclonus has recently been the focus of considerable interest for it has been found to respond to 5-hydroxytryptophan therapy. Unfortunately, other forms of myclonus usually do not benefit from this treatment.

[For a review of the difficult and somewhat confused topic of myoclonus see Halliday (1967).]

Tics, Stereotypies and Mannerisms

A number of disorders of movement may occur in psychiatric patients who have not been given drugs. Abnormal movements are seen most commonly in those with a psychotic schizophreniform psychosis. Characteristically, they consist of repetitive, purposeless, stereotyped movements which resemble ordinary volitional behaviour, can be suppressed by the patient by an effort of will, and can be mimicked by the observer. True dyskinesias of choreiform, dystonic or myoclonic nature are probably never seen in primary psychotic illnesses not due to reorganized organic neurological disease, unless neuroleptic drugs are given. The tics, stereotypies and mannerisms of primary psychotic illness represent motor accompaniments of the abnormal mental state, which may include rituals of both thought and motor expression.

One particular disease of interest is Gilles de la Tourette's syndrome in which repetitive body tics similar to an exaggerated startle response are accompanied by explosive and often foul utterances (coprolalia). This curious condition occurs almost exclusively in childhood and usually resolves spontaneously by adult life. Its cause is unknown, but there are recent suggestions that it may respond to those drugs used to control other dyskinesias.

Another motor accompaniment of psychiatric illness is the motor retardation that occurs in severe depressive illness. The sad expressionless faces, flexed posture and immobility of the melancholic are reminiscent of Parkinsonism, but true rigidity and tremor probably do not occur.

The relationship of disorders of movement in psychiatric disease to the dyskinesias or Parkinsonism of neurological patients, or to the behavioural stereotypes and akinesia produced by many drugs in animals, is not clear. However, one cannot fail to be impressed by the frequent clinical similarities between psychiatric and neurological patients, which pose many theoretical questions. Is the physical motor retardation of the severely depressed individual due to functional and reversible changes in striatal amine activity? Conversely, is the strikingly high incidence of depression in Parkinson's disease more than just a reaction to physical incapacity; might it represent the effect of altered cerebral amine metabolism on mood? Are the abnormal movements seen in some untreated psychotic patients due to changes in cerebral neurotransmitter function that are responsible for the psychotic behaviour? Questions such as these expose the similarity of interests of both neurologists and psychiatrists in this general area and highlight the significance of biochemical pharmacological research to both disciplines.

(For a recent review of abnormal motor behaviour in psychiatric illness see Marsden *et al.,* 1975.)

The Role of Biochemical Pharmacology in Relation to the Dyskinesias

To establish how levodopa and neuroleptic drugs cause dyskinesias in man may shed light on the aetiology of the many unexplained spontaneous illnesses characterized by abnormal involuntary movements.

Levodopa-induced dyskinesias

Levodopa-induced dyskinesias are due to an effect within the brain, for they occur with equal or greater frequency when the drug is combined with a selective extracerebral decarboxylase inhibitor. It is most likely that they are due to stimulation of cerebral dopaminergic receptors, for they are ameliorated by drugs believed to block such receptors, e.g. pimozide, and are enhanced or even produced by dopamine receptor agonists such as piribedil or bromocriptine (see Calne, this volume, Chapter 3). While the possibility that they are caused by some minor metabolite of levodopa, or by a complex effect on many neurotransmitters, cannot be excluded, the most parsimonious explanation for levodopa dyskinesias is that they result from stimulation of striatal dopamine receptors.

However, such a conclusion poses as many questions as it answers. For example, there are clinical hints that the dopamine receptors stimulated by levodopa therapy to improve Parkinsonian disability may not be the same as those responsible for levodopa dyskinesias. Thus some patients may develop dyskinesia despite little or no improvement in their Parkinsonism, while others may obtain considerable clinical benefit without ever manifesting dyskinesias. This possibility, elaborated upon by Klawans (1973) and Marsden (1975), has the implication that it might prove possible to tailor a dopaminergic agonist to selectively stimulate the critical receptors defunct in Parkinson's disease, while avoiding dyskinesias. Such an approach to selective dopamine agonists also might successfully avoid other unwanted effects of therapy, such as nausea and vomiting and the various psychiatric complications of treatment.

There is, too, the question as to whether levodopa-induced dyskinesias are the result of not only drug administration, but also brain receptor abnormality. Certainly they occur with great frequency in patients with Parkinson's disease, and rarely if at all in normal subjects given similar doses of the drug. This may be interpreted as indicating dopamine receptor 'supersensitivity' due to extensive nigral damage in Parkinson's disease. But whether such supersensitivity, if it exists, is due to denervation is not certain. One might have expected levodopa dyskinesias to decrease with the duration of therapy, as dopamine replacement corrected such supersensitivity, but actually the reverse occurs. The whole question of denervation supersensitivity in the brain in Parkinson's disease is central to the development of concepts of dyskinesias, but is not fully understood (see Arbuthnott, this volume, Chapter 4).

Neuroleptic-induced dyskinesias

Neuroleptic-induced dyskinesias are related, in some way, to striatal dopamine activity. It is a current hypothesis that neuroleptic drugs possess antipsychotic activity because they block cerebral dopamine receptors, not only in the striatum but also in areas of the mesolimbic system, brainstem and hypothalamus. The unwanted extrapyramidal effects of neuroleptic therapy are attributed to striatal dopamine receptor blockade, but it is generally accepted that they are not necessary for antipsychotic activity. The latter may depend on the effect of the drugs on dopamine receptors elsewhere, for example, the mesolimbic system. If this assumption is correct, it may prove possible to create neuroleptics devoid of striatal dopamine antagonism and, hence, of extrapyramidal unwanted actions.

Such selective neuroleptic drugs have yet to be developed, but, even so, these drugs possess very variable propensities for causing unwanted extra pyramidal effects. Miller & Hiley (1974) have suggested that this is due to variations in the degree of inherent anticholinergic properties of neuroleptic drugs. Thus while all may block cerebral dopamine receptors, some such as trifluoperazine and spiroperidol possess little anticholinergic activity, while others such as thioridazine and clozapine are potent anticholinergics (see Iversen & Miller, this volume, Chapter 5). Since anticholinergic drugs are valuable in controlling neuroleptic-induced Parkinsonism and acute dystonic reactions (and possibly akathisia), drugs such as thioridazine and clozapine might be expected to be characterized by a low incidence of these unwanted actions. Clinical experience appears to bear this out, and suggests that the best neuroleptics might be those designed to contain a high inherent anticholinergic potency.

However, a real quandary exists when we turn to the other neuroleptic-induced dyskinesia, namely the tardive orofacial dyskinesia syndrome, which, tragically, may persist indefinitely, despite withdrawal of drugs. Here, the clinical evidence suggests that anticholinergic drugs actually make tardive dyskinesia worse. while cholinergic drugs make it better (see Marsden, 1975, for review). In view of this, one might predict that tardive dyskinesias would be a bigger problem with those neuroleptic drugs possessing strong anticholinergic properties, than with those that do not. Whether this is true or not in practice is uncertain, for so few studies have clearly separated tardive dyskinesias from other extrapyramidal complications of treatment, or have been undertaken for long enough to judge the true incidence of this later complication of treatment. So far the evidence does not suggest that long-term treatment with, say, thioridazine or clozapine causes more tardive dyskinesia than similar threapy with, say, trifluoroperazine. The theory may be wrong, or the evidence incorrect.

There is an alternative proposition, namely that the presence of a high inherent anticholinergic activity may actually protect against tardive dyskinesias. It is generally supposed that this often permanent complication of therapy is due to irreversible damage or change in the striatal dopamine receptors on which the neuroleptics act (see Klawans, 1973). Such damaged receptors are held to become 'supersensitive' to endogenous dopamine. This situation certainly has been produced in experimental animals by continuous neuroleptic therapy (Tarsy & Baldessarini, 1974), which leads to excessive response to dopamine agonists when the neuroleptics are stopped. This mirrors the clinical situation when neuroleptic-induced tardive dyskinesia gets worse, or even appears for the first time, when the offending drug is withdrawn. However, neuroleptic-induced tardive dyskinesias are so similar to levodopa-induced dyskinesias as to suggest that they, too, are due to excessive striatal dopaminergic stimulation. Neuroleptic drugs, which block dopamine receptors, are known to provoke a compensatory increase in nigrostriatal dopamine synthesis and probably release (see Fry & Sharman, this volume, Chapter 7). This could lead to excessive dopaminergic stimulation of some population of striatal receptors inadequately blocked by the neuroleptic (Marsden, 1975). Such a mechanism might explain why tardive dyskinesias appear during therapy, but cannot account for their persistence after the drug is withdrawn, when presumably, this compensatory increase in dopaminergic turnover returns to normal after a while. However, it may be that prolonged bombardment of dopamine receptors by dopamine itself renders them supersensitive, or even causes permanent damage. Such a possibility looks unlikely at first sight, but Klawans *et al.* (1975) have shown that chronic amphetamine treatment of guinea-pigs does cause a form of 'innervation supersensitivity' to subsequent administration of dopamine agonists.

If the compensatory changes in striatal dopamine synthesis and turnover caused by neuroleptic drugs are relevant to the tardive dyskinesias they produce, then neuroleptics with high inherent anticholinergic activity might be protected from causing tardive dyskinesias, for drugs such as atropine antagonist the compensatory effects of neuroleptics on dopamine turnover.

Thus, it is possible, theoretically, to suggest reasons why a neuroleptic designed to possess strong anticholinergic potency should cause a high or a low incidence of chronic tardive dyskinesia. Obviously, careful clinical observation to establish the true state of affairs will tell us a great deal about the mechanism of tardive dyskinesias.

Finally, one must draw attention to another quandary posed by these drugs. If it is argued that tardive dyskinesia is due to permanent damage of dopamine receptors in the striatum by neuroleptics, such damage could well occur elsewhere in the brain where the drugs act. In particular, it might occur at mesolimbic dopamine receptors, with possible profound consequences for higher mental function and emotional behaviour. Although such a course of events is not apparent, it must be recalled that it took many years to appreciate that the tardive dyskinesias seen in psychotic patients treated with neuroleptics over a long period were actually due to the drugs given. An urgent solution to the problem of how these agents cause tardive dyskinesias may remove the thought that they may carry risks of causing damage to other parts of the brain, or may lead to better drugs which avoid such risks.

References

Barbeau, A., Chase, T. N. & and Paulson, G. W. (1973) *Advan. Neurol.* **1**, 1–826

Bruyn, G. W. (1968) in *Handbook of Clinical Neurology* (Vinken, P. J. & Bruyn, G. W., eds.), vol. 6, pp. 298–378, North Holland Publishing Co., Amsterdam

Halliday, A. M. (1967) in *Modern Trends in Neurology* (Williams, D., ed.), vol. 4, pp. 69–105, Butterworths, London

Hornykiewicz, O. (1973) *Brit. Med. Bull.* **29** 172–178

Klawans, H. L. Jr. (1973) *The Pharmacology of Extrapyramidal Movement Disorders,* Karger, Basel

Klawans, J. L. Jr., Crossett, P. & Dana, N. (1975) *Advan. Neurol.* **9**, 105–112

Marsden, C. D. (1975) in *Modern Trends in Neurology* (Williams, D., ed.), vol. 6, pp. 141–165, Butterworths, London

Marsden, C. D. & Harrison, M. J. G. (1974) *Brain* **97**, 793–810

Marsden, C. D. & Parkes, J. D. (1973) *Brit. J. Hosp. Med.* **10**, 428–450

Marsden, C. D., Tarsy, D. & Baldessarini, R. J. (1975) in *Psychiatric Aspects of Neurologic Disease* (Benson, D. F. & Blumer, D., eds.), chap. 12, Grune and Stratton, New York

Miller, R. J. & Hiley, C. R. (1974) *Nature (London)* **248**, 596–597

Tarsy, D. & Baldessarini, R. J. (1974) *Neuropharmacology* **13**, 927–940

Chapter 2

Biochemical Aspects of the Dyskinesias

By G. CURZON

Department of Neurochemistry, Institute of Neurology, Queen Square, London WC1 3BG, U.K.

Introduction

Professor Marsden (this volume, Chapter 1) has dealt with the range of dyskinesias as seen by the clinician. This chapter concerns the biochemical and pharmacological aspects of some of these disorders and is introduced by a brief comment on the main methods used in studying them.

Methods of study

A commonly used biochemical method is to determine certain transmitter amine metabolites in lumbar cerebrospinal fluid (CSF). Thus homovanillic acid determination gives some measure of brain dopamine turnover, 5-hydroxyindoleacetic acid of 5-hydroxytryptamine turnover and 3-methoxyhydroxyphenylglycol of noradrenaline turnover. Limitations involved in deducing brain amine turnover from these measurements have been discussed in detail elsewhere (Garelis *et al.*, 1974; Curzon, 1975). The principal essential limitation is that values obtained do not on the whole give clear information on specific brain regions—they represent summations of brain amine metabolism probably distorted by disproportionate contributions from periventricular regions. Also, with the exception of homovanillic acid values they are influenced by metabolism in spinal neurons. However, even taking these and other problems into account CSF amine metabolite determination remains the only way by which we can at present readily obtain any information on amine metabolism in the human brain during life. It is useful, for example, in the study of whether drug-induced dyskinesias are obviously related to effects on transmitter amine metabolism.

Direct determinations on human brain autopsy material provide better evidence on transmitter abnormalities in that regional studies can be made, and not only transmitter metabolites may be determined but also transmitters, precursors and enzymes required for synthesis and catabolism. Since the initial report 15 years ago of defective brain dopamine metabolism in Parkinson's disease there have been only two comprehensive studies reported on brain transmitters in human movement disorders—the account of the work of Hornykiewicz's group on Parkinsonian states (Bernheimer *et al.*, 1973) and the paper by Bird & Iversen (1974) on Huntington's chorea.

More detailed information on the location of transmitter amines in tracts can be obtained by histochemical methods. In principal, the fluorescence method of Falck and Hillarp is ideal for work on human brain amine pathology and has been used to study noradrenaline terminals in biopsy material (Nystrom *et al.*, 1972) but rapid diffusion after death has limited its application in the past. However, as neuronal uptake mechanisms remain active tracts can be visualized up to 7 hours after death by incubating brain slices in α-methylnoradrenaline which is taken up by catecholamine terminals and gives a fluorescence reaction. 5-Hydroxytryptamine-containing tracts can be seen by incubating slices with 5-hydroxytryptamine after destruction of catecholamine fibres by incubation with 6-hydroxydopamine (Olson *et al.*, 1973). This type of method may be widely applicable to postmortem material but has as yet been used relatively little.

An omnipresent problem in transmitter amine studies is that of distinguishing between molecules that have been destroyed intraneuronally without ever having been released to receptors and those which have been released into the synaptic cleft before metabolism and are therefore functionally more relevant. There is some indication from animal experiments (Roffler-Tarlov *et al.*, 1971) that brain homovanillic acid is largely derived from dopamine metabolized in the cleft, but in the case of 5-hydroxytryptamine, for example, there is no readily determinable index of the functional fraction of its metabolism—certainly not one that is applicable to human studies.

In the past few years means of investigating postsynaptic receptor function have improved as receptor agonists and antagonists have become available. Within ethical limits these are available for the detection of receptor abnormalities in man but can be applied most systematically in laboratory animals, for example when using a model thought to have some analogy to human dyskinesias—the rat with unilaterally destroyed dopaminergic tracts (Ungerstedt, 1971). Another approach to postsynaptic receptor response is to look at cyclic AMP synthesis in nerve-ending rich regions. This seems to be involved in receptor response. For example, drugs which are agonists at dopamine receptors increase rat striatal adenyl cyclase activity *in vitro*, while antagonists have opposite effects (Iversen & Miller, this volume, Chapter 5).

Drug-induced Dyskinesias

Drug-induced dyskinesias were almost unknown before the introduction of drugs such as phenothiazines, butyrophenones and L-dopa which interact in various ways with transmitter amine metabolism and function. Though there is evidence that L-dopa, for example, can at very high concentrations cause dyskinesias in normal animals which are similar to those in human Parkinsonian patients on L-dopa (Mones, 1973), it is important to remember that these drugs tend to be given to subjects who already have a brain amine disturbance—perhaps of a complex nature. This is clearly so when L-dopa is given in the treatment of Parkinson's disease and has varying degrees of probability when phenothiazines and butyrophenones are used in the treatment of psychiatric illness. Very complex situations with many possible interactions can therefore arise, and it is perhaps hardly surprising that drug dyskinesias and other harmful effects are so common.

L-*Dopa dyskinesia*

Though it may well be that treating Parkinsonian patients with L-dopa involves a replacement to some degree of the absent striatal dopamine, this is not all that occurs. A considerable fraction of the L-dopa reaching the brain appears to be metabolized to dopamine by L-aromatic amino acid decarboxylase of other regions rich in noradrenaline (Lloyd *et al.*, 1973). Even the fraction converted to striatal dopamine is not functionally comparable to normal brain dopamine as machinery for its uptake and release is defective. Some of this dopamine may be made not in surviving dopamine neurons but in 5-hydroxytryptamine neurons by the action of L-aromatic amino acid decarboxylase (Ng *et al.*, 1970). These changes outside striatal dopaminergic neurons could possibly contribute to the dyskinesia. Also, various metabolites other than dopamine have been suggested to be involved in both the therapeutic and harmful actions of dopa, e.g. 3-*O*-methyldopa, which gradually accumulates in considerable amount. However, in a study of the absorption and metabolism of a standard oral dose of L-dopa in 42 patients with Parkinsonism (Bergmann *et al.*, 1974) we did not find a significant correlation between plasma 3-*O*-methyldopa or CSF homovanillic acid and abnormal movements. Ericsson *et al.* (1971) have claimed that various *O*-methylated dopa metabolites caused chorea when given to reserpinized rats. The lack of effect of 3-*O*-methyldopamine on striatal cyclic AMP synthesis (Iversen *et al.*, 1975) suggests that it is not acting at dopamine receptors. Also, direct intrastriatal injection of relatively large amounts in nialamide-treated guinea-pigs did not cause dyskinesia although the same amount of dopamine caused oral dyskinesia in 9/10 animals (Costall & Naylor, 1975).

L-Dopa is a very reactive substance and is given in large amounts. Therefore many interesting minor metabolites are detectable, e.g. ring trihydroxylated derivatives (Wada & Fellman, 1973) and tetrahydroisoquinoline alkaloids (Sandler *et al.*, 1973). Though the latter substances are relatively inactive in the above two test systems (Costall & Naylor, 1975; Iversen *et al.*, 1975) the possibility cannot be excluded that some quantitatively minor but highly active metabolite is involved in L-dopa dyskinesia in man. However, it seems more likely that these symptoms are more closely related to the fact that the administered L-dopa is interacting with a nervous system with abnormalities of transmitter function. This is consistent with the much easier provocation of dopa dyskinesia in Parkinsonian than in non-Parkinsonian subjects (Chase *et al.*, 1973).

It is therefore important to consider the various abnormalities in the Parkinsonian brain. Dopaminergic neurons may be present at various levels of functional effectiveness. Residual intact neurons releasing dopamine in the normal way to normal receptors may coexist with dead neighbours and with compensatory changes. Early stages of the disease process apparently lead to compensatory changes so that surviving neurons become functionally more effective. This is consistent with the markedly low striatal dopamine even in subjects in whom only mild symptoms had developed (Bernheimer *et al.*, 1973). Compensatory changes may be of two kinds. Firstly, dopamine turnover is increased in surviving neurons, which is consistent with their at least partially performing the tasks of their dead neighbours. The increased turnover is indicated by the fall of the dopamine metabolite, homovanillic acid being less striking in the Parkinsonian striatum than that of dopamine itself (Bernheimer *et al.*, 1973). Sharman *et al.* (1967) reported similar findings in monkeys in an experimental analogy of the Parkinsonian situation, i.e. after making lesions in dopamine cell bodies in the substantia nigra. Also, Agid *et al.* (1973) found that partial unilateral destruction of the substantia nigra with 6-hydroxydopamine led to acceleration of the rate-limiting step of dopamine synthesis (the ring hydroxylation of tyrosine) in surviving dopamine neurons of the damaged side of the brain.

A second kind of compensatory change also occurs, i.e. increased sensitivity of associated dopamine receptors. The classical experiments suggesting this are those of Ungerstedt (1971), in which unilateral destruction of striatal dopaminergic tracts in the rat led to increased sensitivity of receptors on the damaged side so that dopamine agonists such as apomorphine caused circling movements away from the lesioned side.

We know little about the relative importances of these compensatory changes in Parkinson's disease or about their relationships and time courses. It is also not clear whether the receptors fall into two populations with normal and maximal sensitivities or whether a continuous range of sensitivities can occur. What is clear is that in the overt disease compensatory changes have failed. Thus when L-dopa is given, dopamine becomes available to receptors with sensitivities which are not necessarily appropriate to the concentrations of dopamine to which they are now exposed. Therefore L-dopa at dosage adequate for normal action at some receptors may be sufficient for an exaggerated response at others and hence for dyskinesia.

The dopamine resulting from L-dopa administration does not appear to eventually cause increased receptor sensitivity to fall back towards an appropriate level. If this occurred then dopa dyskinesia might disappear during prolonged treatment. Instead it tends to gradually become manifest. This may reflect the development of underlying disease processes or even an effect of the treatment itself on receptor sensitivity. An apparent analogy is the effect of chronic treatment of guinea-pigs with amphetamine (Klawans *et al.*, 1975). This leads to increased dopamine release to receptors and also (somewhat surprisingly) to increased receptor response to amphetamine itself or to the agonist apomorphine as shown by stereotyped movements. Other consequences of the degeneration of dopamine neurons could also be involved in the development of dyskinesia. For example, the invasion of damaged areas by different kinds of terminals (Raisman & Field, 1973) might be important.

A related possibility is that the defective synthesis of transmitters other than dopamine for which there is evidence in Parkinson's disease plays some part, e.g. noradrenaline and 5-hydroxytryptamine (Hornykiewicz, 1964), γ-aminobutyric acid (Lloyd *et al.*, 1973; McGeer *et al.*, 1973). While we do not know what is the relationship between these various transmitter deficiencies, many kinds of animal experiment show that behavioural effects of drugs which alter one transmitter are strongly dependent on the functional state of other transmitters (e.g. Mabry & Campbell, 1973; Everitt *et al.*, 1974; Grabowska & Michaluk,

1974; Green & Grahame-Smith, 1974). Therefore, different relationships between the transmitter deficiencies in different loci important for movement control could be involved in dopa dyskinesia. For example, experiments on the relationship between increased locomotion due to apomorphine and 5-hydroxytryptamine depletion (Grabowska & Michaluk, 1974) suggest that dopamine might have a particularly powerful action at certain receptors if there was a localized 5-hydroxytryptamine deficiency.

Dyskinesias caused by phenothiazines and butyrophenones

Drugs of the phenothiazine and butyrophenone classes which provoke tardive dyskinesia (among other movement disturbances) have little effect on brain amine levels in animals. They behave as if they block receptors and cause an increase of dopamine turnover (Fry & Sharman, this volume, Chapter 7) with increased but functionally ineffective release of amines into the synaptic cleft and elevated concentrations of dopamine metabolites such as homovanillic acid. The receptor blockade can lead to an increased sensitivity of dopamine receptors. This is demonstrable in animals after withdrawing chronic haloperidol treatment by giving the agonist apomorphine (Gianutsos *et al.*, 1974; Moore and Thornburg, 1975). The role of increased receptor sensitivity here may be analogous to that in dopa-provoked dyskinesia. Tardive dyskinesias may be treated with the amine depleter tetrabenazine (Kazamatsuri *et al.*, 1972*a*). They also respond to partial reblockage of the hypersensitive dopamine receptors by haloperidol (Kazamatsuri *et al.*, 1972*b*) though it is obvious that this treatment would eventually be harmful if it led to a further increase in sensitivity.

Huntington's Chorea

Huntington's chorea provides a good example of the importance of relationships between different transmitters in the production of dyskinesia. Initially the demonstration of a dopamine defect in Parkinson's disease suggested that in Huntington's chorea, where to some extent the symptoms represent an opposite pole, there would be an opposite metabolic disturbance, i.e. abnormally high dopamine synthesis. However, brain dopamine is not excessive (Bird & Iversen, 1974). On the contrary, in one study striatal dopamine and homovanillic acid concentrations were found to be moderately decreased with more marked changes in the caudate than in the putamen (Bernheimer *et al.*, 1973). Atrophy of the caudate nucleus and putamen with little or no nigral change is typical of Huntington's chorea. Thus the normal or moderately low striatal concentrations togethèr with the low weights of striatal regions lead to low net *content* of dopamine and homovanillic acid at autopsy. This is consistent with the decrease of lumbar CSF homovanillic acid with increasing severity of symptoms (Curzon *et al.*, 1972). CSF 5-hydroxyindolylacetic acid did not decrease. Very low concentrations of both amine metabolites are reported in lumbar CSF of two patients with the akinetic rigid form of the disease (Curzon *et al.*, 1972; Johansson & Roos, 1974). This is apparently consistent with the Parkinson-like symptoms but, unlike in Parkinson's disease, brain dopamine content is not strikingly low, being somewhat higher than in ordinary Huntingtonians (Bird & Iversen, 1974).

Striatal shrinkage in Huntington's chorea probably involves a major decrease of non-dopamine-containing neurons. Therefore although net dopamine synthesis is not abnormally great there is probably a dopaminergic predominance. This is consistent with the exacerbation of symptoms by L-dopa and the therapeutic effect of amine depleters such as tetrabenazine.

Perry *et al.* (1973) in a study of brain amino acid concentrations obtained clear evidence of significantly low concentration of the transmitter γ-aminobutyric acid and of its peptide derivative homocarnosine. Furthermore, although results suggested moderate γ-aminobutyric acid deficiency throughout the brain it was most striking in the substantia nigra and other basal ganglia regions. The low values did not simply reflect cerebral oedema or non-specific changes as 34 other amino acids or related substances were present at normal concentration.

Bird & Iversen (1974) showed that the γ-aminobutyric acid deficiency occurs together with even more strikingly low values in various parts of the basal ganglia of glutamic acid decarboxylase. This enzyme is responsible for γ-aminobutyric acid synthesis and occurs mainly in inhibitory neurons which release γ-aminobutyric acid as their transmitter. The distribution of γ-aminobutyric acid in the human substantia nigra (Kanazawa *et al.*, 1973) suggests its presence in striatonigral neutrons with terminals mainly in the pars reticulata of the substantia nigra bordering the zone containing the cell bodies of the nigrostriatal dopamine neurons, i.e. the pars compacta. The findings in Huntington's chorea therefore indicate degeneration of such inhibitory neurons in the basal ganglia.

Choline acetyltransferase activity/g tissue—an index of cholinergic neurons—was strikingly low in striata of patients with particularly marked deficits of glutamic acid decarboxylase, although activity was normal in autopsy material from some subjects who had exhibited typically choreiform movements. However, there may also be damage to acetylcholine receptors (Hiley, this volume, Chapter 11). Acetylcholine imbalance could have a role in the disease as Klawans & Rubovits (1972) found that the anticholinesterase physostigmine and the anticholinergic drug benztropine alleviated and exacerbated symptoms respectively.

As reduced γ-aminobutyric acid synthesis also occurs in the basal ganglia in Parkinson's disease (Lloyd *et al.*, 1973; McGeer *et al.*, 1973) it appears that chorea results not from an absolute γ-aminobutyric acid deficiency but from an imbalance between γ-aminobutyric acid and dopamine neurons so that the reduced inhibitory activity of the former neurons leads to an exaggerated response to dopamine. Both the precipitation of chorea by dopa in Huntingtonians (Klawans *et al.*, 1972) and the choreiform movements which can occur when Parkinsonian patients are given dopa are consistent with this relationship. In untreated Parkinsonians the functional deficit in dopamine predominates so that rigidity and akinesia occur. Such a generalization is in keeping with the relative deficits of the two transmitters demonstrable in autopsy material from subjects with the two disorders. Increased responsiveness of dopamine receptors may also lead to the chorea of occasional hyperthyroid patients. Thus haloperidol, a dopamine receptor antagonist, alleviated hyperthyroid chorea in one patient (Klawans & Shenker, 1972) while guinea-pigs made hyperthyroid by thyroxine showed increased responsiveness to the dopamine receptor agonist apomorphine (Klawans *et al.*, 1973*a*).

As with the degeneration of dopamine neurons in Parkinson's disease we can only speculate about the origin of the neuronal degeneration in Huntington's chorea. It suggests, however, that the development of drugs increasing brain γ-aminobutyric acid could prove valuable. Present useful drug treatments decrease functional concentrations of catecholamines and 5-hydroxytryptamine.

Myoclonus, etc.

Excessive 5-hydroxytryptamine synthesis or receptor sensitivity has been suggested to occur in infantile myoclonus, as injection of the precursor 5-hydroxytryptophan caused spasms when given to some patients with Down's syndrome (Coleman, 1971) or to young guinea-pigs (Klawans *et al.*, 1973*b*). However, the relevance of these observations to the biochemical changes in spontaneous myoclonus is unclear as animal experiments indicate that 5-hydroxytryptophan only penetrates to 5-hydroxytryptamine neurons to a limited extent and is also decarboxylated to it in cells where 5-hydroxytryptamine does not normally occur (Fuxe *et al.*, 1971). Also, the failure of *p*-chlorophenylalanine or methysergide to alleviate spontaneous spasms (Coleman *et al.*, 1971) suggests that a 5-hydroxytryptamine excess may not have a primary role in the non-drug-induced symptom. Unlike the above gross spasms, spontaneous oral mycoclonic activity in the rat is inhibited by the 5-hydroxytryptamine receptor blocker methysergide (Bieger *et al.*, 1972). Therefore 5-hydroxytryptamine might be important in the irreversible facial dyskinesias occurring in some subjects after phenothiazine or butyrophenone treatment (Crane, 1973). Pharmacological evidence in the rat experiments of Bieger *et al.* (1972) suggest that the oral myoclonus depends on both dopaminergic and serotoninergic activity. CSF 5-hydroxyindoleacetic and homovanillic acid are both normal in a group of patients with oral dyskinesias of various origins (Curzon, 1973). CSF findings after probenecid treatment (Pind & Faurbye, 1970) appear to be in agreement.

In complete contrast to the above findings, 5-hydroxytryptophan either alone (Lhermitte *et al.*, 1972) or with a peripheral decarboxylase inhibitor (Van Woert & Sethy, 1975) had a beneficial effect on a different form of myoclonus—the intention or action myoclonus following an anoxic episode due to cardiac or respiratory arrest. In the two patients on whom measurements were made CSF 5-hydroxyindoleacetic and homovanillic acids were both low before treatment (Van Woert & Sethy, 1975). Tryptophan did not benefit the single patient to whom it was given. The findings are consistent with degeneration of 5-hydroxytryptamine neurons so that tryptophan hydroxylase is absent at certain sites and therefore 5-HT cannot be effectively made from tryptophan at these sites but can be made from 5-hydroxytryptophan through the action of decarboxylase at other sites in their proximity. Degeneration of dopamine neurons is consistent with the low CSF homovanillic acid, and indeed dopa treatment gave some benefit though considerably less than that obtained with 5-hydroxytryptophan.

References

Agid, Y., Javoy, F. & Glowinski, J. (1973) *Nature* (*London*) **245**, 15–151

Bergmann, S., Curzon, G., Friedel, J., Godwin-Austen, R. B., Marsden, C. D. & Parkes, J. D. (1974) *Brit. J. Clin. Pharmacol.* **1**, 417–424

Bernheimer, H., Birkmayer, W., Hornkiewicz, O., Jellinger, K. & Seitelberger, F. (1973) *J. Neurol. Sci.* **20**, 415–455.

Bieger, D., Larochelle, L. & Hornykiewicz, O. (1972) *Eur. J. Pharmacol.* **18**, 128–136

Bird, E. D. & Iversen, L. L. (1974) *Brain* **97** 457–472

Chase, T. N., Holden, E. & Brody, J. A. (1973) *Arch. Neurol.* **29**, 328–330

Coleman, M. (1971) *Neurology* **21**, 911–919

Coleman, M., Bouillon, D. & Davis, M. (1971) *Neurology* **21**, 421.

Costall, B. & Naylor, R. J. (1975) *Advan. Neurol.* **9**, 285–297

Crane, G. E. (1973) *Brit. J. Psychiat.* **122**, 395–405

Curzon, G. (1975) *Advan. Neurol.* **9**, 349–357

Curzon, G. (1973) *Proc. Roy. Soc. Med.* **66**, 873–876

Curzon, G., Gumpert, J. & Sharpe, D. (1972) *J. Neurol. Neurosurg. Psychiat.* **35**, 514–519

Ericsson, A. D., Wertman, B. G. & Duffy, K. M. (1971) *Neurology* **21**, 1023–1029

Everitt, B. J., Fuxe, K. & Hokfelt, T. (1974) *Eur. J. Pharmacol.* **29**, 187–191

Fuxe, K., Butcher, L. L. & Engel, G. (1971) *J. Pharm. Pharmacol.* **23**, 520–524

Garelis, E., Young, S. N., Lal, S. & Sourkes, T. L. (1974) *Brain Res.* **79**, 1–8

Gianutsos, G., Drawbaugh, R. B., Hynes, M. D. & Lal, H. (1974) *Life Sci.* **14**, 887–898

Grabowska, M. & Michaluk, J. (1974) *Pharmacol. Biochem. Behav.* **2**, 263–266

Green, A. R. & Grahame-Smith, D. G. (1974) *Neuropharmacology* **13**, 949–959

Hornykiewicz, O. (1964) *Proc. Int. Pharmacol. Meet. 2nd,* pp. 57–68

Iversen, L. L., Horn, A. S. & Miller, R. J. (1975) *Advan. Neurol.* **9**, 197–212

Johansson, B. & Roos, B. E. (1974) *Eur. Neurol.* **11**, 37–45

Kanazawa, L., Miyata, Y., Toyokura, Y. & Otsuka, M. (1973) *Brain Res.* **51**, 363–365

Kazamatsuri, H., Chien, C. P. & Cole, J. (1972*a*) *Arch. Gen. Psychiat.* **27**, 95–99

Kazamatsuri, H., Chien, C. P. & Cole, J. (1972*b*) *Arch. Gen. Psychiat.* **27**, 100–103

Klawans, H. L. & Rubovits, R. (1972) *Neurology* **22**, 107–116

Klawans, H. L. & Shenker, D. M. (1972) *J. Neurol. Trans.* **33**, 73–81

Klawans, H. L. & Goetz, C. & Weiner, W. J. (1973*a*) *J. Neurol. Trans.* **34**, 187–193

Klawans, H. L., Goetz, C. & Weiner, W. J. (1973*b*) *Neurology* **23**, 1234–1240

Klawans, H. L., Crossett, P. & Dana, N. (1975) *Advan. Neurol.* **9**, 105–112

Klawans, H. L., Paulson, G. W., Ringel, S. P. & Barbeau, A. (1972) *New Engl. J. Med.* **286**, 1332–1334

Lhermitte, F., Marteau, R. & Degos, C. F. (1972) *Rev. Neurol.* (*Paris*) **126**, 107–114

Lloyd, K. G., Davidson, L. & Hornykiewicz, O. (1973) *Advan. Neurol.* **3**, 173–188

Mabry, P. D. & Campbell, B. A. (1973) *Brain Res.* **49**, 381–391

McGeer, P. L., McGeer, E. C. & Fibiger, H. C. (1973) *Lancet* ii, 623–624

Mones, R. J. (1973) *Mt. Sinai J. Med.* **39**, 197–201

Moore, K. E. & Thornburg, J. E. (1975) *Advan. Neurol.* **9**, 93–104

Ng K. Y., Chase, T. N., Colburn, R. W. & Kopin, I. J. (1970) *Science* **170**, 76–77

Nystrom, B., Olson, L. & Ungerstedt, U. (1972) *Science* **176**, 924–926

Olson, L., Nystrom, B. & Seiger, A. (1973) *Brain. Res.* **63**, 231–247

Perry, T. L., Hansen, S. & Kloster, B. (1973) *New Engl. J. Med.* **288**, 337–342

Pind, K. & Faurbye, A. (1970). *Acta Psychiat. Scand.* **46**, 323–326

Raisman, G. & Field, P. M. (1973) *Brain Res.* **50**, 241–264

Roffler-Tarlov, S., Sharman, D. F. & Tegerdine, P. (1971) *Brit. J. Pharmacol.* **42**, 343–351

Sandler, M., Bonham-Carter, S., Hunter, K. R. & Stern, G. M. (1973) *Nature* (*London*) **241**, 439–443

Sharman, D. F., Poirier, L. J., Murphy, G. F. & Sourkes, T. L. (1967) *Can. J. Physiol.* **45**, 57–62

Ungerstedt, U. (1971) *Acta Physiol. Scand. suppl.* **367**, 49– 7

Van Woert, M. H. & Sethy, V. H. (1975) *Neurology* **25**, 135–140

Wada, G. H. & Fellman, J. H. (1973) *Biochemistry* **12**, 5212–5217

Chapter 3

Clinical Pharmacology of Dopaminergic Effects in Parkinsonism

By DONALD B. CALNE

National Institute of Neurological and Communicative Disorders and Stroke, National Institutes of Health, Department of Health, Education and Welfare, Bethesda, Maryland 20014, *U.S.A.*

Introduction

Extensive clinical experience of the therapeutic and adverse effects of levodopa is now available (Barbeau *et al.*, 1971). Many pharmacological investigations have been undertaken in man in an attempt to elucidate the actions of this drug. The use of dopaminergic agonists, dopaminergic antagonists, phosphodiesterase inhibitors, monoamine oxidase inhibitors and extracerebral decarboxylase inhibitors will be discussed. The therapeutic implications of these studies will be reviewed.

Dopaminergic Agonists

Agonists of dopamine may be anticipated to be useful in the treatment of Parkinsonism provided they readily cross the blood–brain barrier. Since the enzyme responsible for converting levodopa to dopamine (L-aromatic amino acid decarboxylase) is depleted in the brain of Parkinsonian patients (Lloyd *et al.*, 1973), dopaminergic agonists, which do not require this enzyme, have a theoretical advantage over levodopa. Furthermore, individual dopaminergic agonists can be predicted to have a range of diverse profiles of activity on different areas of the brain, depending on accessibility and receptor affinity. By analogy with other transmitters such as acetylcholine and noradrenaline, it is possible that more than one type of dopaminergic receptor exists. By playing 'pharmacological roulette' with different agonists there is a chance that a molecular configuration will be found that has a relatively selective action upon those striatal dopaminergic synapses which are defective in Parkinsonism.

The first two dopaminergic agonists to be studied, apomorphine (Cotzias *et al.*, 1972) and pirebedil (Vakil *et al.*, 1973; Chase *et al.*, 1974; Sweet *et al.*, 1974), have been abandoned. Apormorphine causes prominent emesis and high oral doses lead to azotaemia. Pirebedil induces prominent psychiatric adverse reactions.

The most recent dopaminergic agonist to be investigated is an ergot derivative (Corrodi *et al.*, 1973; Calne *et al.*, 1974*a*, *b*). Bromocriptine is made up of a tripeptide moiety and a lysergic acid diethylamide (LSD) residue. The pharmacologically active component is likely to be the LSD residue, since LSD has recently been shown to act as a dopaminergic agonist *in vivo* and *in vitro*

Hungen *et al.,* 1974; Pieri *et al.,* 1974). Bromocriptine has therapeutic actions similar to levodopa, ameliorating akinesia, rigidity and tremor; if no unwanted toxic effect emerge from prolonged administration it is likely to assume a role in the routine treatment of Parkinsonism. The observation that dopaminergic agonists improve Parkinsonism militates against the suggestion (Sourkes, 1971) that the beneficial consequences of administering levodopa may stem from formation of metabolites other than dopamine. One finding of interest is that patients with more severe neurological disability derive greater benefit from bromocriptine than patients with milder deficits. This may reflect the presence of more severe depletion of L-aromatic amino acid decarboxylase in the brains of patients with advanced Parkinsonism. Alternatively, there may be denervation supersensitivity leading to extension of the dopaminergic receptor area to regions of the postsynaptic membrane remote from the presynaptic enzyme mechanisms responsible for converting levodopa to dopamine. A third possible explanation could be based upon a higher therapeutic index—an increase in the ratio of wanted to unwanted actions, which would allow higher doses to be tolerated by those patients who require more treatment because of the severity of their Parkinsonism.

Apomorphine, piribedil and bromocriptine all cause a wide range of dose-dependent reversible adverse reactions similar to those encountered with levodopa. They include nausea, dyskinesia, confusion and hypotension. We have recently found that 'on–off' phenomena also occur with bromocriptine. The implication of these findings is that dopaminergic activation induces all of the major adverse effects which limit the dose of levodopa. This conclusion is of importance since it has been suggested that metabolites of levodopa other than dopamine may play a role in producing certain adverse reactions (Hornykiewicz, 1973; Dougan *et al.,* 1975). While it must be admitted that no drugs are completely specific, it seems unlikely that dopaminergic agonists of such varied structure as apomorphine, piribedil and bromocriptine share activity at noradrenaline, serotonin, γ-aminobutyric acid and acetylcholine receptors.

Dopaminergic Antagonists

Drugs which block dopaminergic receptors include phenothiazines (such as chlorpromazine), butyrophenones (such as haloperidol), and diphenylbutylpiperidines (such as pimozide). All these drugs induce Parkinsonism; several also produce dyskinesia. Current views on the pharmacology of extrapyramidal disorders infer that Parkinsonism results from inadequate dopaminergic transmission while dyskinesia is produced by excessive dopaminergic function (Klawans *et al.,* 1970; Chase, 1972). This hypothesis appears to be the best currently available in view of the firm evidence that drugs which drive dopaminergic synapses alleviate Parkinsonism (levodopa, apomorphine, piribedil and bromocriptine) and are all capable of causing dyskinesia. Conversely, drugs which decrease dopaminergic transmission by depleting brain dopamine induce Parkinsonism (reserpine and tetrabenazine). Drugs which block dopaminergic receptors occupy a special position since they can induce both Parkinsonism and dyskinesia. The paradoxical finding, in the case of dopaminergic receptor blockers, that the same drugs can produce both Parkinsonism and dyskinesia, may be resolved by postulating that these medications induce denervation supersensitivity. Administration of such drugs commonly results in a situation where factors conducive to Parkinsonism are in approximate balance with mechanisms promoting dyskinesia. The balance is frequently disturbed by alterations in drug dosage because Parkinsonism tends to be rapidly reversible by reducing the dose of causal drug whereas dyskinesia is relatively persistent.

If it is accepted that Parkinsonism and dyskinesia represent opposite poles in a range of pharmacological responses from inadequate (Parkinsonism) to excessive (dyskinesia) dopaminergic drive, one problem which must be considered is the occasional simultaneous occurrence of Parkinsonism and dyskinesia in different parts of the body. While any explanation must be regarded as speculative, it would seem reasonable to interpret this unusual finding in terms of certain brain regions manifesting increased dopaminergic function while at the same time others display decreased dopaminergic transmission. Evidence in support of this concept derives from analogous experience occasionally encountered in the treatment of myasthenia gravis with anticholinesterase agents. It is well recognized that 'brittle myasthenics' may have one group of muscles paretic because of a deficiency of acetylcholine at their neuromuscular junctions (myasthenic crisis) while simultaneously another group of muscles is weak owing to the presence of excessive quantities of acetylcholine at their neuromuscular junctions (cholinergic crisis) (Calne, 1975). The therapeutic dilemma posed by the problem of managing these patients is considerable. It might be appropriate to coin the term 'brittle Parkinsonian' for the dopaminergic counterpart to the 'brittle myasthenic'.

Phosphodiesterase Inhibitors

The recent finding that the striatal dopaminergic receptor is closely associated with an adenylate cyclase (Kebabian *et al.,* 1972; Iversen *et al.,* 1975), provides a useful '*in vitro*' screen for dopaminergic agonists and antagonists. In addition, this observation poses a new question. If the actions of dopamine are mediated by an adenylate cyclase which produces cyclic adenosine monophosphate (cyclic AMP), can Parkinsonian patients be helped by the administration of a phosphodiesterase inhibitor which obstructs the degradation of cyclic AMP? The most readily available phosphodiesterase inhibitor is caffeine; this drug has no action when administered alone to Parkinsonian patients (R. C. Duvoisin, personal communication). Similarly, no benefit has been detected when caffeine is given, in maximum tolerated dosage, in combination with the dopaminergic agonists piribedil (T. N. Chase, personal communication) or bromocriptine (R. Kartzinel & D. B. Calne, unpublished work). These results do not refute the essential role now assigned to cyclic AMP in mediating the therapeutic and adverse consequences of augmenting dopaminergic drive in Parkinsonian patients. However, any further attempt to potentiate anti-Parkinsonian therapy by blocking the degradation of cyclic AMP should involve the pursuit of a drug with more selective actions on striatal phosphodiesterase. If greater specificity can be achieved, this approach may be developed to enhance the wanted as opposed to the deleterious effects of dopaminergic drive.

Monoamine Oxidase Inhibitors

Cotzias *et al.* (1974) have recently reported evidence indicating that monoamine oxidase in the cerebral capillary endothelium plays a major role in blocking the entry to the brain of dopamine, including that formed by L-aromatic amino acid decarboxylase within the endothelial cell. We have recently studied the effect of giving levodopa to patients who have had monoamine oxidase blocked with tranylcypromine (P. F. Teychenne, D. B. Calne, P. J. Lewis & L. J. Findley, unpublished work). Previous investigations have shown that the combination of levodopa with a monoamine oxidase inhibitor (MAOI) results in a potentially dangerous rise in blood pressure (Hunter *et al.*, 1970). Animal studies (Robson, 1971; Calne & Reid, 1973) have indicated that this pressor response is mediated at the periphery and can be suppressed by administration of an extracerebral decarboxylase inhibitor. We have therefore administered the extracerebral decarboxylase inhibitor carbidopa, in addition to levodopa and tranylcypromine. Our results indicate that while this combination of drugs clearly allows rapid therapeutic effects to be achieved with very small doses of levodopa, adverse reactions are also produced and there is no improvement in the ratio of wanted to unwanted responses. Furthermore, there is considerable variation between patients in the dose of carbodopa required to inhibit the pressor reaction to levodopa plus tranylcypromine (the usual dose of carbidopa employed to block extracerbral decarboxylase is 100 mg daily—in one of our patients a marked rise in blood pressure occurred even after administration of 400 mg daily). It may be concluded that this approach to improving the treatment of Parkinsonism is unlikely to prove acceptable as routine therapy.

Extracerebral Decarboxylase Inhibitors

By preventing the peripheral formation of catecholamines, extracerebral decarboxylase inhibitors (DCI) such as carbidopa reduce the risk of levodopa-induced cardiac arrhythmias (Parks *et al.,* 1970). DCI have also been employed to analyse the mechanism of hypotension induced by levodopa. In animals (Dhasmana & Spilker, 1973) and man (Reid *et al.,* 1971), it has been found that levodopa reduces the set of the blood pressure by an action which is uninfluenced by extracerebral inhibition of decarboxylation; this must be mediated through the central nervous system. In contrast, levodopa impairs the baroceptor reflexes by a mechanism which can be blocked by DCI; this must be achieved at the periphery. The peripheral impairment of the baroceptor reflexes may involve the dopaminergic inhibitory interneurons which have been found in the sympathetic ganglia (Greengard & Kebabian, 1974).

In practical therapeutics, the major advantage of DCI is reduction of the emesis induced by levodopa (Yahr, 1974). This effect probably results from penetration of DCI into the area postrema, where the blood–brain barrier is relatively permeable. By blocking the formation of catecholamines in this region of the brain, which is located close to the 'emetic centre', DCI would decrease the central actions of levodopa responsible for anorexia, nausea and vomiting.

Conclusion

The clinical pharmacology of dopaminergic mechanisms in Parkinsonism has been reviewed. The salient problem remains the task of separating therapeutic from adverse effects. The actions of dopaminergic agonists, phosphodiesterase inhibitors, monoamine oxidase inhibitors and extracerebral decarboxylase inhibitors have been discussed in the context of analysing the pharmacology of unwanted actions of levodopa and attempting to reduce their impact on therapy.

Discussion

L. L. Iversen: The idea of using caffeine with bromocriptine is very interesting. However, the difficulty biochemically is that brain phosphodiesterase is only weakly inhibited by caffeine and theophylline. I do not think we have a suitable phosphodiesterase inhibitor available for clinical use yet.

D. B. Calne: Can I ask you for more information on this? Are there different phosphodiesterases in different parts of the brain?

L. L. Iversen: This is a very complicated issue and a lot of work still needs to be done on this subject, but certainly it is a very heterogeneous enzyme.

D. B. Calne: What I cannot do is to discount completely the effects of caffeine on the brain, for clearly it caused considerable cerebral excitation, presumably by influencing phosphodiesterase in the brain, but I would agree with you perhaps not the right phosphodiesterase. Perhaps we need an inhibitor more specific for striatal dopaminergic phosphodiesterase.

L. L. Iversen: I would agree. We should not take the result with caffeine as negative evidence for involvement of cyclic AMP.

G. W. Arbuthnott: I would add to the discussion on phosphodiesterase inhibition that when we tried to use phosphodiesterase inhibitors on the turning animal, the only ones that worked were caffeine and theophylline; the others didn't.

D. B. Calne: I think the Swedish group found the same thing, suggesting caffeine potentiates striatal phosphodiesterase very much in the rat model [K. Fuxe & U. Ungerstedt (1974) *Med. Biol.* **52**, 48–54].

L. L. Iversen: What sort of doses of caffeine are needed?

G. W. Arbuthnott: Fairly large, about 25 mg/kg.

L. L. Iversen: That's not very large.

G. W. Arbuthnott: It was large enough to make the rats hard to handle.

L. L. Iversen: My personal opinion is that if you take total phosphodiesterase activity in brain extract, it is not powerfully inhibited by caffeine or theophylline in very high concentrations.

S. R. Nahorski: Perhaps this indicates that the ability of caffeine to cause turning in unilaterally lesioned rats is not related to its phosphodiesterase inhibiting property. Thus caffeine appears to be a very weak inhibitor of cerebral phosphodiesterase, while other compounds which are much more potent, and which enhance the action of dopamine on cyclic AMP accumulation in cerebral slices, do not enhance turning.

You really need about 20 mg/kg of caffeine by itself to get turning.

G. M. Woodruff: We have studied the behavioural actions of some other ergot alkaloids, including ergometrine. Ergometrine stimulates dopamine receptors and causes strong locomotor stimulation in rats. This locomotor stimulation is strongly potentiated by caffeine [G. N. Woodruff *et al.* (1974) *J. Pharm. Pharmacol.* **26**, 455–456].

D. A. Brown: I think that caffeine has effects on neurotransmission other than inhibition of phosphodiesterase. These other effects may cause the central stimulation. The doses used may not be enough to affect striatal phosphodiesterase.

References

Barbeau, A., Wars, H. & Gillo-Joffroy, L. (1971) *Recent Advances in Parkinson's Disease* (McDowell, F. H. & Warkham, C. H., eds), p. 203, Davis Co., Philadelphia

Calne, D. B. (1975) *Therapeutics in Neurology*, pp. 328, Blackwell, Oxford

Calne, D. B. & Reid, J. L. (1973) *Brit. J. Pharmacol.* **48**, 194

Calne, D. B., Teychenne, P. F., Claveria, L. E., Eastman, R., Greenacre, J. K. & Petrie, A. (1974*a*) *Brit. Med. J.* **iv**, 442

Calne, D. B., Teychenne, P. F., Leigh, P. N., Bamji, A. N. & Greenacre, J. K. (1974*b*) *Lancet* **ii**, 1355

Chase, T. N. (1972) *Res. Publ. Ass. Res. Nerv. Ment. Dis.* **50**, 448

Chase, T. N., Woods, A. C. Glaubiger, G. A. (1974) *Arch. Neurol.* **30**, 383

Corrodi, H., Fuxe, K., Hokfelt, T., Lidbrink, P. & Ungerstedt, U. (1973) *J. Pharm. Pharmacol.* **25**, 409.

Cotzias, G. C., Lawrence, W. H., Papavasilious, P. S., Duby, S. E., Ginos, J. Z. & Mena, I. (1972) *Trans. Am. Neurol. Ass.* **97**, 156

Cotzias, G. C., Tang, L. C. & Ginos, J. Z. (1974) *Proc. Nat. Acad. Sci. U.S.A.* **71**, 2715

Dhasmana, K. M., and Spilker, B. A. (1973) *Brit. J. Pharmacol.* **47**, 437

Dougan, D., Wade, D. & Mearrick, P. (1975) *Nature (London)* **254**, 70

Greengard, P. & Kebabian, J. W. (1974) *Fed. Proc. Fed. Am. Soc. Exp. Biol.* **33**, 1063

Hornykiewicz, O. (1973) *Fed. Proc. Fed. Am. Soc. Exp. Biol.* **32**, 183

Hungen, K., Roberts, S. & Hill, D. F. (1974) *Nature (London)* **252**, 588

Hunter, K. R., Boakes, A. J., Laurence, D. R. & Stern, G. M. (1970) *Brit. Med. J.* **iii**, 388

Iversen, L. L., Horn, A. S. & Miller, R. J. (1975) *Advan. Neurol.* **9**, 197.

Kebabian, J. W., Petzold, G. L. & Greengard, P. (1972) *Proc. Nat. Acad. Sci. U.S.A.* **69**, 2145

Klawans, H. L., Ilahi, M. M. & Shenker, D. (1970) *Acta Neurol. Scand.* **46**, 409

Lloyd, K. G., Davidson, L. & Hornykiewicz, O. (1973) *Advan. Neurol.* **9**, 173

Parks, L. C., Watanabe, A. M. & Kopia, I. J. (1970) *Lancet* **ii**, 1014

Pieri, L., Pieri, M., Haefely, W. (1974) *Nature (London)* **252**, 256

Reid, J. L., Calne, D. B., George, C. F., Pallis, C. & Vakil, S. D. (1971) *Clin. Sci.* **41**, 63

Robson, R. D. (1971) *Circulation Res.* **28**, 662.

Sourkes, T. L. (1971) *Nature (London)* **229**, 413

Sweet, R. D., Wasterlain, C. G. & McDowell, F. H. (1974) *Clin. Pharmacol. Ther.* **16**, 1077.

Yahr, M. D. (1974) *Current Concepts in the Treatment of Parkinsonism*, pp. 257, Raven Press, New York

Vakil, S. D., Calne, D. B., Reid, J. L. & Seymour, C. A. (1973) *Advan. Neurol.* **3**, 121

Chapter 4

Supersensitivity of Dopamine Receptors

By G. W. ARBUTHNOTT

Medical Research Council Brain Metabolism Unit, University Department of Pharmacology, 1 George Square, Edinburgh EH8 9JZ, U.K.

Current interest in the suggestion that dopamine receptors might become supersensitive when denervated, just as peripheral receptors were known to do, originated from a technical advance. Earlier studies reviewed by Cannon & Rosenbleuth (1949), Stavraky (1961) and Sharpless (1964) all concluded that receptors in the central nervous system do become supersensitive after denervation but all were hampered by lesions which destroyed both excitatory and inhibitory systems impinging on the area being studied. Although this objection can be partly overcome by the use of specific receptor agonists, nevertheless it casts considerable doubt on the interpretation of the effects seen.

In 1968 Ungerstedt published the first paper on a technique for the making of discrete lesions in an identified pathway which has proved more specific than any of the previously known methods. The neurotoxin 6-hydroxy-dopamine (6-OH-DA) is concentrated in catecholamine-containing cells by the membrane pump which is normally responsible for the uptake of the transmitter substance. A microinjection of 6-OH-DA into the region of the anterior substantia nigra will produce widespread damage to the nigrostriatal dopamine-containing system without major damage to any other system, with the possible exception of the noradrenaline-containing axons which pass close to the injection site (Iversen, 1974).

Hökfelt & Ungerstedt (1973) were able to show that non-amine-containing cells in the substantia nigra remain undamaged in electron micrographs which show destruction of the neighbouring amine-containing cells. This technique allowed Ungerstedt (1971*a*) to progress further in the analysis of the behavioural characteristics of rats with unilateral damage to the nigrostriatal system. Andén (1966) had demonstrated by the use of conventional lesioning techniques that unilateral damage to the nigrostriatal system results in a postural asymmetry which could be exaggerated into a continuous circling locomotion towards the lesioned side by drugs which were known to act on dopaminergic transmission.

The more specific lesioning techniques revealed that this ipsiversive circling was obtained only with drugs thought to act by releasing amines; amphetamine was the example most studied by Ungerstedt. On the other hand, apomorphine, which is presumed to act on the postsynaptic receptors for dopamine (Andén *et al.*, 1967; Ernst, 1967), made the animals circle in a direction opposite to that caused by the amine-releasing drugs. There is evidence that turning is a result of a unilateral action on striatal dopamine receptors (Arbuthnott & Ungerstedt, 1969, 1975; Ungerstedt *et al.*, 1969; Arbuthnott & Crow, 1971; Christie *et al.*, 1973), and so the action of apomorphine seemed to be on the side opposite to the one on which amphetamine acted. The simplest explanation is that amphetamine caused turning by releasing dopamine on the normal side but that apomorphine had its major effect on the receptors of the denervated side which had become supersensitive. The preparation was the more interesting because L-dopa had an action of the apomorphine type suggesting a direct action on the dopamine receptors of the denervated side. Since the effect could be inhibited by RO 4-4602, a potent dopa decarboxylase inhibitor, Ungerstedt (1971*a*) concluded that the action was not a result of reduced specificity of the denervated receptors allowing dopa a direct action. He suggested that the dopamine synthesized from L-dopa acted predominantly on the supersensitive receptors of the lesioned side.

The suggestion soon followed that this was how L-dopa could be effective in Parkinson's disease (Ungerstedt, 1971*b*) and moreover that here was an animal model which would be valuable in the testing of anti-Parkinsonian drugs with a mechanism of action similar to L-dopa. There are now two drugs, which have received and are receiving clinical assessment, which were discovered and originally tested on this model. The first, piribedil (Corrodi *et al.*, 1971; Corrodi *et al.*, 1972), seems to have proved less useful than L-dopa (Vakil *et al.*, 1973; Chase & Shoulson, 1975; McDowell & Sweet, 1975). The ergot derivatives, ergocornine and 2-Br-α-ergocryptine (Fuxe *et al.*, 1974), are still to be fully tested (Claveria *et al.*, 1975).

The value of such a model does not necessarily rest on the assumptions which generate it. The drugs it produces will be useful in Parkinson's disease or they will not—and the test system stands or falls by its results, Nevertheless the suggestion that supersensitive receptors exist on the cells of the striatum, after a lesion of the dopamine system, is itself an interesting hypothesis worthy of test.

The turning rat model is not a suitable test system for this hypothesis. To be able to show receptor supersensitivity we need to be able to show that the operation has increased a normal response, but turning away from the lesioned side has no 'normal' equivalent. One way round this problem is to look at the details of the striatal response to dopamine. If it is possible to identify the normal effect of dopamine on some aspect of striatal function, then after denervation an increase in the response for the same dose of agonist would be evidence for supersensitivity.

Tests on Single Striatal Units

Like Bloom *et al.* (1965) most authors find the most striking effect of the microiontophoretic application of dopamine to striatal cells to be inhibitory (McLennan & York, 1967; Herz & Zieglgansberger, 1968; Connor, 1970; Feltz & de Champlain, 1972*b*). York (1970) draws attention to an excitatory action in the putamen, but although there are reports of excitatory responses in striatal cells after nigral stimulation (Frigyesi & Purpura, 1967; Feltz & Wackenzie, 1969), Feltz (1969) and Connor (1970) conclude that only the inhibitory effects of nigral stimulation are mimicked by dopamine application.

With this background it is disappointing to find that the first direct attempt to look at the effect of the removal of the nigrostriatal pathway (Ohye *et al.*, 1970) showed L-dopa to have an excitatory effect which was obvious only on cells from the lesioned side. The cells in the putamen on the lesioned side had a faster firing frequency however, suggesting that the lesion had removed an inhibitory influence. Electrolytic lesions were made in this early study and the L-dopa was applied intravenously.

Feltz & de Champlain (1972*b*) were able to show that the depression in firing rate which they found after iontophoretic dopamine was changed in cats which had been injected with 6-OH-DA intraventricularly. An ejection current equivalent to a dose of drug which was able to reduce the firing rate of only some 29% of the striatal cells in normal cats, was 100% effective in the lesioned animals. They did not comment on any increase in firing rate of the cells in the lesioned animals. Feltz & de Champlain (1972*a*) did find other changes in the cells but these lesions only affected an area close to the ventricular fluid and were very variable between animals.

By accepting only rats which show behaviour typical of animals with more than a 95% reduction of DA concentration in the striatum, we have recently been able to show (Arbuthnott, 1975) that cells on the lesioned side do make larger changes in firing rate than do cells on the control side in response to DA receptor agonists. Figure 1 shows that the size of the reduction of the mean firing frequency of cells on the lesioned side is significant after a dose of 0.05 mg/kg apomorphine while the smaller reduction in the control side is not.

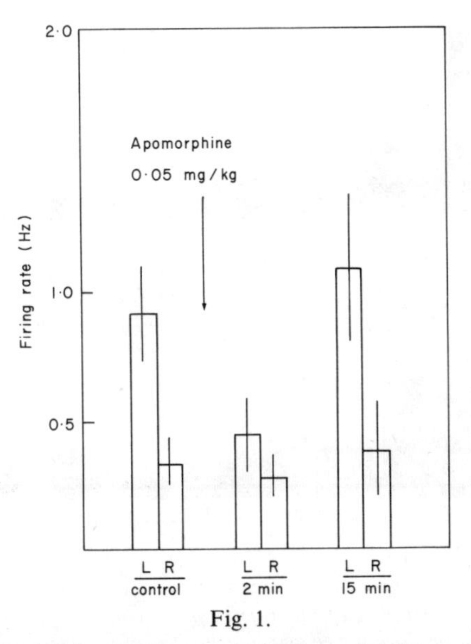

Fig. 1.

The columns represent mean firing rate (±S.E.M.) of groups of cells recorded in the striatum of the two sides of lesioned rats. The left side (L) is the lesioned and the firing rate in this striatum is higher in the control state. Two minutes after 0.05 mg/kg apomorphine intravenously the mean firing rate is reduced—significantly ($P < 0.02$) on the lesioned side but not on the right side. The rates have returned to normal 15 min after the drug injection.

This increased response to apomorphine on the lesioned side is accompanied by a raised frequency of spontaneous activity in the cells there (Arbuthnott, 1974) (Fig. 2). In our rats the firing rate of cells in the striatum of the control side was not significantly different from the firing rate found in normal animals. Hull *et al.* (1974) suggested that lesions in the region of the nigrostriatal bundle reduce the firing rate on the control side while that on the operated side remains within the normal range. Hull *et al.* (1974) used cats and their lesions did not need to involve the nigrostriatal system in order to produce this effect.

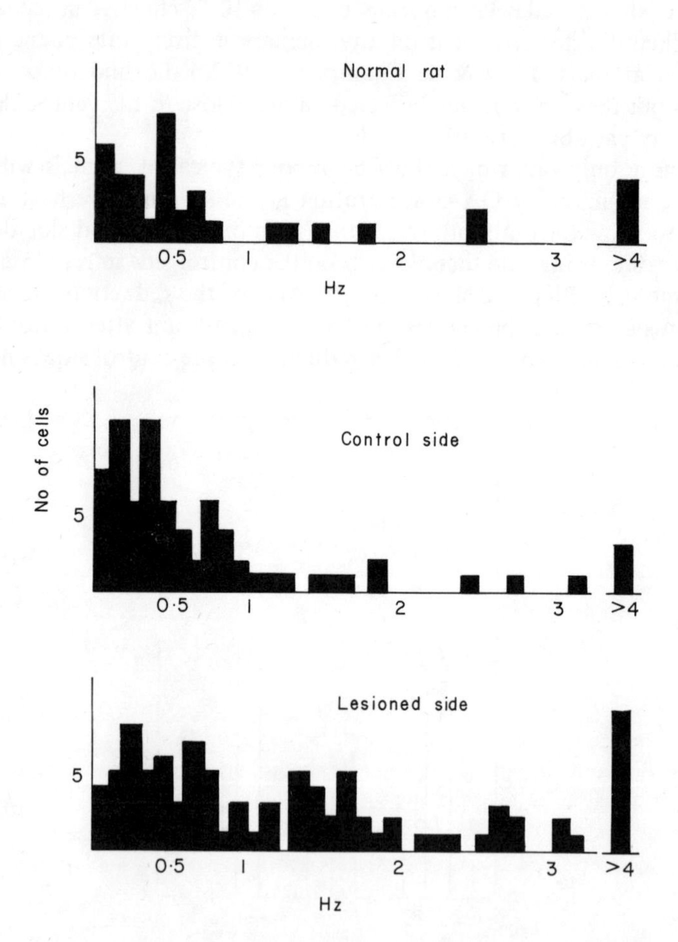

Fig. 2. *The distribution of observed cell firing rate in all the cells sampled in the striatum of normal rats and on the lesioned and control sides of the operated rats*

Although the distributions are not normal, more of the cells sampled fire faster than 1.0 Hz on the lesioned side than in control striata. Logarithmic transformations of the data make the distributions more nearly normal and then the firing rate in the lesioned side is significantly higher than that on the control side ($P < 0.001$) and also than that in the normal animals ($P < 0.005$).

The increased response to apomorphine in Fig. 1 may be a result of the increase in firing rate of the cells on the lesioned side seen in Fig. 2. Although Siggins *et al.* (1974) obtain similar results, Ungerstedt *et al.* (1975) draw attention to another parameter which is also altered on the lesioned side. Figure 3, which

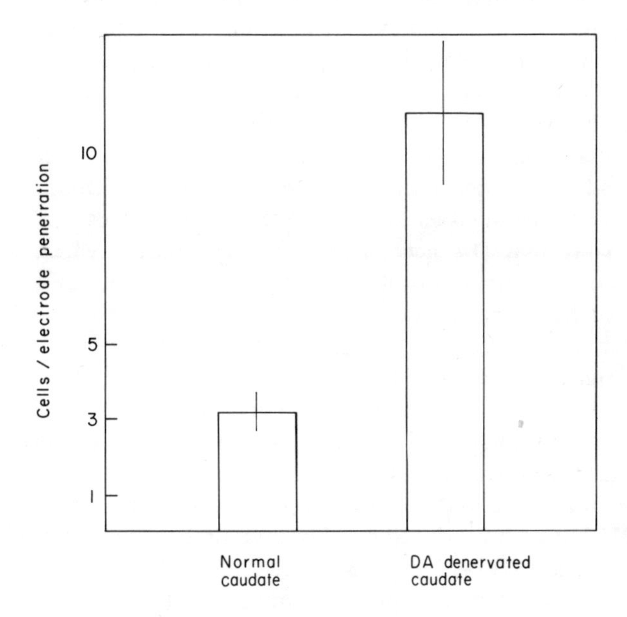

Fig. 3. *Mean number of spontaneously active cells* (±S.E.M.) *found in the caudate nucleus during electrode penetration through its entire depth*

The number of cells is greater on the lesioned side. Reproduced from Ungerstedt *et al.* (1975).

comes from the latter paper, shows that in otherwise identical tracks through the two striata more cells are encountered on the lesioned side. The distribution of these spontaneously active cells within the caudate is also different (Ungerstedt *et al.,* 1975). Not only are the cells on the lesioned side firing faster, but they may be a sample of a population of cells which is normally silent in the striatum.

The cells which are sampled on the control side, and in normal animals, are sensitive to apomorphine and L-dopa either administered iontophoretically (Siggins *et al.,* 1974) or intravenously (Arbuthnott, 1975) but the cells of the lesioned side appear to be more responsive. From this work we have not been able to prove whether the greater responses on the lesioned side derived from the reaction of receptors to denervation, or from the removal of a tonic depression of sensitivity which accompanied the removal of the tonic inhibition from the DA cells.

Tests on Striatal Biochemistry

Siggins *et al.* (1974) tested a number of other substances besides apomorphine and dopamine by microiontophoretic application to the striatal cells. The cells were also slowed by cyclic AMP. Kebabian *et al.* (1972) had already been able to isolate a dopamine-stimulated adenylate cyclase from the striatum. The similarity of the sensitivity of this enzyme to drugs and the sensitivity of dopamine receptors in other test systems suggest that cyclic AMP is part of the receptor complex. In that case it might be expected that more of the cyclase could be isolated from the denervated striatum, or that the enzyme would be more sensitive to dopamine. The noradrenaline-sensitive adenylate cyclase of cerebral cortex responds in just such a manner after damage to the noradrenergic input to the cortex (Eccleston, 1973). Although Makman *et al.* (1975) report some preliminary results which suggest that the striatal adenylate cyclase activity is modified by denervation, Iversen (1975) comments that this result has not been obtained in either his own or other laboratories.

It seems likely that DA acts on cells in the striatum which have acetylcholine (ACh) as their transmitter. Drugs which inhibit or stimulate DA receptors have reciprocal effects on ACh release (Stadler *et al.*, 1973). Anticholinergic agents are useful in the treatment of Parkinson's disease. A recent report by Fibiger & Grewaal (1974) shows that although 6-OH-DA lesions do not alter ACh levels in the striatum the effect of apomorphine is exaggerated on the lesioned side, a finding which supports more directly the interpretation of the behavioural data as the result of supersensitivity.

Intraventricular injections, bilateral lesions or bilateral intracerebral injections of 6-OH-DA also produced clear behavioural effects in rats (Ungerstedt, 1971*a*, 1974; Iversen, 1974; Iversen & Creese, 1975). These treatments produce animals whose psychomotor response to amphetamine is abolished and whose response to apomorphine now follows lower doses. Iversen & Creese (1975) were able to show a supersensitivity to apomorphine in all of their preparations with reduced dopamine concentrations in the striatum. These experiments have the advantage of being able to show the occurrence of a normal response to the drugs but with a dose response curve shifted towards lower doses.

Drug-induced Supersensitivity

In earlier experiments Dominic & Moore (1969) showed that an increased response to adrenergic stimulating drugs followed long-term treatment with α-methyltyrosine, and in unilateral lesioned animals Ungerstedt (1971*a*) was able to show that reserpine in a single dose caused a prolonged supersensitivity to apomorphine. Although this was always less than the supersensitivity caused by denervation it points the way to experiments which may have even greater clinical relevance.

Studies by Moore & Thornburg (1975) after α-methyltyrosine, and by Ungerstedt *et al.* (1975) after reserpine and after spiroperidol, suggest that the prolonged hyperactivity which follows even a single dose of these drugs may reflect a prolonged change in the sensitivity of the striatal receptors for dopamine. These experiments suggest that the loss of dopaminergic transmission—even for a short period—may cause a near-permanent change in the behavioural sensitivity to dopamine or its analogues. Perhaps to the feedback control of dopamine synthesis (Andén *et al.*, 1970), which has a fairly short time course, we must add a control of the sensitivity to the transmitter which has a very long time scale.

In summary, there seems to be considerable behavioural, biochemical and physiological evidence that removal of dopaminergic transmission causes an increased responsiveness to dopamine receptor agonists. Whether this increase is due to a receptor supersensitivity which is the result of mechanisms similar to those acting in the periphery, or to changes in the postsynaptic cell systems which themselves lead to an increase sensitivity, must still remain an open question. So far, no experiments seem to rule out the possibility that dopamine normally acts to inhibit a system of cells whose function in the brain is evidenced only after removal of this inhibition, but it seems a much less attractive idea than the simple extension of the results of experiments on peripheral synapses to the central nervous system.

Acknowledgements

Many of the ideas in this paper originated in discussions with Dr. U. Ungerstedt over the five years since we first watched rats chase their tails. I am grateful for his continuing co-operation and to the European Training Programme in Brain and Behaviour Research for the Twinning grant which is financing it at present.

I am indebted to Mr. Michael Martin for technical assistance and Miss Lorna Guthrie for secretarial assistance.

References

Andén, N. E. (1966) in *Mechanisms of Release of Biogenic Amines* (von Euler, U. S., Rosell, S. & Uvnas, B., eds.), pp. 357–359, Pergamon Press, Oxford
Andén, N. E., Rubenson, A., Fuxe, K. & Hokfelt, T. (1967) *J. Pharm. Pharmacol.* **19**, 627–629
Andén, N. E., Butcher, S. G., Corrodi, H., Fuxe, K. & Ungerstedt, U. (1970) *Eur. J. Pharmacol.* **11**, 303–314
Arbuthnott, G. W. (1974) *J. Physiol. (London)* **239**, 121–122P
Arbuthnott, G. W. (1975) *Brit. J. Pharmacol.* in press.
Arbuthnott, G. W. & Crow, T. J. (1971) *Exp. Neurol.* **30**, 484–491
Arbuthnott, G. W. & Ungerstedt, U. (1969) *Acta Physiol. Scand.* **77** suppl. 330, p. 117
Arbuthnott, G. W. & Ungerstedt, U. (1975) *Exp. Neurol.* **47**, 162–172
Bloom, F. E., Costa, E. & Salmoiraghi, G. C. (1965) *J. Pharmacol. Exp. Ther.* **150**, 244–252
Cannon, W. B. & Rosenbleuth, A. (1949) *The Supersensitivity of Denervated Structures,* Macmillan, New York
Chase, T. N. & Shoulson, I. (1975) *Advan. Neurol.* **9**, 359–366
Christie, J., Ljungberg, T. & Ungerstedt, U. (1973) *J. Physiol. (London)* **234**, 80P
Claveria, L. E., Teychenne, P. F., Calne, D. B., Petrie, A. & Bassendine, M. F. (1975) *Advan. Neurol.* **9**, 393–397
Connor, J. D. (1970) *J. Physiol. (London)* **208**, 691–703
Corrodi, H., Fuxe, K. & Ungerstedt, U. (1971) *J. Pharm. Pharmacol.* **23**, 989–991
Corrodi, H., Farnebo, L. O., Fuxe, K., Gamberger, B. & Ungerstedt, U. (1972) *Eur. J. Pharmacol.* **20**, 195–204
Dominic, J. A. & Moore, K. D. (1969) *Psychopharmacologia,* **15**, 96–101
Eccleston, D. (1973) *Biochem. Soc. Spec. Publ.* **1**, 121–126
Ernst, A. M. (1967) *Psychopharmacologia* **10**, 316–323
Feltz, P. (1969) *J. Physiol. (London)* **205**, 8–9P
Feltz, P. & de Champlain, J. (1972*a*) *Brain Res.* **43**, 595–600
Feltz, P. & de Champlain, J. (1972*b*) *Brain Res.* **43**, 601–605

Feltz, P. & Wakenzie, J. S. (1969) *Brain Res.* **13**, 612–616
Fibiger, H. C. & Grewaal, D. S. (1974) *Life Sci.* **15**, 57–63
Frigyesi, T. L. & Purpura, D. P. (1967) *Brain. Res.* **6**, 440–456
Fuxe, K., Corrodi, H., Hökfelt, T., Lidbrink, P. & Ungerstedt, U. (1974) *Med. Biol.* **52**, 121–132
Herz, A. & Zieglgansberger, W. (1968) *Int. J. Neuropharmacol.* **7**, 221–230
Hökfelt, T. & Ungerstedt, U. (1973) *Brain Res.* **60**, 269–297.
Hull, C. D., Levine, M. S., Buchwald, N. A., Heller, A. & Browning, R. A. (1974) *Brain Res.* **73**, 241–262
Iversen, L. L. (1975) *Advan. Neurology* **9**, 415–418
Iversen, S. D. (1974) in *The Neurosciences Third Study Programme* (Schmitt, F. O. & Warden, F. G., eds.), pp. 705–711, MIT Press, Cambridge, Mass.
Iversen, S. D. & Creese, I. (1975) *Advan. Neurol.* **9**, 81–92
Kebabian, J. W., Petzold, G. L. & Greengard, P. (1972) *Proc. Nat. Acad. Sci. U.S.A.* **69**, 2145–2149
McDowell, F. H. & Sweet, R. (1975) *Advan. Neurol.* **9**, 367–372
McLennan, H. & York, D. H. (1967) *J. Physiol.* (*London*) **189**, 393–402
Makman, M. H., Mishra, R. K. & Brown, J. H. (1975) *Advan. Neurol.* **9**, 213–222
Moore, K. E. & Thornburg, J. E. (1975) *Advan. Neurol.* **9**, 93–104
Ohye, C., Bouchard, R., Boucher, R. & Poirier, L. J. (1970) *J. Pharmacol. Exp. Ther.* **175**, 700–708
Sharpless, S. K. (1964) *Annu. Rev. Physiol.* **26**, 357–388
Siggins, G. R., Hoffer, B. J. & Ungerstedt, U. (1974) *Life Sci.* **15**, 779–792
Stadler, H., Lloyd, K. E., Gadae-Ciria, M., Bartholini, G. (1973) *Brain Res.* **55**, 476–480
Stavraky, G. W. (1961) *Supersensitivity Following Lesions of the Nervous System*, Univ. of Toronto Press, Toronto
Ungerstedt, U. (1968) *Eur. J. Pharmacol.* **5**, 107–110
Ungerstedt, U. (1971*a*) *Acta Physiol. Scand. Suppl.* **367**, 69–93
Ungerstedt, U. (1971*b*) in *Monoamines and the Central Grey Nuclei* (de Ajuriaquerra, J., ed.), Proceedings of the Bel Air Symposium
Ungerstedt, U. (1974) in *The Neurosciences Third Study Programme* (Schmitt, F. O. & Warden, F. G., eds.), pp. 695–703, MIT Press, Cambridge, Mass.
Ungerstedt, U., Butcher, L. L. Butcher, S. G., Anden, N. E. & Fuxe, K. (1969) *Brain Res.* **14**, 461–471
Ungerstedt, U., Ljungberg, T., Hoffer, B. J. & Siggin,s G. R. (1975) *Advan. Neurol.* **8**, 57–65
Vakil, S. D., Calne, D. B., Reid, J. L. & Seymour, C. A. (1973) *Advan. Neurol.* **3**, 121–125
York, D. H. (1970) *Brain Res.* **20**, 233–249

Chapter 5

Biochemical Pharmacology of Dopamine Receptors in Mammalian Central Nervous System

By L. L. IVERSEN and R. J. MILLER

Medical Research Council Neurochemical Pharmacology Unit, Department of Pharmacology, University of Cambridge Medical School, Cambridge CB2 2QD, U.K.

Introduction

Studies of the detailed pharmacological properties of dopamine receptors in the central nervous system have proved difficult, owing to the absence of any simple model for such receptor sites in the peripheral nervous system. Recently, however, two useful model systems have emerged. One is based on the remarkable similarity between the CNS effects of dopaminergic drugs and their effects on the renal artery, a system which responds to dopamine and related agonists by vasodilation (Goldberg *et al.*, 1968; Goldberg, 1974). This has proved to be a most valuable peripheral model for studies of dopaminergic agonists and antagonists. The second model arose from the finding that low concentrations of dopamine stimulate the formation of cyclic AMP in bovine superior cervical ganglia (Greengard *et al.*, 1972), rat and bovine retina (Brown & Makman, 1972, 1973) and in cell-free homogenates of rat basal ganglia (Kebabian *et al.*, 1972) and other dopamine-rich areas of brain such as olfactory tubercle and nucleus accumbens (Horn *et al.*, 1974; Clement-Cormier *et al.*, 1974). The effects of dopamine in some of these systems are antagonized by neuroleptic drugs such as chlorpromazine and haloperidol, and are mimicked by apomorphine (Kebabian *et al.*, 1972; Brown & Makman, 1973). The dopamine-sensitive adenylate cyclase is absent from brain regions, such as cerebellum, which lack dopamine-containing nerve terminals, and it is not potently affected by α- or β-adrenoceptor agonists or antagonists. The enzyme appears to be located predominantly on postsynaptic cells in the striatum, since the dopamine-stimulated activity persists unchanged or is even increased after destruction of the nigrostriatal dopaminergic terminals (von Voightlander *et al.*, 1973; Mishra *et al.*, 1974). The existence of a dopamine-type adenylate cyclase is consistent with the hypothesis that many of the postsynaptic actions of catecholamines are mediated by cyclic AMP production in adrenergically innervated cells (Weiss & Costa, 1968; Greengard *et al.*, 1972; Hoffer *et al.*, 1972). In our own work we have used the dopamine-sensitive adenylate cyclase of the rat striatum as a biochemical model in an attempt to define some of the structural requirements for dopamine receptor agonists and antagonists (Miller & Iversen, 1974*a*, *b*; Miller *et al.*, 1974*a*, *b*, 1975*a*).

Actions of Agonist Drugs on Cyclic AMP Formation

Method

The method was as described by Kebabian *et al.* (1972). Rat striata or other regions were dissected according to the procedure described by Glowinski & Iversen (1966), and homogenized with a motor-driven teflon-glass homogenizer in approximately 25 volumes of 2 mM tris–maleate buffer pH 7.4 containing 2 mM EGTA. Aliquots of this homogenate were transferred to a solution containing 80 mM tris–maleate buffer pH 7.4 with 2 mM magnesium sulphate 0.2 mM EGTA and 10 mM theophylline, and various drugs. The incubation tubes were kept at 0°C while ATP was added to a final concentration of 0.5 mM and they were then incubated at 30°C for 2.5 min. The reaction was stopped by heating at 100°C and the contents of the tubes were then assayed for cyclic AMP using the protein-binding method of Brown *et al.* (1972).

Effects of dopamine and related β-phenylethylamines

Addition of dopamine to rat striatal homogenates increased cyclic AMP production by approximately 100% during a brief incubation *in vitro*. Half-maximum stimulation was produced by about 2 μM dopamine. Of the various other phenylethylamines tested (Table 1), the only one that was as active as

Table 1. *Stimulation of cyclic AMP formation by β-phenylethylamines*

Compound	Maximum stimulation* % of dopamine	EC_{50}–μMolar†
Dopamine (DA)	100	2.0
Epinine	100	1.5
N-DimethylDA	48	>1000.0
N-TrimethylDA	30	>1000.0
L-Noradrenaline	97	40.0
(±)-α-MethylDA	58	850.0

Compounds inactive at 10^{-3} M: *m*-tyramine; *p*-tyramine; 3,4-dihydroxyphenylbutylamine; 3,4-dihydroxyphenylpropylamine; 3,4-dihydroxyphenylbenzylamine; D-noradrenaline; 3-methoxy, 4-hydroxyphenylethylamine; 3 hydroxy, 4-methoxyphenylethylamine; L-dopa; (±)-amphetamine; amantadine.

* Maximum stimulation above basal cyclic AMP production is expressed as a percentage of that obtained with 100 μM dopamine, with 1 mM drug concentration.

† EC_{50} is the drug concentration needed to produce 50% of the maximum effect observed with 100 μM dopamine.

dopamine was the *N*-methyl analogue, epinine. The *N*-dimethyl and *N*-trimethyl analogues were less potent agonists. Compounds with 1, 3 or 4 carbon side chains, instead of the two carbon chain of dopamine, were completely inactive, as were non-catecholamines, such as *m*-tyramine, amphetamine and *O*-methoxylated dopamine metabolites (Table 1). (±)-α-Methyldopamine was at least 400 times less potent than the parent compound, and L-noradrenaline was 20 times less potent. D-Noradrenaline was inactive at concentrations up to 1 mM, as was L-dopa.

Naphthalene and tetrahydroisoquinoline derivatives

The compound 2-amino-6,7-dihydroxy-1,2,3,4-tetrahydronaphthalene (ADTN), in which the side chain of dopamine is incorporated into a second ring system in the fully extended form, was equipotent with dopamine. The non-catechol analogue, however, was completely inactive (Table 2). When the

Table 2. *Effect of β-naphthylamine analogues on rat striatal cyclic AMP production*

Compound	Maximum stimulation % of dopamine	EC_{50}—μMolar†
2-Amino-6,7-dihydroxy-(1,2,3,4)-tetrahydro-naphthalene	115	4.0
2-Amino-1,2,3,4-tetra-hydronaphthalene	—	—

† EC_{50} is the drug concentration needed to produce 50% of the maximum effect observed with 100 μM dopamine.

dopamine side chain was incorporated into a ring system in its non-extended form, as in 6,7-dihydroxytetrahydroisoquinoline, the compound was much less active than ADTN (Miller *et al.*, 1974*b*). Tetrahydropapaveroline and emetine were completely inactive.

Piribedil and its metabolite S-584

Piribedil is a non-catechol analogue of dopamine, and was found to be inactive in stimulating cyclic AMP formation in concentrations up to 0.1 mM. The catechol metabolite of this drug, however, S-584, was highly effective in the test system, and was approximately equipotent with dopamine (Miller & Iversen, 1974*b*). This finding suggests that the long-lasting dopamine-like effects of piribedil observed *in vivo* may be mediated through the production of active catechol metabolites such as S-584, which is known to be an important urinary metabolite of the parent drug in man and rat (Jenner *et al.*, 1973; Campbell *et al.*, 1974).

(—)-*Apomorphine and related compounds*

The action of (—)-apomorphine and related compounds has been examined (Kelly *et al.*, 1975). It has previously been reported (Kebabian *et al.*, 1972; Miller *et al.*, 1974*b*) that (—)-apomorphine stimulates striatal adenylate cyclase at low concentrations but is inhibitory at higher concentrations. Among ten other aporphine alkaloids tested only (—)-*N*-*n*-propyl-norapomorphine possessed a stimulatory activity comparable to that of (—)-apomorphine. Several of the compounds tested (10^{-5} M), however, had some ability to inhibit the stimulation of adenylate cyclase produced by dopamine. (±)-Bulbocapnine was the most potent compound in this respect, although the K_i value of $1{\cdot}6 \times 10^{-7}$ M indicates that bulbocapnine is not as potent a dopaminergic antagonist as some neuroleptic drugs, and in accordance with this it is somewhat less potent as a cataleptic agent (Costall & Naylor, 1973).

Topography of the dopamine receptor

Sheppard & Burghardt (1974) and Sheppard (1975) also tested a number of dopamine analogues on the striatal adenylate cyclase and have reported findings similar to those described here. It is of interest that in two other dopamine-sensitive preparations a similar spectrum of agonist activity has also been observed, namely in the renal artery of the dog (Goldberg *et al.*, 1968) and in neurons of the snail *Helix aspersa* (Woodruff & Walker, 1969). In all of these systems the structure–activity relations for agonists are quite distinct from those of classical α- or β-adrenoceptors. In each case epinine was found to be equipotent with dopamine. For agonists there is an absolute requirement for a catechol grouping and a two carbon side chain attached to an amino group. Among rigid analogues of dopamine, in which the side chain is incorporated into a second ring system, the compounds with the greatest potency are those in which the dopamine side chain is in the fully extended form, as in ADTN and apomorphine. This suggests that the fully extended *trans* form of the dopamine side chain is the preferred conformation of the molecule on interaction with the dopamine-sensitive adenylate cyclase (Rekker *et al.*, 1972; Cannon, 1974; Miller *et al.*, 1974*b*; Sheppard, 1975). This suggestion also supports the hypothesis of Horn & Snyder (1971) that the preferred confrontation of dopamine at its receptor site is the fully extended *trans* form, which can be superimposed on the X-ray crystallographic structure of chloropromazine, hence accounting possibly for the receptor blocking activity of the latter compound.

Neuroleptic Drugs and Other Antagonists of Dopamine-sensitive Cyclic AMP Formation in Striatal Homogenates

A number of neuroleptics and other drugs have been examined as antagonists of the dopamine-stimulated adenylate cyclase in rat striatal homogenates (Table 3). In these studies a constant concentration of 100 μM dopamine was used to ensure maximum stimulation. Assuming competitive antagonism as the prevalent mode of inhibition (Clement-Cormier *et al.*, 1974; Miller *et al.*, 1974*a*) it was possible to calculate K_i values for each compound from the graphically determined IC_{50} values (drug concentration producing 50% inhibition of dopamine-stimulated cyclic AMP formation) as described by Clement-Cormier *et al.* (1974) in their similar studies. The results reported by Clement-Cormier *et al.* (1974), by Brown & Makman (1973) and by Karobath & Leitich (1974) from similar experiments are largely in accordance with our own.

Among the phenothiazines tested there was good agreement between the potency of drugs as inhibitors of the dopamine-stimulated adenylate cyclase and their known *in vivo* potencies as neuroleptics. Thus, for example, among the phenothiazines the potent neuroleptics fluphenazine and trifluoperazine were most active, while promazine and promethazine, which have only weak neuroleptic activity, were several orders of magnitude less potent as antagonists of the dopamine-sensitive adenylate cyclase (Table 3). The 7-hydroxy and *N*-desmethyl metabolites of chloropromazine retained substantial inhibitory activity in the cyclase test system, while the *N*-oxide and sulphoxide were inactive

Table 3. *Inhibition of dopamine-stimulated adenylate cyclase by neuroleptic drugs*

Drug	Inhibition constant K_i (nM)*
α-Flupenthixol	1.0
αβ-Flupenthixol	3.5
Fluphenazine	4.3
(+)-Butaclamol	8.8
α-Clopenthixol	16.0
Trifluoperazine	19.0
α-Chlorprothixene	37.0
Chlorpromazine	48.0
Spiroperidol	95.0
Prochlorperazine	100.0
Thioridazine	130.0
Pimozide	140.0
(+)-Bulbocapnine	160.0
Chlorimipramine	420.0
β-Chlorprothixene	950.0

Compounds lacking neuroleptic activity and with K_i values > 1000 nM:† promazine, β-clopenthixol, morphine, β-flupenthixol, (−)-butaclamol, promethazine, benztropine, desipramine, ethopromazine, diethazine, mezapine, fenethazine, chlorpromazine sulphoxide, pyrathiazine, diphenhydramine, methdilazine, pentolamine, propanalol, prostaglandin E_1, DL-amphetamine, amantadine.

* K_i values calculated from IC_{50} values determined graphically. Values from Miller *et al.* (1974*a*).
† Data from Brown & Makman (1973), Clement-Cormier *et al.* (1974), Karobath & Leitich (1974), Miller *et al.* (1974*a*).

(Miller & Iversen, 1974*a*). These findings are in agreement with the activity of these compounds in accelerating dopamine turnover in the intact brain (Sedvall, 1975) (Table 4).

Table 4. *Inhibition of dopamine-sensitive adenylate cyclase by metabolites of chloropromazine*

Metabolite	K_i (nM)
Chlorpromazine	48
Desmethyl chlorpromazine	270
Bisdesmethyl chlorpromazine	510
7-Hydroxy chlorpromazine	600
Chlorpromazine nitroxide	2220
Chlorpromazine sulphoxide	7500

The thioxanthenes are an interesting group of neuroleptic drugs in which a double bond connects the side chain to the hererocyclic nucleus. Due to the presence of a 2-substituent, they therefore exhibit geometric *cis*/*trans* isomerism. The relative effects of the different isomeric forms of flupenthixol, clopenthixol and chlorprothixene were assessed. In each case the α-isomer was considerably more potent than the β-isomer in antagonizing the effects of dopamine on adenylate cyclase activity. This was particularly marked with α- and β-flupenthixol. The activity of flupenthixol in the mixed αβ-form used clinically appears to be entirely due to the α-isomer. These findings agree well with the reported neuropharmacological properties of these drugs. Moller-Nielsen *et al.* (1973)

showed that α-flupenthixol was very considerably more potent than β-flupenthixol in various animal tests for neuroleptic activity, and more potent than clopenthixol and chlorprothixene. It is known from X-ray (Dunitz *et al.*, 1964; Schaefer, 1967; Post *et al.*, 1974) and nuclear magnetic resonance (n.m.r.) analysis (Kaiser *et al.*, 1974) that the pharmacologically more active α-isomers of the thioxanthenes have the *cis* configuration, i.e. the 2-substituent and the amine side chain are on the same side of the double bond linking the side chain to the ring system.

The results obtained with the biochemical test system are consistent with the known structure–activity rules for neuroleptic activity in both phenothiazines and thioxanthenes, (Petersen & Moller-Nielsen, 1964; Zirkle & Kaiser, 1970). A 2-substituent is of critical importance, and the most potent compounds (fluphenazine, trifluoperazine and flupenthixol) are those having a CF_3 substituent in this position. The highest potency both in *in vivo* tests and in the present system was also found in compounds having a β-hydroxyethylpiperazinyl side chain (fluphenazine, flupenthixol).

Neuroleptic drugs of other chemical classes, such as clozapine, the butyrophenones spiroperidol and haloperidol and pimozide were all active in the present test system, although the potencies of the latter drugs in the biochemical experiments were low in view of the very high potencies reported for these compounds as dopaminergic antagonists in whole animal experiments and clinically. Thus pimozide and haloperidol, which are many times *more* potent than chlorpromazine *in vivo*, were two to three times *less* active than chlorpromazine in the *in vitro* tests. This appears to be the most serious discrepancy at the moment between results obtained using the dopamine-sensitive adenylate cyclase and those obtained from *in vitro* studies. This discrepancy cannot be explained, although there are various possible reasons for it which should be carefully examined before rejecting the hypothesis that neuroleptic potency correlates closely with antidopaminergic potency as measured by the present test system. The high *in vivo* potency of these drugs may be related to their differential distribution in the CNS after administration *in vivo*. In the case of pimozide a selective concentration of the drug occurs in the caudate nucleus after systemic administration (Soudijn & Van Wijngaarden, 1972) and this could account for the high potency of the drug *in vivo*.

That the effects of drugs as inhibitors of the dopamine-sensitive adenylate cyclase do have predictive value in assessing neuroleptic activity, however, is shown by more recent findings with a newly described neuroleptic drug, butaclamol (Lippmann *et al.*, 1975). This compound is unique among the neuroleptics in possessing asymmetric carbon atoms, and thus exhibiting stereoisomerism. In animal tests only the (+)-enantiomer possesses neuroleptic activity, and when the enantiomers were tested at concentrations up to 10 μM only the (+) form was active as an inhibitor of the dopamine-stimulated formation of cyclic AMP (Lippmann *et al.*, 1975; Miller *et al.*, 1975*b*). The (+)-enantiomer proved to be an effective competitive inhibitor and is one of the four most potent compounds examined so far (Table 3). The active and inactive enantiomers of butaclamol are clearly of considerable pharmacological interest as research tools for studies of dopaminergic systems in brain.

Anticholinergic Actions of Neuroleptic Drugs

Certain neuroleptic drugs act as antagonists at muscarinic cholinergic receptors and this property may contribute importantly to the neuropharmacological profiles of such drugs (Schelkunov, 1967; Andén & Bedard, 1971; Andén, 1972, 1973). Miller & Hiley (1974) and Snyder *et al.* (1974) tested various neuroleptic drugs as muscarinic antagonists, using recently developed assays *in vitro* which depend on measurements of the binding of radioactively labelled muscarinic receptor ligands. In the studies of Miller & Hiley (1974) the atropipe-sensitive component of the binding of [^{3}H]-*N*-propyl-benzylcholine mustard to membrane fragments in homogenates of rat cerebral cortex was used as the test system (Burgen *et al.,* 1974). The results with neuroleptic drugs and related substances (Table 5) showed that neuroleptics exhibit a wide range of antimuscarinic potencies. Very similar results were reported by Snyder *et al.* (1974). The most potent compounds were thioridazine and clozapine, followed by chlorpromazine, whereas flupenthixol, trifluoperazine, pimozide and spiroperidol were relatively weak anticholinergics. If one expresses the results in terms of the ratio of anticholinergic to antidopaminergic potencies, the neuroleptics span a wide range from those such as thioridazine and clozapine which are considerably more potent as anticholinergics than as antidopaminergics, to flupenthixol and spiroperidol which show the converse properties. We believe that these results

Table 5. *Relative potencies of drugs as muscarinic or dopamine antagonists*

Compound	Dissociation constant for binding to receptor sites (nM)		Ratio of cholinergic: dopaminergic potency
	Muscarinic	Dopaminergic	
Atropine	0.5	—	
Benztropine	1.3	—	
Ethopromazine	10.0	—	
Thioridazine	25.0	130	5.2
Clozapine	55.0	170	3.1
Pimozide	160.0	140	0.87
Chlorpromazine	350.0	48	0.14
α-Flupenthixol	2200.0	1	0.0005
Trifluoperazine	4000.0	19	0.005
Spiroperidol	12000.0	95	0.008

may have an important bearing on the neuropharmacological properties of individual neuroleptic drugs. In particular the low incidence of Parkinson-like effects seen with drugs such as clozapine and thioridazine (Cole & Clyde, 1961; Shader and Di Mascio, 1970; Angst *et al*., 1971) could be related to the fact that such compounds carry a built-in anticholinergic activity, which may prevent them from manifesting extrapyramidal side effects as a consequence of their antidopamine effects (Schelkunov, 1967). On the other hand, potent antidopamine agents which are weak anticholinergics, such as spiroperidol, haloperidol, flupenthixol and fluphenazine, are known to induce extrapyramidal side effects much more frequently. The anticholinergic properties of clozapine and thioridazine may also obscure the effects of these compounds as dopamine antagonists in many of the animal behavioural and biochemical tests used to assess neuroleptic agents. For example, thioridazine and clozapine do not antagonize the behavioural effects of amphetamine, either in intact animals or animals with unilateral lesions of the nigrostriatal pathways. This has led some authors to suggest that these drugs lack dopamine antagonist properties, and that they are thus exceptions to the 'dopamine antagonist' hypothesis of neuroleptic activity (Stille & Hippius, 1971; Crow & Gilbe, 1973). Miller & Sahakian (1974), however, were able to show that clozapine and thioridazine can act as antagonists of the locomotor stimulation induced by amphetamine if tested in 11-day-old rats; in such animals the cholinergic systems of the basal ganglia are not fully developed (McGeer *et al.*, 1971) so that the anticholinergic properties of the drugs do not obscure their antidopamine properties. Thioridazine and clozapine clearly do have antidopamine properties when assessed in an *in vitro* biochemical test system, and it seems more parsimonious to propose that their anomalous behaviour in standard *in vivo* tests for antidopaminergic properties points to weaknesses in the design of such tests, rather than to any weakness in the hypothesis concerning their mode of action as neuroleptics.

Effects of Intracerebral Injection of Cholera Toxin

The enterotoxin from *Vibro cholerae* (choleragen) has been shown to have the property of activating the enzyme adenylate cyclase in all cell systems in which it has been tested, and in consequence it has been shown to activate cyclic AMP mediated processes in all these systems (Finkelstein, 1973). Progress has been made in identifying the receptor for the toxin, which is probably the GM_1 ganglioside on the cell surface (Cuatrecasas, 1973). After binding of the toxin to the cell surface there is a characteristic lag period lasting several hours before adenylate cyclase activation occurs. In order to examine the hypothesis that dopaminergic transmission in the CNS is mediated by cyclic AMP we have attempted to mimic some behavioural effects of dopamine receptor stimulation with cholera toxin.

It is known that bilateral injections of dopamine into the nucleus accumbens area in the rat brain causes a transient stimulation of locomotor activity (Pijnenburg & Van Rossum, 1973). Miller & Kelly (1975) injected choleragen (1 μg) bilaterally into the nucleus accumbens of rat and observed its effect on locomotor activity. During the first two hours after the injection rats treated with vehicle or toxin showed no difference in their locomotor activity. During the third and fourth hour the toxin-treated animals showed increased locomotor activity. This stimulaion became even more pronounced over the subsequent time period, and at 12 hours toxin-treated animals showed an approximately tenfold increase in locomotor activity. This is comparable to the stimulation seen on injecting dopamine into the same area. However, in contrast to the effects of dopamine the effects of choleragen were very long lasting. Increased locomotor activity was still observed in the toxin-treated animals up to 12 days after the injection. Assay of the adenylate cyclase activity in the nucleus accumbens area of toxin-treated rats showed that there was also a lag period before this became activated. There was no significant activation at 90 min but there was an approximately fourfold increase after 5 hours. The enzyme was even further increased at 22 and 48 hours. These preliminary findings suggest that choleragen may be a useful tool for investigation of cyclic AMP mediated processes in the central nervous system.

Discussion

B. E. Leonard: It is suggested that the dopamine-sensitive adenylate cyclase of the striatum and the limbic cortex respond to the agonists in the same way. Clozapine and the butyrophenones are not as effective in inhibiting the striatal dopamine adenylate cyclase as would be expected from their clinical effects. Could it be that the drugs have a different effect on the dopamine-sensitive adenylate cyclase in the limbic area from that on striatal adenylate cyclase? Regarding clozapine, there is evidence that it does not affect dopamine metabolism in the striatum in the same way as the phenothiazine type of neuroleptic drugs.

L. L. Iversen: This is a very good suggestion. We have made some tentative starts in that direction, although we really haven't done very much yet. We have been able to show that in nucleus accumbens and olfactory tubercle homogenates, the dopamine-sensitive adenylate cyclase certainly does exist, and it does appear to have pharmacological properties very similar to the system in striatum. Greengard and his collaborators at Yale have done a more systematic study of quite a number of neuroleptic drugs, and measured the actual K_i values in the olfactory tubercle and striatal homogenates, and the answers for the drugs that they tested appear to be identical [Kebabian, J. W. *et al.* (1972) *Proc. Nat. Acad. Sci. U.S.A.* **69**, 2145–2149]. So at the moment there is no real evidence pharmacologically or biochemically that dopamine receptors in the limbic forebrain or the striatum are different. We would prefer to rationalize the effects of clozapine, or the lack of effects of clozapine, on the striatal dopamine turnover in a different way. We would say that there is another property of the drug, namely, that it is a powerful anticholinergic. The anticholinergic property would tend to cancel out the antidopamine property, especially in the striatum, where we know this mutual antagonism exists, and perhaps not in the limbic forebrain system. But that is just one way of looking at it.

S. R. Nahorski: It was perhaps a bit worrying to note that recent observations [Blumberg *et al.* (1975) *J. Pharm. Pharmacol.* **27**, 128] have shown that the so-called specific dopamine antagonist pimozide is a very potent antagonist of noradrenaline-stimulated cyclic AMP formation in rat limbic forebrain slices. Have you looked at the effect of pimozide on noradrenaline-sensitive adenylate cyclase in the striatum or limbic brain areas?

L. L. Iversen: We don't have the answers to this question, but these data are indeed very interesting. These workers have found that haloperidol and pimozide block noradrenaline-stimulated adenylate cyclase in the forebrain slices at very low drug concentrations and we are attempting to reproduce their findings, but I can't tell you the answer.

References

Andén, N. E. (1972) *J. Pharm. Pharmacol.* **24**, 905–906
Andén, N. E. (1973) *J. Pharm. Pharmacol.* **25**, 346–348
Andén, N. E. & Bedard, P. (1971) *J. Pharm. Pharmacol.* **23**, 460–462
Angst, J., Bente, D., Berner, P., Heiman, H., Helmchen, H. & Hippius, H. (1971) *Pharmakopsychiat. Neuro-Psychophamarkol.* **4**, 201–211
Brown, B. L., Ekins, R. D. & Albano, J. D. M. (1972) in *Advances in Cyclic Nucleotide Research* (Greengard, P. & Robinson, G. A., eds.), vol. 2, pp. 25–40, Raven Press, New York
Brown, J. H. & Makman, M. H. (1972) *Proc. Nat. Acad. Sci. U.S.A.* **69**, 539–543
Brown, J. H. & Makman, M. H. (1973) *J. Neurochem.* **21** 477–479
Burgen, A. S. V., Hiley, C. R. & Young, J. M. (1974) *Brit. J. Pharmacol.* **51**, 279–285
Campbell, D. B., Jenner, P. & Taylor, A. R. (1974) *Advan. Neurol.* **3**, 199–213
Cannon, J. G. (1974) *Advan. Neurol.* **9**, 177–184
Clement-Cormier, Y. C., Kebabian, J. W., Petzold, G. L. & Greengard, P. (1974) *Proc. Nat. Acad. Sci. U.S.A.* **71**, 1113–1117
Cole, J. O. & Clyde, D. J. (1961) *Rev. Can. Biol.* **10**, 565–574
Costall, B. & Naylor, R. J. (1973) *Psychopharmacologia* **32**, 161–170
Crow, T. J. & Gillbe, C. (1973) *Nature* (*London*) **245**, 27–28
Cuatrecasas, P. (1973) *Biochemistry* (*N.Y.*) **12**, 3547–3558
Dunitz, J. D., Eser, H. & Strickler, P. (1964) *Helv. Chim. Acta* **47**, 1897–1902
Finkelstein, R. A. (1973) *C.R.C. Crit. Rev. Microbiol.* **2**, 553–623
Glowinski, J. & Iversen, L. L. (1966) *J. Neurochem.* **13**, 655–669
Goldberg, L. I. (1974) *Advan. Neurol.* **9**, 53–56
Goldberg, L. I., Sonneville, P. F. & McNay, J. L. (1968) *J. Pharmacol. Exp. Therm.* **163**, 188–197
Greengard, P., McAfee, D. A. & Kebabian, J. W. (1972) in *Advances in Cyclic Nucleotide Research* (Greengard, P. & Robinson, G. A., eds.), vol. 1, pp. 337–357, Raven Press, New York
Hoffer, B. J., Siggins, G. R., Oliver, A. P. & Bloom, F. E. (1972) in *Advances in Cyclic Nucleotide Research* (Greengard, P. & Robinson, G. A., eds.), pp. 411–423, Raven Press, New York
Horn, A. S. & Snyder, S. H. (1971) *Proc. Nat. Acad. Sci. U.S.A.* **68**, 2325–2328
Horn, A. S., Cuello, A. C. & Miller, R. J. (1974) *J. Neurochem.* **22**, 265–270
Jenner, P., Taylor, A. R. & Campbell, D. B. (1973) *J. Pharm. Pharmacol.*. **25**, 749–750
Kaiser, C., Warren, R. J. & Zirkle, C. L. (1974) *J. Med. Chem.* **17**, 131–133
Karobath, M. & Leitich, H. (1974) *Proc. Nat. Acad. Sci. U.S.A.* **71**, 2915–2918
Kebabian, J. W., Petzold, G. L. & Greengard, P. (1972) *Proc. Nat. Acad. Sci. U.S.A.* **69**, 2145–2149
Kelly, P. H., Miller, R. J. & Neumeyer, J. L. (1975) *Brit. J. Pharmacol.* **54**, 271P
Lippmann, W., Pugsley, T. & Merker, J. (1975) *Life Sci.* **16**, 213–224
McGeer, E. C., Fibiger, H. C. & Wickson, V. (1971) *Brain Res.* **32**, 433–440
Miller, R. J. & Iversen, L. L. (1974*a*) *J. Pharm. Pharmacol.* **26**, 142–144
Miller, R. J. & Iversen, L. L. (1974*b*) *Naunyn-Schmiedebergs Arch. Exp. Pathol. Pharmakol.* **282**, 213–216
Miller, R. J., Horn, A. S. & Iversen, L. L. (1974*a*) *Mol. Pharmacol.* **10**, 759–766
Miller. R. J.. Horn, A. S., Iversen, L. L. & Pinder, R. M. (1974*b*) *Nature* (*London*) **259**, 238–241
Miller, R. J., Horn, A. S. & Iversen, L. L. (1975*a*) *in Proc. Congr. Coll. Int. Neuropsychopharmacol. 9th,* Excerpta Medical Congress Series.
Miller, R. J., Horn, A. S. & Iversen, L. L. (1975*b*) *J. Pharm. Pharmacol.* **27**, 212–213
Miller, R. J. & Hiley, C. R. (1974) *Nature* (*London*) **248**, 596–597
Miller, R. J. & Kelly, P. (1975) *Nature* (*London*) **255**, 163–167
Miller, R. J. & Sahakian, B. J. (1974) *Brain Res.* **81**, 387–392
Mishra, R. K., Gardner, E. L., Katzman, R. & Makman, M. H. (1974) *Proc. Nat. Acad. Sci. U.S.A.* **71**, 3883–3887
Moller-Nielsen, I., Pedersen, V., Nymark, M., Franc, K. F., Boek, U., Fjallan, B. & Christiansen, A. V. (1973) *Acta Pharmacol. Toxicol.* **33**, 353–362
Petersen, P. V. & Moller-Nielsen, I. (1964) in *Psychopharmacological Agents* (Gordon, M., ed.), vol. 1, pp. 301–324, Academic Press, New York & London
Pijnenburg, A. J. J. & Van Rossum, J. M. (1973) *J. Pharm. Pharmacol.* **25**, 1003–1005
Post, M. L., Kennard, O. & Horn, A. S. (1974) *Acta Crystal.* **B30**, 1644–1646
Rekker, R. F., Engel, D. J. C. & Nys, G. G. (1972) *J. Pharm. Pharmacol.* **24**, 589–591
Schaefer, J. P. (1967) *Chem. Commun.* 743–744
Schelkunov, E. L. (1967) *Activitas nervosa superior* **9**, 207–217

Sedvall, G. (1975) in *Handbook of Psychopharmacology* (Iversen, L. L., Iversen, S. & Snyder, S., eds.), vol. 5, Plenum, New York
Shader, R. I. and Di Maschio, A. (1970) *Psychotropic Drug Side Effects,* Williams and Wilkins, Baltimore
Sheppard, H. (1975) *in Proc. Congr. Coll. Int. Neuropsychopharmacol.* 9th, Excerpta Medical Congress Series (in press)
Sheppard, H. & Burghardt, C. R. (1974) *Res. Commun. Chem. Pathol. Pharmacol.* **8**, 527–534
Snyder, S. H. Greenberg, D. & Yamamura, H. (1974) *Arch. Gen. Psychiat.* **31**, 58–61
Soudijn, W. & van Wijngaarden, I. (1972) *J. Pharm. Pharmacol.* **24**, 773–780
Stille, G. & Hippius, H. (1971) *Neuropsychopharmakologie* **4**, 182–191
Voigtlander, P. F. von, Boukma, S. J. & Johnson, G. A. (1973) *Neuropharmacology* **12**, 1081–1086
Weiss, R. & Costa, E. (1968) *J. Pharmacol. Exp. Ther.* **161** 310–319
Woodruff, G. N. & Walker, R. J. (1969) *Int. J. Neuropharmacol.* **8**, 279–286
Zirkle, C. L. & Kaiser, C. (1970) in *Medicinal Chemistry* (Burger, A., ed.), vol. 2, pp. 1420–1469, Wiley Interscience, New York

Chapter 6

Clinical Aspects of Tardive Dyskinesia

By J. D. PARKES

University Department of Neurology, Institute of Psychiatry and King's College Hospital, London SE5 8AF, U.K.

Introduction

Antipsychotic drugs are a unique advance in psychiatry and have revolutionized the treatment of schizophrenia. Two-hundred and fifty million prescriptions for these drugs, antidepressants and minor tranquillizers, are issued yearly in the United States where 20% of the population (twice as many women as men) use them on one or more occasion each year. The pattern of drug usage is probably similar in Great Britain. As compared to previously used sedative and hypnotic drugs, these antipsychotic agents have proved remarkably safe compounds, the incidence of side effects in one large patient group being less than 3% (National Institute of Mental Health, 1964). Nevertheless, certain of these drugs do occasionally cause blood dyscrasias and jaundice, as well as less dangerous effects which include growth changes, galactorrhoea, amenorrhoea, skin pigmentation, somnolence or alternatively delirium, and various motor disorders including Parkinsonism, acute dystonia, akathisia and tardive dyskinesia (Shepherd *et al.*, 1972).

These hormonal and movement disorders may result from the action of antipsychotic drugs on monoamine systems in the brain. Most of these side effects occur early during treatment and are reversed when this is stopped, although tardive dyskinesias are late in onset and may be permanent. The important distinction between the early onset of Parkinsonism and the late appearance of tardive dyskinesias, and between the reversible nature of the first and permanent character of the second, has not been satisfactorily accounted for. As with tardive dyskinesias, pigmentation of the eye and skin as a result of phenothiazine treatment may be late in onset and irreversible.

Extrapyramidal Drug Reactions

Movement disorders due to antipsychotic drugs have been comprehensively reviewed by Marsden *et al.* (1975). Drug-induced Parkinsonism was first described shortly after the introduction of reserpine in the 1930s. This is in many respects similar to idiopathic Parkinson's disease with akinesia in both conditions, although a rhythmic resting tremor, which is often the first symptom of the naturally occurring disease, is less common in the drug-induced variety. The condition is common, as shown by the finding of Dynes (1968) that over half of 500 patients on major tranquillizers had some symptoms of Parkinsonism, although there is a marked variation in susceptibility of patients.

Parkinsonism first appears comparatively early during treatment: in 50–75% of patients within 4 weeks, and in 90% within 3 months of starting phenothiazine, butyrophenone, or reserpine treatment in conventional dosage. There is a considerable difference in the antipsychotic potency of these different drugs and their ability to cause Parkinsonism. In two controlled trials involving trifluperidol, trifluoperazine and chlorpromazine, no relationship was found between the degree of improvement of the mental state and the severity of extrapyramidal symptoms (Tétreault *et al.*, 1968) and Shepherd *et al.* (1972) considered that the therapeutic potency of antipsychotic drugs and their ability to cause Parkinsonism were not related. However, the two may be associated with some but not all drugs: thus piperazine derivatives (e.g. trifluoperazine) have a high therapeutic potency and frequently cause Parkinsonism, while piperidine derivatives (e.g. thioridizine) have an equal therapeutic action but rarely cause Parkinsonism (Marsden *et al.*, 1975). Signs of Parkinsonism can occasionally persist for up to a year after stopping treatment, although they usually disappear within a few weeks or months. Many different anti-Parkinsonian drugs seem to have little or no effect in controlling drug-induced Parkinsonism.

Acute dystonic reactions are sudden in onset and can occur within hours of taking a single dose of antipsychotic drug. The muscles of the face, mouth and neck are most frequently affected and those of the trunk and limbs less often. Depending on which muscle groups are involved, dysarthria, mutism, trismus, dysphagia, torticollis, dystonic postures of hands and feet, or respiratory distress may occur. Blepharospasm and oculogyric crises are not uncommon although less frequent than lower facial dystonia. These acute dystonic reactions occur more often in young people than in older subjects, and are frequently due to piperazine phenothiazines or butyrophenones. They are not caused by reserpine. Dystonia is rapidly abolished by many different drugs including anticholinergics, barbiturates, diazepam, apomorphine and methylphenidate.

Restlessness, often accompanied by a subjective feeling of motor discomfort, is a common side effect of antipsychotic drugs. An extrapyramidal origin for this symptom has not been established. All classes of drugs discussed will cause akathisia, which often first occurs within a week of starting phenothiazines and may subsequently slightly increase in severity. Similar symptoms occur in a minority of untreated patients with Parkinsonism, and may be made worse and not better by levodopa treatment. Drug-induced Parkinsonism, akathisia and acute dyskinesias may occur simultaneously and result in different combinations of movement disorders.

Tardive dyskinesia

Ey *et al.* (1956) described the persistence of involuntary movements of the lower face after stopping chlorpromazine, and subsequently Uhrbrand & Faurbye (1960) reported the delayed appearance of this dyskinesia after starting antipsychotic drug treatment. This condition of chronic or tardive dyskinesia is a relatively well-defined clinical entity although occasionally it is confused with acute drug reactions. Tardive dyskinesia frequency does not appear until after one to two years antipsychotic drug treatment, and paradoxically may first appear after dosage reduction or drug withdrawal (Degwitz *et al.*, 1967; Crane, 1968).

It is difficult to assess the incidence of tardive dyskinesia with any accuracy: in a study of 750 patients on chronic major tranquillizer treatment, 12.3% of the males and 20.2% of the females had some involuntary movements, usually of the face or mouth (Heinrich *et al.*, 1968). Similar movements are present in between 5–30% of geriatric patients, many of whom have diffuse brain diseases, in long-stay hospitals (Brandon *et al.*, 1971) and in between 1–2% of younger general medical patients, without brain disease, on long-term phenothiazine treatment. The condition also occurs spontaneously in up to 2% of an elderly population never treated with neuroleptics (Heinrich *et al.*, 1968). It is probably more readily provoked by drugs in patients with pre-existing organic brain disease than in those without, although the data is conflicting (Greenblatt *et al.*, 1968; Edwards, 1970). It has been suggested that tardive dyskinesia is especially common in females and East European Jews (Simpson, 1973) as well as in patients over 40, but these facts are difficult to substantiate. It is not known in what proportion of patients the condition is permanent, and Kline (1968) has questioned whether most of these dyskinesias are irreversible. However, Shepherd *et al.* (1972) conclude that the balance of evidence is in favour of this view.

The first sign of tardive dyskinesia to appear is often a vermicular movement of the tongue, followed by repetitive movements of the lips, jaw and cheeks, which result in constant sucking, licking, chewing, blowing and tonguing. There is little or no regular rhythm to these movements which almost always affect the lower face more than the upper, although the periorbital muscles may be involved. Many patients also have choreic movements of the extremities and swaying or rocking of the trunk, torsion of the axial skeleton and particularly the head and neck, legs and disturbances of respiratory rhythms. When extreme, the appearance is characteristic and bizarre although surprisingly tardive dyskinesia may cause little distress to the patient. In some but not all patients there are signs of Parkinsonism, and tremor, not always identical to the rhythmic rest tremor of this condition, occurs in between 15–50% of cases (Crane, 1973). As with most involuntary movements, those of tardive dyskinesia are increased by tension and disappear during sleep.

A syndrome resembling tardive dyskinesia occurs in children on antipsychotic drug treatment (American College of Neuropsychopharmacology Report, 1973). In contrast to adults, mouth and jaw movements are less prominent in children, who, however, develop marked chorea of the limbs. As with the adult syndrome, movements may only first appear when treatment is stopped. In none of these children was the condition permanent. The foetus may also be susceptible to phenothiazines; tremor, muscle spasms and failure to thrive have been described in two neonates after maternal ingestion of phenothiazines in pregnancy (Tamer *et al.*, 1969).

Drugs causing tardive dyskinesia

No effective antipsychotic drug in common use has not been implicated as causing tardive dyskinesia (Schiele *et al.*, 1973), although this is most commonly due to long-term phenothiazine or butyrophenone treatment and is extremely rare with reserpine. Marsden *et al.* (1975) reported only three cases of reserpine being associated with chronic dyskinesias and found no evidence that this drug ever produces acute dyskinesias. With the exception of reserpine, the drugs which regularly result in Parkinsonism are the same as those which after prolonged use cause tardive dyskinesia.

Piperazine derivatives may be especially liable to cause tardive dyskinesia (Uhrbrand & Faurbye, 1960), but many different phenothiazine compounds including the commonly used chlorpromazine and trifluoperazine, the thioxanthenes thiothixene and chlorprothixene, and haloperidol and droperidol may be responsible. Paulson (1968) considered that patients who develop tardive dyskinesia had invariably received large amounts of medication (commonly over 1000 g of chlorpromazine, or equivalent drug). However, Trurek *et al.* (1972) found there was little relationship between the dosage of antipsychotic drug and the severity of tardive dyskinesia, and Briones (1973) described a woman of 46 who developed irreversible orofacial dyskinesia after only 300 mg chlorpromazine daily for 2 months. The ultimate sojourn of phenothiazines in the body is exceedingly long, and various metabolites and even free chlorpromazine can be detected in the urine of hospitalized mental patients for 6–18 months after stopping treatment (Goodman & Gilman, 1970).

Levodopa and, much less commonly, other anti-Parkinsonian drugs cause involuntary movements in patients with Parkinson's disease very similar to the lingual, facial, buccal and choreic movements of patients on long-term phenothiazines. In contrast to tardive dyskinesia, levodopa-induced movements appear earlier during treatment (within days of starting combined therapy with dopa decarboxylase inhibitors) and are almost always abolished on reduction of dosage or levodopa withdrawal. As is the case with tardive dyskinesia, perioral movements are often the first to appear followed by chorea, and also the older patients and those with the longest standing disease may be especially liable to levodopa dyskinesias (Mones *et al.*, 1971). Other anti-Parkinsonian drugs may rarely cause orofacial dyskinesia or increase the severity of levodopa dyskinesias: amantadine (Pearce, 1971); amphetamines (Tarsy *et al.*, 1975) and the dopamine agonist, bromocriptine (Calne *et al.*, 1974). Anticholinergic drugs given as a single treatment do not cause acute or chronic orofacial dyskinesias, although toxic doses of hyoscine can result in chorea.

α-Methyldopa treatment is occasionally associated with extrapyramidal symptoms, and Yamadori & Albert (1972) reported acute orofacial dyskinesia with chorea in a hypertensive patient with cerebrovascular disease treated with this drug. These symptoms were reversed on drug withdrawal and although acute dystonic, choreic and other movements are occasionally due to a wide variety of other drugs, the nature, time of onset and duration of these involuntary movements are distinct from tardive dyskinesia. Thus the myoclonus, chorea, and choreathetosis that may result from treatment with oral contraceptives, isoniazid and perhaps lithium salts (Hamilton & Mahapatra, 1972), and choreic movements of the face and limbs as well as myoclonus in a 2-year-old boy caused by imipramine poisoning (Burks *et al.*, 1973) have a different clinical presentation from tardive dyskinesia. However, tardive as well as permanent dyskinesias have been described in three young drug addicts taking narcotics and marihuana. Involuntary movements involved the upper rather than the lower face as well as the limbs (Marshall, 1972). It is notoriously difficult to obtain full drug histories from many subjects, and these patients may conceivably have taken many different drugs.

Differential Diagnosis

There is an impressive number of conditions which may be accompanied by orofacial and choreic limb movements, and Greenhouse (1966) includes, as well as Huntington's chorea, Addison's disease, ataxia telangiectasia, beri-beri, cerebrovascular disease, hypocalcaemia, hypoglycaemia, hypomagnesaemia, hypoparathyroidism, kernicterus, measles and mumps encephalitis, neurosyphilis, senile chorea, Pick's disease and thalassaemia. The movement disorder in most of these conditions has little resemblance to that of tardive dyskinesia, which Baker (1969) and Crane (1973) considered was a specific and unique clinical entity, the predominance of lower facial movements not occurring in other known diseases of the nervous system. Despite this generalization, in some patients the involuntary movements may be indistinguishable from those of Huntington's chorea, encephalitis lethargica, Wilson's disease or torsion dystonia (Hunter *et al.,* 1964). Older subjects with tardive dyskinesia may have more orofacial and less limb choreic movements than younger subjects, and age may determine the topographical localization of dyskinesia (Marsden *et al.,* 1975). However, the relative involvement of different parts of the body in Huntington's chorea does not change with age, although the severity of choreic movements may increase.

The tics, stereotopies and mannerisms which occur in schizophrenia are different from the facial, limb and trunk movements of tardive dyskinesia. The highly repetitive movements of the psychiatric disorder seem behaviourally more complex and often resemble normal activity. However, amphetamines can apparently cause both kinds of movement disorder, although possibly in different groups. Amphetamine addicts occasionally develop seemingly compulsive tearing, picking, and 'pounding' behaviour (Rylander, 1972). As an example of this, subjects may repetitively dismantle and then repair motors, or behave like Penelope at the loom. In hyperkinetic children, some with slight degrees of brain damage, and in patients with narcolepsy on amphetamines for long periods, acute (and not tardive or permanent) choreic and orofacial movements occasionally result.

Treatment of Tardive Dyskinesia

Tardive dyskinesia is made worse and not better by the anti-Parkinsonian drugs levodopa and amantadine. In 12 of 40 patients studied by Hippius & Logeman (1970) the severity of involuntary movements increased following 100 mg levodopa i.v., and Klawans & McKendall (1971) described a striking deterioration in a 73-year-old woman with chlorpromazine-induced orofacial dyskinesia who was given oral levodopa. Anticholinergic drugs are customarily given with many antipsychotic drugs to protect against the emergence of Parkinsonism although they may possibly potentiate the occurrence of tardive dyskinesia (Klawans 1973). Thus the anticholinergic drug trihexiphenidyl worsened and the centrally acting anticholinesterase physostigmine improved dyskinesia in the patient reported by Klawans & McKendall (1971). However, Tarsy *et al.* (1974) showed that physostigmine caused no significant change in most patients with tardive dyskinesia.

The results of manipulating brain serotonin have also been investigated in patients with tardive dyskinesia, on the whole with little change although serotonin precursors may increase the severity of involuntary movements (Chase *et al.*, 1972).

In contrast to anti-Parkinsonian drugs, phenothiazines and haloperidol suppress the movements of tardive dyskinesia as is also the case with reserpine, tetrabenazine and α-methyl-*p*-tyrosine. In some, but not all, patients improvement in involuntary movements occurs at the expense of increasing Parkinsonism; however, orofacial dyskinesia may improve before any increase in hypokinesia is obvious. If allowed by the patient's psychotic condition, drug withdrawal is the preferred treatment; in those whose psychosis worsens, the choice between two evils is not clear. In patients in whom orofacial dyskinesia is masked by the dosage of antipsychotic drugs taken and only appears on drug withdrawal, replacement in full dosage may be the only satisfactory treatment.

Aetiology of Tardive Dyskinesia

The permanent nature of tardive dyskinesia may be due to drug-induced structural changes in the central nervous system although these have not been convincingly demonstrated in man. Hunter *et al.* (1968), after expert and very detailed examination of the brain, could find no significant neuropathological change in three patients with tardive dyskinesia. Neuronal damage was found by Christensen *et al.* (1970) in the brain of 27 of 28 adults with chronic dyskinesias, taking antipsychotic drugs. However, similar changes with gliosis of the substantia nigra were also present in five controls, and it is possible that neuronal or glial changes in old age or disease may increase susceptibility to tardive dyskinesia rather than its being the direct result of antipsychotic drug treatment.

A pigment related to melanin is produced in the exposed skin, cornea and lens after relatively high doses of phenothiazines, and chronic epithelial changes in the cornea have been reported in schizophrenics on chlorpromazine. As with tardive dyskinesia these changes may be permanent despite stopping treatment (Johnson & Buffaloe, 1968). In an important review, Shepherd *et al.* (1972) discuss the mechanism of ocular pigmentation. The total drug ingested, duration of treatment, hormonal and racial factors are all important. Pigmentation may be related to the ability of chlorpromazine, when exposed to ultraviolet irradiation, to form free radicals and then combine with melanin. Chlorpromazine pigmentation apparently does not occur in albino humans (Howard *et al.*, 1969). The relevance of these changes to possible effects of phenothiazines on neuromelanin is doubtful; in one series of 97 patients no correlation was found between ocular pigmentation, and extrapyramidal symptoms including dyskinesia (Wheeler *et al.*, 1968).

Differences in cerebral metabolic responses to antipsychotic drugs may account for the appearance of extrapyramidal disorders in some patients but not in others. Chase *et al.* (1970) found that patients on antipsychotic medication with extrapyramidal symptoms (Parkinsonism, tardive dyskinesia, or a combination of the two) had significantly lower levels of homovanillic acid (HVA) in the cerebrospinal fluid than patients who did not have movement disorders. There is normally an increase in HVA concentration in the cerebrospinal fluid after phenothiazine or butyrophenone treatment, and it was suggested that this compensatory increase in monoamine formation in response to these antipsychotic drugs was impaired in some subjects, who were more liable to extrapyramidal drug reactions. However, Pind & Faurbye (1970) found the cerebrospinal fluid HVA concentration after probenecid block was normal in eight patients with tardive dyskinesia.

The drugs which cause tardive dyskinesia have the pharmacological properties of blocking dopamine receptors, and chronic blockade may eventually cause irreversible receptor damage. The clinical findings indicate there is a state of relative dopaminergic hypersensitivity in patients with tardive dyskinesia Klawans, 1972, 1973; Marsden *et al.*, 1975). Thus very small doses of levodopa increase the severity of tardive dyskinesia, and this condition is reversed at least temporarily by drugs which block dopamine receptors or synthesis. The mechanism of this state of dopaminergic hypersensitivity is uncertain, and although the effects of phenothiazines on cell membranes and cellular respiration are well known, it has not been established whether chronic treatment with these drugs causes changes in dopamine-sensitive adenylate cyclase in patients with tardive dyskinesia. Dopamine receptors are present in different regions of the human nervous system, and phenothiazines cause blockade of these in medullary centres concerned with vomiting, and hypothalamic regions governing hormonal release mechanisms, as well as in the nigrostriatal system. Signs of Parkinsonism are present in many but not all patients with tardive dyskinesia; hormonal changes have not been established. It is unlikely therefore that tardive dyskinesia results from universal chronic dopamine receptor blockade in the nervous system. There are, however, two populations of neurons in the caudate nucleus, one inhibited and the other facilitated by dopamine (McLennan & York, 1967) and other regional differences in the nature of dopamine receptors may exist, analogous to those of the adrenergic nervous system.

Discussion

D. F. Sharman: How often do acute dystonias, particularly the oral-buccal-lingual syndrome, occur after neuroleptic drugs in humans?

J. D. Parkes: Very commonly. I have seen two in the last month: one was a prisoner, who, on discharge from Brixton, was given a shot of prochlorperazine in his bottom, and was admitted akinetic and mute; and the second was a schoolgirl, who after a dental procedure, went totally mute and anarthric.

Figures vary enormously for different drugs. With some drugs they occur as often as in 10–20% of cases. In others it is rare.

L. L. Iversen: Can I ask you a question about the drug use and tardive dyskinesias? The question is about reversibility of the syndrome. Is it reversible?

J. D. Parkes: Again the evidence is not all that good. Undoubtedly the consensus of evidence is that in some patients this is irreversible, although others deny this, and say it is reversible over perhaps 18 months or two years.

L. L. Iversen: The neuropathological study that you referred to, was that done on patients that showed drug-induced tardive dyskinesias of long standing?

J. D. Parkes: Yes.

References

American College of Neuropsychopharmacology—Food and Drug Administration Task Force (1973) *New Engl. J. Med.* **289**, 20–22

Baker, A. B. (1969) in *Psychotropic Drugs and Dysfunctions of the Basal Ganglia* (Crane, G. E. & Gardner, R. Jr. eds.), p. 30, Public Health Service Publication 1938, U.S. Govt. Print. Off., Washington, D.C.

Brandon, S., McClelland, H. A. & Protheroe, C. (1971) *Brit J. Psychiat* **118**, 171–184

Briones, R. V. (1973) *Brit. J. Psychiat.* **122**, 493

Burks, J., Walker, J., Ott, J. E. & Rumarck, B. (1973) *Neurology (Minneapolis)* **23**, 393

Calne, D. B., Teychenne, P. F., Claveria, L. E., Eastman, R., Greenacre, J. K. & Petrie, A. (1974) *Brit. med. J.* **iv**, 442–444

Chase, T. N., Schnur, J. A. & Gordon, E. K. (1970) *Neuropharmacology* **9**, 265–268

Chase, T. N., Watanabe, A. M., Brodie, K. H. & Donnelly, E. S. (1972) *Arch. Neurol.* **26**, 282–284

Christensen, E., Møller, J. E. & Faurbye, A. (1970) *Acta Psychiat. Scand.* **46**, 14–23

Crane, G. E. (1968) *Am. J. Psychiat.* **124** suppl., 40–48

Crane, G. E. (1973) *Advan. Neurol.* **1**, 115–122

Degwitz, R., Binsack, K. F. & Herkert, H. (1967) *Nervenarzt* **38**, 170–174

Dynes, J. B. (1968) *Virginia Med. Mth.* **95**, 746

Edwards, H. (1970) *Brit. J. Psychiat.* **116**, 271–275

Ey, H., Faure, H. & Rappard, P. (1956) *Encephale* **45**, 790–796

Goodman, L. S. & Gilman, A. (1970) in *The Pharmacological Basis of Therapeutics*, p. 163, Macmillan, London

Greenblatt, D. L., Shader, R. I., Stotsky, B. A. & Di Mascio, A. (1968) *J. Am. Geriat. Soc.* **16**, 27–34

Greenhouse, A. M. (1966) *Arch. Intern. Med.* **117**, 389–393

Hamilton, M. & Mahapatra, S. B. (1972) in *Side Effects of Drugs* (Meyer, L. & Herxheimer, A., eds.), vol. 1, p. 32, Excerpta Medica, Amsterdam

Heinrich, K., Wegerer, I. & Bender, H. J. (1968) *Pharmakopsychiat. Neuro-Psychopharmakol.* **1**, 169–195

Hippius, V. H. & Logeman, G. (1970) *Arzneimittel-Forsch.* **20**, 894–896

Howard, R. O., McDonald, C. J., Dunn, B. & Creasy, W. A. (1969) *Invest. Ophthalmol* **8**, 413–421

Hunter, R., Earl, C. J. & Thornicroft, S. (1964) *Proc. Roy. Soc. Med.* **57**, 758–762

Hunter, R., Blackwood, W., Smith, M. C. & Cummings, J. N. (1968) *J. Neurol. Sci.* **7**, 263–273

Johnson, A. W. & Buffaloe, W. J. (1968) *S. Med. J.* (*Birmingham Ala.*) **61**, 993–994

Klawans, H. L. Jr. (1972) *J. Neurol. Trans.* **33**, 235–246

Klawans, H. L. Jr. (1973) *Am. J. Psychiat.* **130** 82–86

Klawans, H. L., Jr. & McKendall, R. R. (1971) *J. Neurol. Sci.* **14**, 189–192

Kline, N. S. (1968) *Am. J. Psychiat. Suppl.* **124**, 48–54

McLennan, H. & York, D. H. (1967) *J. Physiol.* **189**, 393–402

Marsden, C. D., Tarsy, D. & Baldessarini, R. J. (1975) in *Psychiatric Aspects of Neurologic Disease* (Benson, F. & Bloomer, D. eds), pp. 219–265, Grune and Stratton, New York

Marshall, M. H. (1972) *J. Am. Med. Ass.* **221**, 86–7

Mones, R. J., Elizan, T. S. & Siegel, G. J. (1971) *J. Neurol. Neurosurg. Psychiat.* **34**, 668–673

National Institute of Mental Health, Psychopharmacology Service Center Collaborative Study Group (1964) *Arch. Gen. Psychiat.* **10**, 246–261

Paulson, G. W. (1968) *Geriatrics* **23**, 105–110
Pearce, J. (1971) *Brit. Med. J.* **iii**, 529
Pind, K. & Faurbye, A. (1970) *Acta Psychiat. Scand.* **46**, 323–326
Rylander, G. (1972) *Psychiat. Neurol. Neurochir.* **75**, 203–212
Schiele, B. C., Gallant, D., Simpson, G. Gardner, A. & Cole, J. O. (1973) *Ann. Intern. Med.* **79**, 99–100
Shepherd, M., Lader, M. & Lader, S. (1972) in *Side Effects of Drugs* (Meyler, L. & Herxheimer, A., eds.), vol. 7, pp. 69–97, Excerpta Medica, Amsterdam
Simpson, G. M. (1973) *Brit. J. Psychiat.* **122**, 618
Tamer, A., McKey, R., Arias, D., Worley, L. & Fogel, B. J. (1969) *J. Pediat.* **75**, 479–480
Tarsy, D., Leopold, N., and Sax, D. S. (1974) *Neurology* (*Minneapolis*) **24**, 28–33
Tarsy, D., Parkes, J. D., Marsden, C. D., Bovill, K. T., Phipps, J. A., Rose, P. & Asselman, P. (1975) *J. Neurol. Neurosurg. Psychiat.* **38**, 331–335
Tétreault, L., Filotto, J. & Bordeleau, J.-M. (1968) *Can. Psychiat. Ass.* **13**, 507–512
Trurek, I., Kurland, A. A., Hanlon, T. E. & Bohm, M. (1972) *Brit. J. Psychiat.* **121**, 605–612
Uhrbrand, L. & Faurbye, A. (1960) *Psychopharmacologia* **1**, 408–18
Wheeler, R. H., Bhalerao, V. R. & Gilkes, M. J. (1968) *Brit. J. Psychiat.* **115**, 687–690
Yamadori, A. & Albert, M. L. (1972) *New Engl. J. Med.* **286**, 610

Chapter 7

The Effects of Tranquillizing Drugs on the Metabolism of Transmitter Substances in the Central Nervous System

By J. P. FRY and D. F. SHARMAN

Agricultural Research Council Institute of Animal Physiology, Babraham, Cambridge CB2 4AT, U.K.

Effects of Major Tranquillizing (Neuroleptic) Drugs on the Metabolism of Monoamines in the Brain

The main pathways for the metabolism of the monoamines, dopamine, noradrenaline and 5-hydroxytryptamine in the brain are illustrated in Fig. 1.

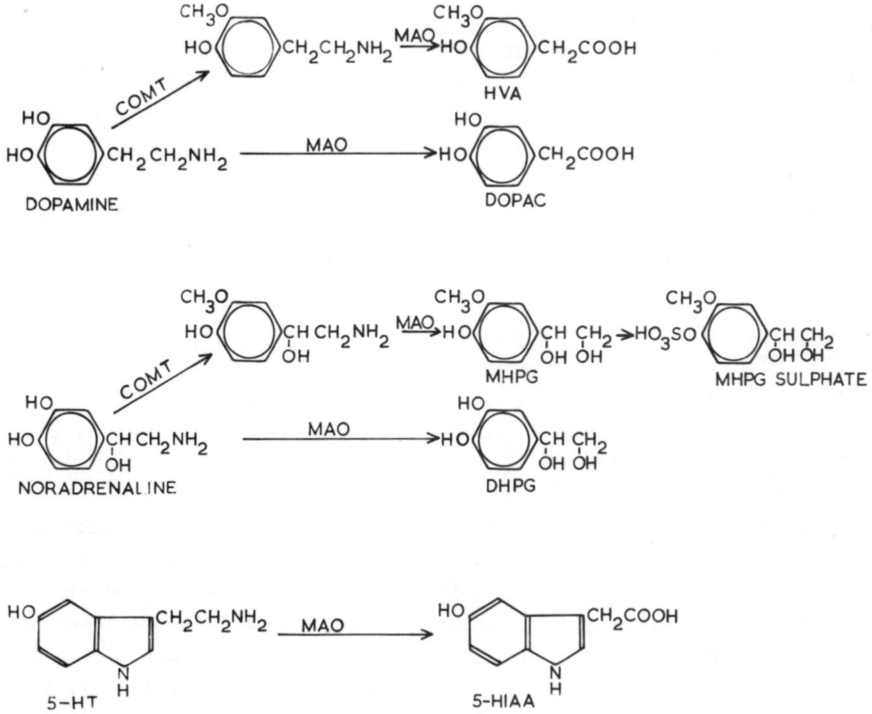

Fig. 1. *The major pathways for the metabolism of dopamine, noradrenaline and 5-hydroxytryptamine in the brain*

Abbreviations: HVA, homovanillic acid; DOPAC, 3,4-dihydroxyphenylacetic acid; MHPG, 1-(4-hydroxy-3-methoxyphenyl)ethan-1,2-diol; DHPG, 1-(3,4-dihydroxyphenyl)ethan-1,2-diol; 5-HT, 5-hydroxytryptamine; 5-HIAA, 5-hydroxyindol-3-ylacetic acid. (The conversion of MHPG to MHPG-sulphate does not occur in the mouse brain to any great extent.)

These monoamines have been localized to separate, specific neuron systems in the brain by fluorescence histology and by chemical analyses. Biochemical, pharmacological and electrophysiological evidence allows us to presume that they serve as transmitter substances between neurons. Studies on the cerebral metabolism of dopamine and noradrenaline (Roffler-Tarlov *et al.*, 1971; Braestrup & Nielson, 1975) have suggested that changes in the concentrations of the deaminated catechol metabolites might reflect changes in the metabolism within the neurons which form the parent catecholamine (intrahomoneuronal metabolism) and that part of the concentrations of the methoxylated deaminated metabolites represents metabolism at a different site (extrahomoneuronally). In the case of 5-hydroxytryptamine (5-HT) there is only one major metabolic product in the brain, 5-hydroxyindol-3-ylacetic acid (5-HIAA) so that a similar distinction in the locus of the metabolism comparable with that suggested for the catecholamines is much more difficult to investigate.

Although a definite action of reserpine on the cerebral metabolism of the three monoamines could be demonstrated by studying their concentrations in the brain and observing that this drug caused a profound depletion of all three monoamines, it was not until methods for the measurement of the metabolites of the biogenic amines had been developed that the effects of other neuroleptic drugs on their metabolism could be clearly demonstrated. Carlsson & Lindqvist (1963) estimated the concentrations of 3-methoxytyramine and normetanephrine in the brains from mice which had been pretreated with a monoamine oxidase inhibiting drug. These authors observed that the increases in the cerebral concentrations of 3-methoxytyramine and normetanephrine following the inhibition of monoamine oxidase were enhanced by small doses of chlorpromazine or haloperidol. The development of methods for the estimation of acidic metabolites of dopamine (Andén *et al.*, 1963; Sharman, 1963) allowed further studies to be made. Andén *et al.* (1964), Laverty & Sharman (1965*a*, *b*), and Roos (1965) showed that the administration of a wide variety of phenothiazine and butyrophenone neuroleptic drugs or reserpine to different species resulted in an increase in the cerebral concentration of 4-hydroxy-3-methoxyphenylacetic acid (homovanillic acid; HVA) and 3,4-dihydroxyphenylacetic acid (DOPAC) with little effect on the concentration of 5-HIAA except in the case of reserpine, when an increase in the concentration of the acidic metabolite of 5-HT was observed (Ashcroft & Sharman, 1962).

Studies on the conversion of radioactivity labelled tyrosine into dopamine in the brain revealed that the increase in the cerebral concentration of the metabolites of dopamine was accompanied by an increase in the rate of formation of dopamine, presumably by an increase in the activity of the enzyme tyrosine hydroxylase (Burkard *et al.*, 1967; Nybäck *et al.*, 1967). Originally, it was proposed that the increased metabolism of dopamine and noradrenaline seen after chlorpromazine and haloperidol was a result of compensatory activation of catecholaminergic neurons as a result of blockade of catecholamine receptors (Carlsson & Lindqvist, 1963). Further evidence for this hypothesis was provided by Aghajanian & Bunney (1973), who demonstrated that the peripheral administration of chlorpromazine and haloperidol caused an increased rate of firing in dopamine-containing cells located in the substantia nigra. The possibilities for

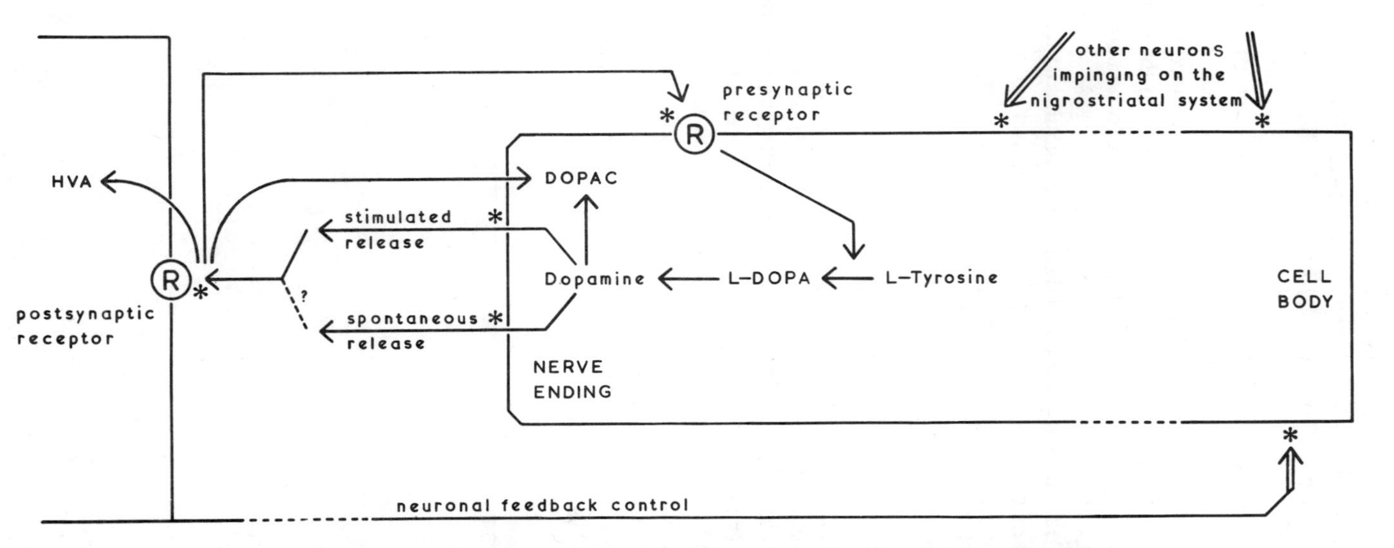

Fig. 2. *Diagrammatic representation of a dopamine-containing nigrostriatal neuron showing possible sites of action of neuroleptic drugs* (*)

modifying the metabolism of dopamine through blockade of dopamine receptors have recently become more complex with the finding that there is a dopamine-sensitive adenylate cyclase system in the striatum which can be inhibited by neuroleptic drugs (Kebabian *et al.*, 1972; Clement-Cormier *et al.*, 1974; Miller & Iversen, 1974). Kebabian *et al.* (1972) have considered that this system might represent cerebral dopamine receptors. In addition Kehr *et al.* (1972) proposed that there is a presynaptic dopamine receptor in the striatum which is concerned in the control of tyrosine hydroxylase activity in the dopaminergic nerve endings. Aghajanian & Bunney (1973) have further demonstrated that the firing rate of cell bodies of dopaminergic neurons in the substantia nigra can be inhibited by the iontophoretic application of dopamine or the dopamine agonist drug apomorphine. Thus there are at least three sites at which neuroleptic drugs might act to modify the metabolism of dopamine (Fig. 2). Some aspects of the effects of tranquillizing drugs on the metabolism of biogenic amines in the brain will now be considered in more detail.

Does the Increase in Homovanillic Acid Represent a Compensation in Respect of the Behavioural Effects of Neuroleptic Drugs?

The effects of neuroleptic drugs on dopamine metabolism are often interpreted as being due to an increased release of dopamine in compensation for postsynaptic receptor blockade. As far as behavioural effects are concerned, the catalepsy produced in laboratory animals after administration of certain neuroleptics is thought to be a result of blockade of transmission at dopaminergic nerve endings in the striatum (Hornykiewicz, 1973). Stille *et al.* (1971) found that the ability of different neuroleptics to increase the concentration of HVA in the rat striatum was roughly paralleled by the potency of the drugs in producing catalepsy in these animals (see Table 1). A similar parallel relation between the ability to

Table 1. *A comparison of the ability of some neuroleptic drugs to cause catalepsy, to increase the striatal concentration of homovanillic acid (HVA) and to antagonize apomorphine-induced stereotypy in the rat*

Drug	Catalepsy ED_{50} s.c.	Anti-apomorphine ED_{50} s.c.	Catalepsy ED_{50} p.o.	Increase in HVA % of control		
				Dose 1	mg/kg 5	p.o. 20
Spiperone	0.2	0.17				
Loxapine			0.1	605	628	
Clothiapine			0.15		595	
Fluphenazine	0.16	0.13	0.28		722	
Haloperidol	0.20	0.20	0.3		774	
Pimozide	0.20	0.17				
Perphenazine	0.31	0.32	0.5		399	
Chlorpromazine	7.5	6.5	3.8		206	
Chlorprothixene	1.8	24	6.4		202	
Levomepromazine	5.0	24	5.0		158	
Perlapine			6.8		100	314
Thioridazine	13	>160	17.0		100	249
Clozapine		>20	Inactive		100	116

Data from Janssen *et al.* (1965, 1968) and Stille *et al.* (1971).

increase the cerebral HVA concentration and the cataleptogenic properties of narcotic analgesics in the rat was described by Ahtee & Kääriäinen (1973).

In the following experiments dose/response curves were constructed for the effects of pimozide, a diphenylbutylamine-type neuroleptic, on the concentrations of dopamine, HVA and DOPAC in the striatum of the mouse. Pimozide was used, since this drug was reported to block dopamine rather than noradrenaline receptors in the central nervous system (Andén *et al.*, 1970). The catalepsy and sedation produced in mice by pimozide were quantified and dose/response curves constructed for these behavioural effects of the drug (J. P. Fry & D. F. Sharman, unpublished work).

Catalepsy was estimated by two different methods. In the rotarod test (Jones & Roberts, 1968) the mice were placed on a rotating rod which increased in speed linearly during the test session. The length of time for which the animals stayed on the rod was measured. The other test for catalepsy involved placing the mice on a vertical wire mesh and measuring the length of time they stayed immobile during a five minute period. Sedation was estimated by placing each mouse in a photocell activity cage for five minutes.

Figure 3 shows the effects of different doses of pimozide on the concentrations

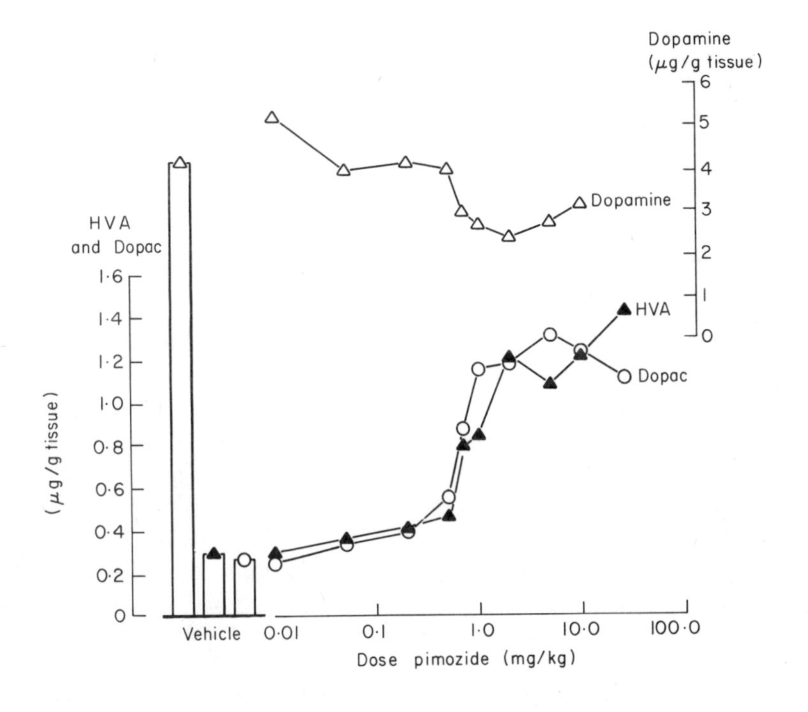

Fig. 3.

The concentration of dopamine (△) and its acidic metabolites HVA (homovanillic acid; ▲) and DOPAC (3,4-dihydroxyphenylacetic acid; ○) in the striatum of the mouse 75 min after the intraperitoneal injection of pimozide. Dopamine was estimated by the method of Laverty & Sharman (1965), HVA and DOPAC by the method of Murphy *et. al.* (1969). Estimates are means from at least four determinations.

of dopamine, HVA and DOPAC in the mouse striatum. The lowest dose of pimozide causes a slight increase in dopamine concentration. An increased dopamine concentration in the mouse striatum was also observed by O'Keeffe *et al.* (1970) after injecting low doses of chlorpromazine or haloperidol. As can be seen in Fig. 3, higher doses of pimozide decrease the concentration of striatal dopamine and increase the concentrations of HVA and DOPAC. Figure 4 shows that a dose/response curve can also be constructed for the catalepsy produced in mice by pimozide.

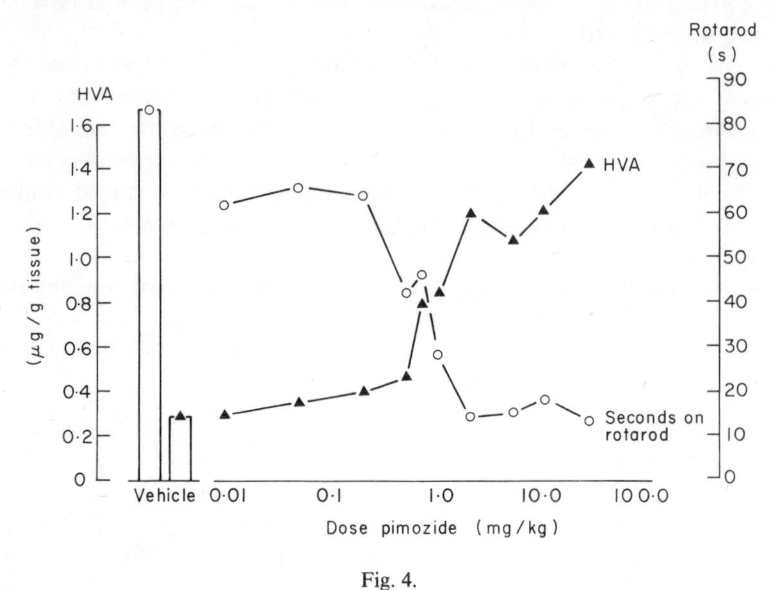

Fig. 4.

Catalepsy (○), as measured in a rotating rod test and the striatal concentration of HVA (▲) in mice, 75 min after the intraperitoneal injection of pimozide.

If the increase in striatal HVA and DOPAC concentrations represented an increased compensatory release at dopaminergic nerve endings then the behavioural effects of pimozide should not become apparent until doses are used which cause nearly maximal increases in HVA and DOPAC concentrations. The dose/response curves that have been obtained, however, show that the dose-dependent increases in HVA and DOPAC concentrations are paralleled by dose-dependent increases in catalepsy and sedation. Thus the increased metabolism of dopamine following intraperitoneal injection of pimozide does not compensate for the catalepsy and sedation produced by the drug. Some tentative explanations can be made for the above results.

Pimozide could produce a blockade of dopaminergic transmission of a type that cannot be overcome by activation of the nigrostriatal neurons. The drug could, for example, irreversibly block postsynaptic receptors or prevent the impulse-coupled release of dopamine (Seeman & Lee, 1975). If this were the correct explanation then for any fractional blockade of dopamine receptors a compensatory mechanism, if present, would increase to maximum activity, since it would not be able to overcome the blockade. The dose-dependent effects of pimozide on dopamine metabolism and behaviour could then be explained by certain receptors or populations of neurons having different sensitivities to the drug.

If the increased striatal metabolism of dopamine following injection of pimozide represents an increased release from nigrostriatal neurons then the possibility arises that this increased release could compensate for behavioural or neurophysiological changes that have not been measured in the present study. This suggestion, however, does not explain why the dose/response curves for catalepsy and sedation run so close to those for changes in the concentrations of dopamine, HVA and DOPAC after pimozide. When chlorpromazine is given intraveneously to cats, an increased release of dopamine can be detected into the perfusate of the caudate nucleus (Lloyd and Bartholini, 1975), presumably because of an increased release of dopamine from nigrostriatal neurons. The possibility remains that this neuroleptic-induced release of dopamine interacts with receptors not blocked by neuroleptic drugs. Such receptors have recently been found by Ben Ari & Kelly (1974) on cells in the putamen and amygdala of the cat. These authors report that the inhibitory actions of dopamine on spontaneously active or glutamate-excited cells are unaffected after intravenous injection of large, cataleptogenic doses of pimozide or α-flupenthixol.

The Mechanism of Action of Neuroleptic Drugs in Causing an Increase in the Concentration of Homovanillic Acid in the Brain

There is evidence that impulse traffic in the dopaminergic neurons is necessary for at least part of the neuroleptic-induced increase in amine turnover (Andén *et al.*, 1971). The intravenous administration of chlorpromazine or haloperidol to rats results in an increase in the firing rate of dopaminergic neuronal cells in the midbrain (Aghajanian & Bunney, 1973). Taken together, these observations suggest that the increases in the acidic metabolites of dopamine are a result of an increased release of dopamine at the nerve terminals. However, the maximum effect on the firing rate of the dopaminergic neuronal cells was achieved with relatively low doses of the neuroleptic drugs. Recently, the presence of a presynaptic dopamine receptor which is involved in the control of the rate of dopamine synthesis in the terminals of dopamine-containing neurons has been demonstrated (Kehr *et al.*, 1972). The presence of such a receptor, which responds to dopamine-agonist drugs to bring about an inhibition of the rate of synthesis of dopamine can also be demonstrated in synaptosome-containing fractions of striatal tissue and in tissue slices (Goldstein *et al.*, 1973; Anagnoste *et al.*, 1974; Kuczenski, 1975; L. L. Iversen, personal communication). Neuroleptic drugs can antagonize this effect and have been reported to increase the rate of formation of dopamine in such preparations. Neuroleptic drugs and dopamine-agonist drugs have been shown to have opposing effects on the properties of the enzyme tyrosine hydroxylase, the rate limiting step in the formation of dopamine (Zivkovic & Guidotti, 1974; Zivkovic *et al.*, 1974), isolated from striatal tissue of animals treated with such drugs. The effects of the neuroleptic drugs appear to be due to a change in the K_m for the pteridine co-factor in the enzyme system. The contribution of this mechanism to the neuroleptic-induced increase in HVA in the brain is difficult to assess, particularly in the light of the observation by Seeman & Lee (1975) that with electrically stimulated slices of brain tissue the stimulus-induced release of dopamine is reduced by neuroleptics whereas the spontaneous efflux of the catecholamine is increased. It is suggested that neuroleptic drugs inhibit the coupling between the nerve impulse and dopamine release from the nerve ending.

At present there is insufficient evidence to enable a description of the locus of the mechanism by which the activation of the dopamine-containing neurons occurs in response to neuroleptic drugs. Although section of the ascending dopamine-containing fibres reduces the effect of neuroleptic drugs at dopamine-containing neuron terminals, section of the descending cholinesterase containing strionigral fibres does not suppress the accelerated dopamine turnover in the striatum in response to haloperidol in the rat (Bedard & Larochelle, 1973). These authors concluded that the effect was a local phenomenon or that other unknown neuronal pathways might be involved.

Is It Possible to Distinguish Actions of Neuroleptic Drugs at Different Sites in the Brain From *in vivo* Biochemical and Behavioural Effects?

Among the tests which have been used to study compounds with neuroleptic activity, inhibition of the stereotyped behaviour caused by apomorphine is thought to represent the effect of neuroleptic drugs on the effector cells innervated by dopamine-containing neurons in the brains of intact animals. It has been suggested that the dopamine-sensitive adenylate cyclase system which can be isolated from brain tissue might represent the dopamine-receptor in the tissue, but since it has been shown that cyclic AMP can stimulate dopamine synthesis in slices of striatal tissue (Anagnoste *et al.*, 1974) this system could reflect a mixed receptor population. It is possible that there is a cholinergic neuronal link in the mechanism that causes the increased metabolism of dopamine following the administration of neuroleptic drugs. O'Keeffe *et al.* (1970) reported that atropine reduced the concentration of HVA in the striatum without changing the concentration of dopamine and also antagonized the increase in HVA caused by chlorpromazine or haloperidol. Bartholini & Pletscher (1971) and Bartholini *et al.* (1973) concluded that there are two cholinergic systems which can affect dopamine metabolism in opposite directions since atropine, injected into the region of the substantia nigra, increased the concentration of HVA in the striatum.

Aghajanian & Bunney (1973) have demonstrated that the dopamine-containing cells in the midbrain respond to iontophoretically applied dopamine with a reduction in their firing rate. This response is not antagonized by chlorpromazine iontophoresed onto the same cell but is antagonized by peripherally administered neuroleptic drugs. With cells in the putamen, which are inhibited by iontophoretically applied dopamine, the local application of a neuroleptic drug antagonizes the response to dopamine but the peripheral administration of the same neuroleptic drug does not modify the response (Ben Ari & Kelly, 1974). Thus it is extremely difficult to decide the site or sites at which neuroleptic drugs might act to bring about an increase in the concentration of HVA in the brain, and it is possible that the increase in the concentration of HVA is a result of the action of the drugs at more than one site. Andén *et al.* (1967) and Ernst (1967) obtained evidence that apomorphine stimulates dopamine receptors in the brain and retards the release of dopamine from dopamine-containing nerve endings. There is little evidence to enable a comparison of the ability of neuroleptics to increase the concentration of HVA in the brain with their ability to antagonize the behavioural effects of apomorphine to be made. Stille & Lauener (1971) and Stille *et al.* (1971) have demonstrated that the time course and intensity of the catalepsy produced in rats by neuroleptic drugs parallel the increase in cerebral HVA. Our experiments described above provide further evidence for this relation.

A direct comparison of the ability to cause catalepsy in rats with the ability to inhibit apomorphine-induced chewing can be found in the work of Janssen *et al.* (1965). Some of the results obtained by these authors are given in Table 1, which compares the relation between the ability to cause catalepsy and the abilities to inhibit apomorphine-induced chewing and to increase the striatal

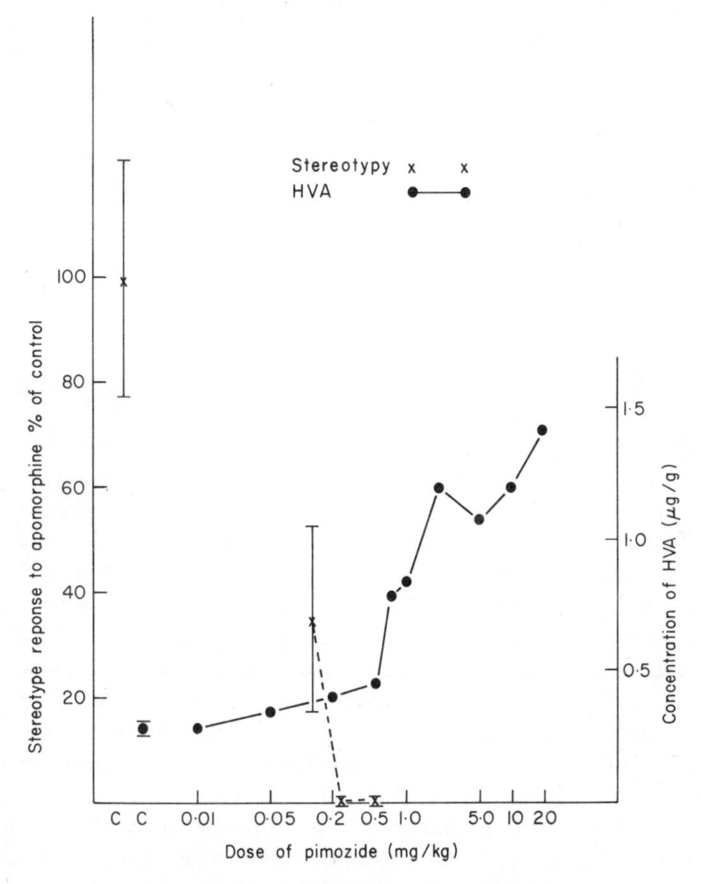

Fig. 5. *The effect of pimozide on apomorphine-induced, stereotyped rearing behaviour and on the striatal concentration of HVA (homovanillic acid) in mice*

concentration of HVA. Although the antagonist activity towards apomorphine appears to be in the same sequence as the ability to cause catalepsy and the effectiveness in producing an increase in the striatal concentration of HVA there is not a linear relation between the dose required to produce catalepsy and the dose required to antagonize the chewing response to apomorphine.

A clear separation between the increase in the concentration of HVA produced in the striatum of the mouse by pimozide and the activity of this drug in antagonizing stereotyped behaviour in the mouse is shown in Fig. 5. The behavioural test used in obtaining these results was a rearing response which appears to be specific for apomorphine, since it was not observed with other drugs which are thought to stimulate cerebral dopamine receptors and which is only observed with certain strains of mice (Hage *et al.*, unpublished work). Thus it would seem that the receptors involved in this response to apomorphine are not the receptors involved in the increase in the cerebral concentration of HVA following treatment with neuroleptic drugs. If more than one set of dopamine receptors are present in the brain, subserving different physiological functions and showing different sensitivity to blockade by drugs, and if the increase in HVA caused by neuroleptic or other drugs represents an increased release of dopamine from the terminals of dopamine-containing neurons then one might expect to see on occasions behavioural reponses after treatment with neuroleptic drugs which resemble those seen after drugs which are thought to stimulate dopamine receptors. Such responses do not seem to have been observed in rodents, but in the domestic pig many neuroleptic drugs are comparatively ineffective in causing catalepsy. We have examined the behaviour of pigs after the administration of apomorphine or neuroleptic drugs and after treatment with metoclopramide, an antiemetic drug which causes catalepsy and an increase in striatal HVA in rodents (Ahtee, 1975) and is an effective antagonist of apomorphine-induced stereotyped behaviour in the rat, and it has been reported to cause dyskinesias in humans but does not exacerbate the symptoms of Parkinsonian patients (Jenner *et al.*, 1975). In addition metoclopramide is devoid of any antipsychotic action.

Over 100 years ago Feser (1874) described the action of apomorphine in the pig and reported that this drug did not cause emesis but that the animals were excited and showed continuous sniffing and rooting accompanied by chewing. In the pigs we have used, apomorphine administered intravenously invariably induced a behaviour which resembled the natural rooting of the animal; the snout was rubbed continuously against the floor or side of the pen or cage. This was accompanied by chewing and salivation. If food was presented to the region of the snout the animal would usually eat but tended to ignore food placed elsewhere in the cage. When pigs were given spiperone or haloperidol, by intramuscular injection, nose rubbing and chewing was often seen. In a series of ten pigs to which metoclopramide was administered intravenously, two showed a behavioural response after 10–15 min that was almost identical to that seen after apomorphine. The movements of the animal after metoclopramide, however, were somewhat slower than those observed after apomorphine. In four litter-mate pigs only one showed the nose-rubbing behavioural response to metoclopramide. When tested on subsequent occasions, the same animal always responded in the same way. Increasing or reducing the dose of metoclopramide did not induce nose-rubbing in animals which did not show this response after the first injection. A similar nose-rubbing chewing response was observed after the intravenous administration of haloperidol or spiperone. This neuroleptic-induced behaviour was not suppressed by the subsequent intravenous injection of acaperone, a butyrophenone derivative used as a sedative in pigs, although the administration of large doses of this latter drug caused unco-ordinated convulsive movements. The pigs treated with metoclopramide which did not show the nose-rubbing response showed cataleptic postures, the head being held motionless for relatively long periods of time. In addition, several of the animals showed a half sitting posture, which was observed in two animals to be the first stage of a rapid jumping response. Similar jumping behaviour was observed in one pig treated with haloperidol.

Ahtee (1974) has described the stereotyped behaviour which occurs in rats in parallel with catalepsy following the chronic administration of methadone. It was suggested that the stereotyped behaviour was due to the stimulation of dopamine receptors by excess dopamine produced by a compensatory mechanism and reflected in the increase in the striatal concentration of HVA which also occurred. We have not yet examined whether there are changes in the metabolism of dopamine in the brains of pigs treated with neuroleptic drugs or with metoclopramide and a possible explanation of the observed behavioural response is that, in pigs, these drugs show agonist activity at dopamine receptors and are not acting through the release of dopamine in the brain. An abnormal compulsive gnawing response to neuroleptics has been observed in the baboon by Meldrum *et al.* (1975). However, the demonstration that in species other than those usually used in laboratory experiments neuroleptic drugs can induce behavioural responses which resemble those caused by a drug thought to activate dopamine receptors indicates a possible mechanism by which neuroleptic drugs can produce dyskinesias in man.

The Effect of Neuroleptic Drugs on the Metabolism of Other Transmitter Substances in the Brain

Noradrenaline

Ehringer *et al.* (1960) demonstrated that chlorpromazine could antagonize the increase in the concentration of noradrenaline in the rat brain caused by monoamine oxidase inhibition. This observation was extended by Dresse & De Meyer (1965) to other phenothiazine and butyrophenone neuroleptics. These authors concluded that the neuroleptics could act at the level of the intraneuronal amine-storage granule or at the level of the cell membrane. The important report of Carlsson & Lindqvist (1963) demonstrated the increased formation of a metabolite of noradrenaline in the brains of mice which had been treated with a monoamine oxidase inhibiting drug together with haloperidol or chlorpromazine. The major metabolite of noradrenaline in the rat brain appears to be the ethereal sulphate conjugate of 1 (4-hydroxy-3-methoxyphenyl)ethan-1,2-diol (MHPG–SO_3H). The development of a method for the estimation of this metabolite (Meek & Neff, 1972) enabled studies to be made on the effect of neuroleptic drugs on the concentration of MHPG–SO_3H in the rat brain (Keller *et al.*, 1973; Berridge & Sharman, 1974). There is a wide range of potency. Methiothepin, haloanisone, azaperone and haloperidol are very active whereas pimozide has a very low potency. The response seems to parallel the ability of the neuroleptic drug to reduce the concentration of noradrenaline in the brain (Bartholini *et al.*, 1973*a*).

Acetylcholine.

As stated above, atropine can reduce the HVA concentration and antagonize the neuroleptic-induced increase in HVA in the brain. The striatum is rich in acetylcholine and the enzymes for its synthesis and degradation. There are many cholinergic nerve endings in this part of the brain and there are also cells from which cholinergic nerve fibres originate. If the dopamine in the striatum is acting to modulate the activity of these cells and nerve endings, changes in the metabolism of acetylcholine might occur secondary to the changes in the metabolism of dopamine. Many of the neuroleptic drugs possess antiacetylcholine properties, and these might also be instrumental in any change in the metabolism of acetylcholine. High doses of chlorpromazine or clozapine increase the turnover of acetylcholine in the striatum of the cat, and this effect can be antagonized by apomorphine (Stadler *et al.*, 1973, 1974; Cheney *et al.*, 1974). Atropine is also capable of increasing the release of acetylcholine from brain tissues (Mitchell, 1963; Portig & Vogt, 1969) and as yet it is not possible to separate completely the two possible components of neuroleptic drug action on the metabolism of acetylcholine.

5-Hydroxytryptamine

Many of the neuroleptic drugs appear to have little effect on the metabolism of 5-hydroxytryptamine (5-HT) to 5-hydroxyindol-3-ylacetic acid (5-HIAA) (Laverty & Sharman, 1965; Sharman, 1966) but chlorpromazine can increase 5-HT turnover (D. A. Bender, personal communication). Those, like reserpine, which release 5-HT from intraneuronal storage sites increase the concentration of 5-HIAA in the brain. Monachon *et al.* (1972) have described the effects of methiothepin on cerebral 5-HT metabolism. This neuroleptic drug blocks 5-HT receptors in the brain in addition to its actions on other neurotransmitter systems and causes an increase in the cerebral concentration of 5-HIAA with little change in that of 5-HT.

γ-Aminobutyric acid

Lahti & Losey (1974) have reported that the administration of aminooxyacetic acid to rats antagonizes the effect of chlorpromazine or morphine on cerebral dopamine metabolism. Aminooxyacetic acid inhibits the metabolism of γ-aminobutyric acid (GABA) and thus causes an increase in the concentration of GABA in the brain. It is thought that GABA might exert an inhibitory influence on the nigrostriatal neurons. Andén (1974) has further examined the action of aminooxyacetic acid on cerebral dopamine metabolism and concluded that GABA acts as an inhibitor on cells in the mesencephalon and that the closely related γ-hydroxybutyric acid, which has a profound effect on dopamine metabolism, stimulates the inhibitory GABA mechanism. However, the precise interaction between GABA and the metabolism of other transmitter substances in the brain remains to be elucidated.

Summary

Most neuroleptic drugs increase the cerebral concentration of the acidic metabolites of dopamine and this is associated with an increase in the turnover of dopamine in the brain.

The increased turnover of dopamine does not compensate the catalepsy which neuroleptics cause in laboratory rodents.

It appears possible to separate the blocking action of neuroleptic drugs, at least at some receptor sites, from the actions on dopamine metabolism. In the pig, it is possible to induce a behavioural response to neuroleptic drugs, or metoclopramide, which closely resembles the response to apomorphine.

Neuroleptics interact with the metabolism of other neurotransmitter substances to varying degrees. At present it is not possible to define clearly the way in which most neuroleptic drugs bring about an increase in the turnover of dopamine in the brain.

Discussion

D. A. Bender: Just a quick piece of information. Chlorpromazine does increase brain serotonin pool size and synthesis rate, and brain tryptophan uptake in rats [unpublished work].

D. F. Sharman: We did it in the cat and didn't get any effect on 5-HIAA. That's all I can say [R. Laverty & D. F. Sharman (1965) *Brit. J. Pharmacol.* **24**, 759–772].

P. V. Taberner: I would like to know whether you have examined the time course of the behavioural effect of pimozide together with the biochemical changes, and, if so, whether there is a temporal correlation between the two.

D. F. Sharman: In the pig, this is a bit difficult because most of the pigs I get to do work on are rather radioactive; I get them between the radioactive laboratory and our slaughterhouse. I have them in the Department for about two hours and a half, and they are usually still showing the behavioural response to pimozide at the time they are slaughtered. We do hope to be able to do a bit of detailed work on this. We don't have dyskinesias in the farmyard, we call them vices!

P. V. Taberner: What about the mouse or rat?

J. P. Fry: We are still doing this in the mouse. We are looking at the time course after pimozide; the catalepsy does wear off sooner than the HVA effect, but the catalepsy correlates with DOPAC far better. About four hours after pimozide the DOPAC will start to go down as catalepsy wears off.

L. L. Iversen: Would you like to comment on the suggestion that some neuroleptics may act on dopamine turnover more potently in the limbic forebrain than they do in the striatum?

D. F. Sharman: Like you, I've read the literature, and somebody ought to do some more experiments. That's the answer!

C. J. Van den Berg: If one thinks that aminooxyacetic acid acts only by inhibiting GABA transaminase and that the subsequent accumulation of GABA acts as a neurotransmitter, one of course then assumes that GABA accumulates presynaptically in GABA neurons. But there is almost no GABA transaminase in the presynaptic terminals. So how does aminooxyacetic acid act? The assumption of a mechanism based on interaction at the neurotransmitter level might only just be a hypothesis.

D. F. Sharman: Oh yes, I agree.

B. E. Leonard: With reference to this apomorphine/GABA story, apomorphine appears to have a complex action on GABA metabolism. We have recently found that the stereotyped behaviour of movement and gnawing produced by apomorphine is associated with an increase in the concentration of GABA in most of the brain regions examined. However, in some apomorphine-treated rats fighting occurred. In these animals the concentration of GABA was profoundly reduced in the septal region, in particular.

D. F. Sharman: We looked at GABA concentrations in the region of the substantia nigra in the mouse under the influence of apomorphine, and we could detect no change there [J. D. M. Pearson & D. F. Sharman (1975) *J. Neurochem.* **24**, 1225–1228].

L. J. Fowler: May I suggest that it would be interesting to discover whether aminooxyacetic acid inhibits dopa decarboxylase as well as GABA transminase, when injected *in vivo* in the concentrations used by Dr. Sharman. Aminooxyacetic acid is a carbonyl trapping agent, and, as such, is likely to inhibit all pyridoxal phosphate dependent enzymes. This might give a simpler explanation of the results obtained rather than postulating a secondary effect of increased GABA concentration on the dopamine system.

D. A. Brown: What is the transmitter of the dopamine-receptive neuron?

D. F. Sharman: I think they are acetylcholine neurons. At least, there is a cholinergic link.

D. A. Brown: Where do the axons go?

D. F. Sharman: One set of the axons, from the striatum that have been investigated, appear to go to the substantia nigra.

References

Aghajanian, G. J. & Bunney, B. S. (1973) in *Frontiers in Catecholamine Research* (Usdin, E. & Snyder, S., eds.), pp. 643–648, Pergamon Press, Oxford

Ahtee, L. (1974) *Eur. J. Pharmacol.* **27**, 221–230

Ahtee, L. (1975) *Brit. J. Pharmacol.* **53**, 460P

Ahtee, L. & Kääriäinen, I. (1973) *Eur. J. Pharmacol.* **22**, 206–208

Anagnoste, B., Shirron, C., Friedman, E. & Goldstein, M. (1974) *J. Pharmacol. Exp. Ther.* **191**, 370–376

Andén, N.-E. (1974) *Acta Pharmacol. Toxicol.* **35**, 16

Andén, N.-E., Roos, B.-E. & Werdinius, B. (1963) *Life Sci.* **2**, 448–458

Andén, N.-E., Roos, B.-E. & Werdinius, B. (1964) *Life Sci.* **3**, 149–158

Andén, N.-E., Rubenson, A., Fuxe, K. & Hökfelt, T. (1967) *J. Pharm. Pharmacol.* **19**, 627–629

Andén, N.-E., Butcher, S. G., Corrodi, H., Fuxe, K. & Ungerstedt, U. (1970) *Eur. J. Pharmacol.* **11**, 303–314

Andén, N.-E., Corrodi, H., Fuxe, K. & Ungerstedt, U. (1971) *Eur. J. Pharmacol.* **15**, 193–919
Ashcroft, G. W. & Sharman, D. F. (1962) *Brit. J. Pharmacol. Chemother*, **19**, 153–160
Bartholini, G. & Pletscher, A. (1971)
Bartholini, G., Keller, H. H. & Pletscher, A. (1973*a*) *Neuropharmacology* **12**, 751–756
Bartholini, G., Stadler, H. & Lloyd, K. (1973*b*) *Advan. Neurol.* **3**, 233–241
Bedard, P. & Larochelle, L. (1973) *Expl. Neurol.* **41**, 413–322
Ben Ari, Y. & Kelly, J. S. (1974) *J. Physiol.* (*Lond.*) **242**, 66–67P
Berridge, T. L. & Sharman, D. F. (1974) *Brit. J. Pharmacol.* **50**, 156–158
Braestrup, C. & Nielsen, M. (1975) *J. Pharm. Pharmac.* **27**, 413–419
Burkard, W. P., Gey, K. F. & Pletscher, A. (1967) *Nature (London)* **213**, 732–733
Carlsson, A. & Lindqvist, M. (1963) *Acta Pharmacol. Toxicol.* **20**, 140–144
Cheney, D. L., Costa, E., Racagni, G. & Trabucchi, M. (1974) *Brit. J. Pharmacol.* **52**, 427–428P
Clement-Cormier, Y. C., Kebabian, J. W., Petzold, G. L. & Greengard, P. (1974) *Proc. Nat. Acad. Sci. U.S.A.* **71**, 1113–1117
Dresse, A. & De Meyer, R. (1965) *Biochem. Pharmacol.* **14**, 1129–1134
Ehringer, H., Hornykiewicz, O. & Lechner, K. (1960) *Naunyn-Schmiedebergs Arch. Exp. Pathol. Pharmakol.* **239**, 507–519
Ernst, A. M. (1967) *Psychopharmacologia* **10**, 316–323
Feser, Prof. (1874) *Z. Prakt. Veterinairwissenschaft* 310–317
Goldstein, M., Anagnoste, B. & Shirron, C. (1973) *J. Pharm. Pharmacol.* **25**, 348–351
Hornykiewicz, O. (1973) *Brit. med. Bull.* **29**, 172–178
Janssen, P. A. J., Niemegeers, C. J. E. & Schellekens, K. H. L. (1965) *Arzneimittel-Forsch.* **15**, 104–117
Janssen, P. A, J., Niemegeers, C. J. E., Schellekens, K. H. L., Dresse, A., Lenaerts, F. M. Pinchard, A., Schaper, W. K. A., Van Nueten, J. M. & Verbruggen, F. J. (1968) *Arzneimittel-Forsch.* **18**, 261–279
Jenner, P., Marden, C. D. & Perringer, E. (1975) *Br. J. Pharmacol.* **54,** 275–276P
Jones, B. J. & Roberts, D. J. (1968) *J. Pharm. Pharmacol.* **20**, 303–304
Kebabian, J. W., Petzold, G. L. & Greengard, P. (1972) *Proc. Nat. Acad. Sci. U.S.A.* **69**, 2145–2149
Kehr, W., Carlsson, A., Lindqvist, M., Magnusson, T. & Atack, C. (1972) *J. Pharm. Pharmacol.* **24**, 744–747
Keller, H. H., Bartholini, G. & Pletscher, A. (1973) *Eur. J. Pharmacol.* **23**, 183–186
Kuczenski, R. (1975) *Neuropharmacology* **14**, 1–10
Lahti, R. A. & Losey, E. G. (1974) *Res. Commun. Chem. Pathol. Pharmacol.* **7**, 31–40
Laverty, R. & Sharman, D. F. (1965*a*) *Brit. J. Pharmacol. Chemother.* **24**, 538–548
Laverty, R. & Sharman, D. F. (1965*b*) *Brit. J. Pharmacol. Chemother.* **24**, 759–772
Lloyd, K. G. & Bartholini, G. (1975) *Experimentia* **31**, 560–561
Meek, J. L. & Neff, N. H. (1972) *Brit. J. Pharmacol.* **45**, 435–441
Meldrum, B. S., Anlezark, G. M. & Trimble, M. (1975) *Eur. J. Pharmac.* **32**, 203–213
Miller, R. J. & Iversen, L. L. (1974) *J. Pharm. Pharmacol.* **26**, 142–144
Mitchell, J. F. (1963) *J. Physiol. (London)* **165**, 98–116
Monachon, M.-A., Burkard, W. P., Jalfre, M. & Haefely, W. (1972) *Naunyn-Schmiedebergs Arch. Exp. Pathol. Pharmakol.* **274**, 192–197
Murphy, G. F., Robinson, D. & Sharman, D. F. (1969 *Brit. J. Pharmacol.* **36**, 107–115
Nybäck, H., Sedvall, G. & Kopin, I. J. (1967) *Life Sci.* **6**, 2307–2312
O'Keeffe, R., Sharman, D. F. & Vogt, M. (1970) *Brit. J. Pharmacol.* **38**, **287**–304
Perringer, E., Jenner, P. & Marsden, C. D. (1975) *J. Pharm. Pharmacol.* **27**, 442–444
Portig, P. J. & Vogt, M. (1969) *J. Physiol. (London)* **204**, 687–715
Roffler-Tarlov, S., Sharman, D. F. & Tegerdine, P. (1971) *Brit. J. Pharmacol.* **42**, 343–351
Roos, B.-E. (1965) *J. Pharm. Pharmacol.* **17**, 820–821
Seeman, P. and Lee, T. (1975) *Science* **188**, 1217–1219
Sharman, D. F. (1963) *Brit. J. Pharmacol. Chemother.* **20**, 204–213
Sharman, D. F. (1966) *Brit. J. Pharmacol. Chemother.* **28**, 153–163
Stadler, H., Lloyd, K. G., Gadea-Ciria, M. & Bartholini, G. (1973) *Brain Res.* **55**, 47 –480
Stadler, H., Lloyd, K. G. & Bartholini, G. (1974) *Naunyn-Schmiedebergs Archs. Exp. Pathol. Pharmakol.* **2L3**, 129–134
Stille, G. & Lauener, H. (1971) *Arzneimittel. Forsch.* **21**, 252–255
Stille, G., Lauener, H. & Eichenberger, E. (1971) *Il Farmaco* **26**, 603–625
Zivkovic, B. & Guidotti, A. (1974) *Brain Res.* **79**, 505–509
Zivkovic, B., Guidotti, A. & Costa, E. (1974) *Mol. Pharmacol.* **10**, 717–725

Chapter 8

Long-term Effects of Dyskinesia-inducing Drugs

By CELIA M. YATES

Medical Research Council Brain Metabolism Unit,
University Department of Pharmacology,
1 George Square, Edinburgh EH8 9JZ, U.K.

Introduction

This brief review will be restricted to the phenothiazines, in particular chlorpromazine, on which there is an extensive literature. The butyrophenones will not be discussed due to lack of information on their long-term effects. Some chronic effects of the phenothiazines which may be related to the long-term action of these drugs are: (1) their patterns of metabolism, (2) pigmentation and binding to melanin, (3) effects on metals, (4) pathological effects and (5) effects on CSF metabolites.

Metabolism of Phenothiazines

The complexity of the metabolism of the phenothiazines is indicated by the fact that at least 35 metabolites have been identified in man (Turano *et al.*, 1974). Some of these metabolites, notably the 7-hydroxy derivative, share with chlorpromazine the ability to block dopamine receptors, whereas others, such as chlorpromazine sulphoxide appear to be pharmacologically inactive (Bunney & Aghajanian, 1974). Considerable variation in the plasma levels of the parent drug and its metabolites is found between patients ostensibly receiving the same dose. Mackay *et al.* (1974) found that, although there was no direct correlation between the level of plasma chlorpromazine and clinical response, schizophrenic patients responding well to the drug had higher plasma levels of the active metabolites than did patients responding poorly to the drug. There is some evidence that the metabolism of chlorpromazine changes with the duration of treatment. For instance, the level of chlorpromazine in plasma from patients given a constant daily dose has been reported to fall several weeks after commencement of therapy (Curry *et al.*, 1972; Stevenson *et al.*, 1972). I have noticed that in rats given daily intraperitoneal injections of 20 mg chlorpromazine/kg the initial signs of extreme prostration and loss of appetite pass off after about 2 days. It is also possible that the metabolism of chlorpromazine, as of other drugs, varies with the dose. This could contribute to the well-documented extremely slow clearance of the drug from the body, chlorpromazine and/or its metabolites being detectable in the urine several months after discontinuation of therapy (Sakalis *et al.*, 1972). These variations in metabolism may be related to the onset of long-term clinical effects such as the dyskinesias, although the time course of the metabolic changes would appear to be shorter than the duration of treatment required to produce dyskinesias.

Pigmentation Caused by Phenothiazines

Another long-term effect of phenothiazines is pigmentation. Marzuli (1968) found that skin pigmentation occurred in about 1% of patients given not less than 800 mg chlorpromazine/day. The pigmentation takes several months to develop, requires light, is related to the dose and is reversible. Pigment is also deposited in the viscera after prolonged treatment with chlorpromazine (Greiner & Nicholson, 1964). The chemical nature of the chlorpromazine pigment, which resembles melanin, is not known; it may be a derivative of chlorpromazine itself possibly coupled to a tyrosine moiety. In addition to producing melanization, chlorpromazine tends to accumulate in melanin-containing tissues such as the skin and the choroid of the eye (Meier-Ruge & Cerletti, 1966). Lindquist & Ullberg (1972), using whole body autoradiography, found that chlorpromazine was very rapidly accumulated by the melanin-bearing tissues of mice and that the drug was found in these areas in high concentrations up to 90 days after injection. Albino animals, on the other hand, showed a very low uptake. Retention of chlorpromazine by mice brain was not demonstrable, possibly because mice, unlike men, have no melanin in the substantia nigra (Marsden, 1961). Recently, Lindquist & Ullberg (1974) have shown that chlorpromazine is selectively accumulated by the pigmented cells of human substantia nigra *in vitro*. An attractive hypothesis regarding the function of neuromelanin was proposed by Marsden (1965) and recently extended by Forrest (1974). They suggested that neuromelanin acts, like peripheral melanin (McGinness & Proctor, 1973), as an electron acceptor. Phenothiazines, which have been shown to be electron donors (Lyons & Mackie, 1963), may attach to neuromelanin converting it to an inert compound incapable of acting as an intraneuronal electron trap. This hypothesis demands that the melanin-containing areas of brain concentrate phenothiazines *in vivo* as Lindquist & Ullberg (1974) have demonstrated they do *in vitro*. As far as I know there is no evidence that such an *in vivo* concentration mechanism exists in man or in dog, a species in which the distribution of chlorpromazine in brain has been studied (De Jaramillo & Guth, 1963).

Chelating Action of Phenothiazines

Of possible relevance to their melanization effects is the ability of the phenothiazines to act as chelating agents (Rajan *et al*., 1974). Curzon (1968) suggested that the increased levels of copper found in brain and plasma following treatment with chlorpromazine might be related to an increased activity of the copper-containing enzyme tyrosinase, which is required for the production of peripheral melanin. Guth & Spirtes (1964) speculated that chlorpromazine might bind and transport copper to the basal ganglia where the copper could be deposited to cause a degeneration similar to that of Wilson's disease.

Pathological Studies

Postmortem studies of tissue from patients on long-term phenothiazines have been few and have so far been restricted to pathological examinations.

There have been two such studies, one, mentioned by Dr. Parkes (this volume, Chapter 6) from the National Institute of Mental Health, in which brains from patients on chronic phenothiazines were found not to differ histologically from brains from age-matched controls. The other group (Christensen *et al.*, 1970) reported degenerative changes in the substantia nigra in 27 out of 28 cases of long-standing dyskinesia following administration of different neuroleptics, mainly chlorpromazine. Only 4 out of 28 age- but not sex-matched controls showed similar changes. These pathological findings suggest that neuroleptic-induced dyskinesia may be related to denervation supersensitivity of the caudate neurons consequent to destruction of the dopaminergic cells of the substantia nigra.

Effect of Neuroleptics on CSF Amine Metabolites

In dogs, the increased turnover of brain dopamine produced by acute administration of neuroleptics has been shown to be mirrored by a rise in the concentration of homovanillic acid, the acid metabolite of dopamine, in ventricular cerebrospinal fluid (CSF) (Guldberg & Yates, 1969). In man, the level of homovanillic acid in lumbar CSF was likewise increased after a single massive dose of haloperidol (Persson & Roos, 1968) and following 12 days treatment with chlorpromazine (Fyrö *et al.*, 1974). Long-term treatment with neuroleptics may not, however, produce such marked effects on dopamine metabolism. Messiha (1974) found no increase in the concentration of homovanillic acid in lumbar CSF taken from monkeys showing lingual-buccal dyskinesias as a result of administration of chlorpromazine over a 12 month period. Reports on the effects of chronic dosage with neuroleptics on the level of homovanillic acid in lumbar CSF in man have also been conflicting (Persson & Roos, 1969; Chase *et al.*, 1970), although the problem of obtaining representative age-matched controls in such studies could contribute to these discrepancies.

Long-term Effects of Chlorpromazine in Rats

In view of the paucity of information on the long-term effects of chlorpromazine we are currently investigating Wistar rats which have been given chlorpromazine equivalent to about 3, 8 and 20 mg/kg/day in their drinking water for over a year. The animals have been divided into two groups for behavioural and biochemical studies, respectively. Our aim is to relate the sensitivity to dopamine stimulation as assessed behaviourally, with the turnover of dopamine in the brain as assessed by the levels of homovanillic acid and tyrosine hydroxylase. We use the peripheral dopa decarboxylase inhibitor RO 4-4062 50 mg/kg plus L-dopa 300 mg/kg given intraperitoneally, as the dopamine challenge. In control rats this treatment produced a rather variable increase in locomotor activity followed within 40 min by a more consistent sniffing and licking of the metal lid of the cage. These activities were measured by placing the rat in a cage having a high frequency input to the stainless steel grid floor and a sensor fitted on each of two adjacent sides of the plastic cage basin. Rats which were still receiving chlorpromazine did not appear to react differently to the dopamine challenge than control animals. The next step is to test the rats after they have been off the drug for several days as it is possible that any supersensitivity effects may be masked by the dopamine-blocking effect of administered chlorpromazine. By contrast, animals given chlorpromazine by intraperitoneal injection in higher doses, viz. 2 or 10 mg/kg twice daily over a shorter period (6 days), showed a potentiation of the behavioural response to RO 4-4062 plus L-dopa while on chlorpromazine and a decline in response when injection of chlorpromazine was stopped; the lower dose producing less potentiation than the higher dose. At no time was there a depression in the response to dopamine stimulation such as could be attributed to block of dopamine receptors. These preliminary results suggest that repeated administration of chlorpromazine can produce a dose-related potentiation of dopaminergic stimulation without any behaviourally demonstrable blockade of dopamine receptors such as has been found after single doses of chlorpromazine (Derkach, *et al.*, 1974; McKenzie & Sadof, 1974).

Discussion

H. S. Bachelard: Were there no movement disorders at all in the chronically chlorpromazine-treated rats?

C. M. Yates: Not that we've noticed yet.

E. D. Bird: Some years ago I gave phenothiazines to rhesus monkeys and, in short-term experiments, found an increased basal ganglia dopamine and manganese concentration. However, the manganese increase was greater in the caudate than in the substantia nigra, which was a little surprising as we might have expected more manganese in a pigmented area, such as subtantia nigra.

References

Bunney, B. S. & Aghajanian, G. K. (1974) *Life Sci.* **15**, 309–318
Chase, T. N., Schnur, J. A. & Gordon, E. K. (1970) *Neuropharmacol.* **9**, 265–268
Christensen, E., Møller, J. E. & Faurbye, A. (1970) *Acta Psychiat. Scand.* **46**, 14–23
Curry, S. H., Lader, M. H., Mould, G. P. & Sakalis, G. (1972) *Brit. J. Pharmacol.* **44**, 370–371P
Curzon, G. (1968) in *Biochemical Aspects of Neurological Disorders* (J. N. Cumings & M. Kremer, eds.), 3rd series, pp. 82–98, Blackwell, Oxford
De Jaramillo, G. A. V. & Guth, P. S. (1963) *Biochem. Pharmacol.* **12**, 525–532
Derkach, P., Larochelle, L., Bieger, D. & Hornykiewicz, O. (1974) *Can. J. Physiol. Pharmacol.* **52**, 114–118
Forrest, F. M. (1974) *Advan. Biochem. Psychopharmacol.* **9**, 255–268
Fyrö, B., Wode-Helgodt, B., Borg, S. & Sedvall, G. (1974) *Psychopharmacologia* **35**, 287–294
Greiner, A. C. & Nicholson, G. A. (1964) *Can. Med. Ass. J.* **91**, 627–635
Guldberg, H. C. & Yates, C. M. (1969) *Brit. J. Pharmacol.* **36**, 535–548
Guth, P. S. & Spirtes, M. A. (1964) *Int. Rev. Neurobiol* **7**, 231–278
Lindquist, N. G. & Ullberg, S. (1972) *Acta Pharmacol. Toxicol. suppl.* **31**, 3–32
Lindquist, N. G. & Ullberg, S. (1974) *Advan. Biochem. Psychopharmacol.* **9**, 413–423
Lyons, L. E. & Mackie, J. C. (1963) *Nature (London)* **197**, 589
McGinness, J. & Proctor, P. (1973) *J. Theor. Biol.* **39**, 677–678
Mackay, A. V. P., Healey, A. F. & Baker, J. (1974) *Brit. J. Clin. Pharmacol.* **1**, 425–430
McKenzie, G. M. & Sadof, M. (1974) *J. Pharm. Pharmacol.* **26**, 280–281
Marsden, C. D. (1961) *J. Anat.* **95**, 256–261

Marsden, C. D. (1965) *Lancet,* **iii** 475–476
Marzuli, F. N. (1968) *Fd Cosmet. Toxicol.* **6**, 221–234
Meier-Ruge, W. & Cerletti, A. (1966) *Ophthalmologica* **151**, 512–533
Messiha, F. S. (1974) *J. Neurol. Sci.* **21**, 39–46
Persson, T. & Roos, B. E. (1968) *Nature (London)* **217**, 854
Persson, T. & Roos, B. E. (1969) *Brit. J. Psychiat.* **115**, 95–98
Rajan, K. S., Marian, A. A., Davis, J. M. & Skripkus, A. (1974) *Advan. Biochem. Psychopharmacol.* **9**, 571–591
Sakalis, G., Curry, S. H., Mould, G. P. & Lader, M. H. (1972) *Clin. Pharmacol. Therap.* **13**, 931–94
Stevenson, I. H., O'Malley, K., Turnbull, M. J. & Ballinger, B. R. (1972) *J. Pharm. Pharmacol.* **24**, 577–578
Turano, P., Turner, W. J. & Donato, D. (1974) *Advan. Biochem. Psychopharmacol.* **9**, 315–322

Discussion of Chapters 1 to 8

C. E. Rowe: Are the increases in sensitivity to dopamine observed after denervation, or after the blocking of receptors by drugs, accompanied by increases in sensitivity of adenylate cyclase to dopamine?

L. L. Iversen: Thank you, that is something I forgot to say. There is evidence of that sort from Dr. Makman and his colleagues in New York [R. K. Michra, E. L. Gardner, R. Katzman & M. H. Makman (1974) *Proc. Nat. Acad. Sci. U.S.A.* 7, 3883–3887], who have shown quite recently that one does get the predicted changes in the dopamine-sensitive adenylate cyclase activity in striatal homogenates after 6-hydroxydopamine or electrolytic lesions of the nigro-striatal pathways. Looking at the striatal tissue they found an increase of approximately two fold in response to dopamine sometime after the lesion, so this fits quite well. We tried to do similar experiments with rats, but we were not able to pick up the expected increase in dopamine sensitivity of adenylate cyclase, so the situation is still a little confusing.

J. D. Parkes: In human Parkinsonism, there is some very interesting evidence to suggest that one aspect of the behaviour of dopamine receptors just may change with prolonged use of levodopa. In patients and in normal controls, after a single oral dopa of levodopa, the level of growth hormone in the blood will go up three- to fivefold (from about 2 ng to about 30 ng or thereabouts). The fasting level of growth hormone isn't affected by chronic dopa treatment. Prolactin, in contrast, does the opposite with a single dose of levodopa. One group of American workers has recently shown that, after chronic dopa treatment of patients with Parkinson's disease for three or four years, there may be a dissociation between these two responses [Malarkey *et al.* (1974) *J. Clin. Endocrinol. Metab.* **39**, 229–235]. The response of growth hormone is totally lost, whereas the response of prolactin continues. Six or seven patients were studied and I think one showed a slight rise in growth hormone, the rest didn't. We've repeated the growth hormone studies, but have been unable to find this effect after three to five years dopa treatment of six patients; we are still getting quite good rises of growth hormone, but we may not have gone on long enough. Certainly the mechanisms of supersensitivity we are talking about occur in weeks, and after a few weeks treatment of levodopa we could get no change at all in the responses of the hormones to a standard dose of dopa.

L. L. Iversen: This is an interesting observation. I have been particularly impressed by the opposite prolactin response that one gets from neuroleptic drugs. All the antidopamine drugs give this response; even drugs like clozapine and thioridazine, which as you have heard are anomalous in some other respects, give a perfectly respectable prolactin response.

J. D. Parkes: There is, of course, another anomaly. As yet, bromocriptine almost certainly does not cause a rise in growth hormone in normals or Parkinsonians although it lowers it in acromegalics.

C. D. Marsden: Can I pursue this question of supersensitivity. All of us use this term, and the theme going through the whole morning has been whether these curious dyskinesias we are talking about are due to supersensitivity of dopamine receptors in the corpus striatum. The nigrostiatal dopamine pathway appears to terminate on an inhibitory striatal synapse. Dopamine released at this synapse can only stop the receptor neuron from firing. If the dopamine receptor becomes supersensitive, the neuron can only stop firing to a lower concentration of transmitter. This is what supersensitivity must imply physiologically. Yet, by all accounts, these striatal neurons are pretty silent anyway. So how can an extrasensitivity to inhibition of already quiet neurons generate abnormal movements?

D. B. Calne: The whole thing is too complicated. Inhibition at any level may eventually lead to excitation.

D. W. Straughan: I think the big paradox is to try and associate what happens at the neuronal level because this will always lead one astray. One always thinks that if neurons are firing less, then behaviourally there should be less movement, but so often the exact opposite is the case. I wouldn't even try to think from one level to the other.

C. D. Marsden: I would accept that; one just has to throw in another inhibitory or excitatory synapse.

R. Morris: I often wonder why people don't look more at the pallidal outflow. Globus pallidus neurons are highly spontaneously active and the removal of inhibitory actions in the striatum may be reflected most clearly here. If the removal of dopaminergic inhibition results in an increased activity in striatal efferents, and these inhibit the globus pallidus, a reduction in pallidal activity may be anticipated. Generally the output of the basal ganglia is expressed primarily via pallidal–thalamic projections and more attention should be focused on this element of the system.

C. D. Marsden: Certainly, but the dopamine receptors that are said to be supersensitive and then cause abnormal movements, are in the striatum.

C. D. Richards: We have heard a great deal about the role of the basal ganglia in dyskinesias, but no evidence that they are involved in motor control. Is there any evidence to suggest that basal ganglia are implicated in motor control?

C. D. Marsden: There is one piece of evidence other than the clinical human evidence that when something goes wrong with the basal ganglia motor performance is compromised. The work of Evart's group [M. R. De Long & P. L. Strick (1974) *Brain Res.* **71**, 327–335] has shown that these cells do discharge preparatory to a voluntary movement in unanaesthetized monkeys.

D. W. Straughan: What happens if you make lesions in these areas, because you can get compensation?

C. D. Marsden: Virtually nothing. That is one of the things that has made the whole subject of basal ganglia physiology so difficult. One can stick electrodes into it, you can stimulate it, you can make holes in it, but very little happens.

L. L. Iversen: I wonder if I can have the prerogative of having the last word on this subject. Even if we look at it in the very vague pharmacological way, it still makes some sense. We know pharmacologically that there is some sort of balance between cholinergic and dopaminergic factors in the motor functions of the basal ganglia. One can upset and produce motor disorders either with dopamine drugs or with cholinergic agents. If you have supersensitivity in one or other system, say the dopamine system, then surely you are upsetting the normal balance and would expect to be in a situation leading to motor abnormality. But that is not getting any nearer the actual neuronal circuitory involved.

Chapter 9

Biochemical Studies on γ-Aminobutyric Acid Metabolism in Huntington's Chorea

By EDWARD D. BIRD

Department of Neurological Surgery and Neurology, Addenbrooke's Hospital, Cambridge,
and
Medical Research Council Neurochemical Pharmacology Unit,
Department of Pharmacology, University of Cambridge Medical School,
Cambridge CB2 2QD, U.K.

Introduction

Huntington's chorea is an autosomal dominant disorder that may have a sex-related modifying factor (Bird *et al.*, 1974). Usually the disorder is manifest during middle life with the onset of involuntary movements that become more severe until these patients are bedridden and die some 15 years after the onset. Dementia also occurs and progresses throughout the course so that in their last few years afflicted cases usually require hospitalization in a mental hospital. The course of the disease is related to the age of onset with a short course in the young and a mild progressive course in the elderly. It is also of some biochemical interest that epilepsy is fairly common when the disorder occurs in young subjects.

The greatest atrophy in the brain is seen in the basal ganglia, a large group of nuclei that control movement through numerous interconnections with both the thalamus and motor cortex. This area of the brain is rich in amine neurotransmitters, and it was in Parkinson's disease, another disorder affecting this region of the brain, that Ehringer & Hornykiewicz (1960) found a decreased concentration of dopamine in postmortem brain. As a result of this investigation L-dopa and the decarboxylase inhibitors have been developed as useful pharmacologic agents for Parkinson's disease.

The observations that the phenothiazine drugs are effective in reducing choreiform movements and that these drugs in high or prolonged doses can produce a Parkinson-like state stimulated hypotheses concerning the possibility that dopamine might be involved in Huntington's chorea. Since there was little evidence that there was either an increased concentration or increased turnover of dopamine in the brain, it was suggested that dopamine receptors were hyperresponsive to dopamine (Klawans, 1970).

The clinical presence of uninhibited movements in chorea also indicated that a neuroinhibitory transmitter, such as γ-aminobutyric acid (GABA), might be deficient in the basal ganglia of the brain. It appears that GABA is an inhibitory neurotransmitter. In the lobster nervous system there is 100 times more GABA present in the inhibitory motor axons than in the excitory axons (Hall *et al.*, 1970).

GABA is synthesized from glutamate by glutamic acid decarboxylase (GAD) and is metabolized to succinic semialdehyde by GABA-glutamate transferase (GABA-T) which is a reversible enzyme, although the amount of GABA formed from succinic semialdehyde is considered to be very small. GABA-T is very unstable in postmortem brain, whereas GAD is fairly stable (McGeer *et al.*, 1971); therefore, there is a postmortem rise in the GABA concentration in brain tissue (Minard & Mushahwar, 1966). For this reason GAD activity in postmortem brain may provide a better reflexion of the state of GABA metabolism prior to death.

Preliminary studies on both animal and human postmortem brain tissue indicated to us that a number of neurotransmitter biosynthetic enzymes could be measured some hours after death, and we therefore initiated a systematic protocol for the collection of postmortem choreic brain tissue throughout England and Wales.

In our early studies we organized the collection of postmortem choreic brain that had been frozen at the time of autopsy (Bird *et al.*, 1973). The dissection of the atrophic choreic brain in the frozen state was difficult, and had to be limited to recognizable areas. In order to provide a broader neurochemical brain profile, we now arrange to collect the brain at the time of autopsy, keep it at ice temperature and dissect ten areas that are of neurochemical interest when the tissue arrives back in Cambridge a few hours after being collected. Dissection of areas such as substantia nigra, hypothalamus, olfactory tubercle and hippocampus were made from microtome sections, and tissues were stored in liquid nitrogen along with aliquots from larger areas, such as caudate nucleus, putamen, globus pallidus and dentate nucleus of the cerebellum, that were dissected directly from the fresh tissue. An equal number of coroner's cases at Addenbrooke's Hospital were used for controls and were handled in a similar way to choreic brains.

Since the basal ganglia of the brain contain the highest concentrations of GABA, dopamine and acetylcholine in the central nervous system it would appear that these neurotransmitters are all related one to another in some way, and we considered it important to examine all three systems in the postmortem choreic brain.

The neurotransmitters measured in the postmortem brain were GABA, dopamine, noradrenaline and their respective biosynthetic enzymes GAD, tyrosine hydroxylase (T-OH), dopamine β-hydroxylase (DBH) and choline acetyltransferase as a marker enzyme for cholinergic neurons. The biochemical assay methods used have been described previously (Bird & Iversen, 1974) except for noradrenaline, which was measured by the method of Cuello *et al.* (1973) and DBH which was measured by a method similar to that described by Molinoff *et al.* (1971).

Glutamic Acid Decarboxylase

GAD activity has now been measured in ten areas of the choreic brain and the values are compared to those in the control group in Table 1. The areas of the control group have been listed in descending order of GAD activity. The

GAD activities in choreic brain may be broadly divided into two groups: those areas at the upper portion of table 1 that have the greatest decrease in GAD activity and those areas at the bottom of the table that reveal no significant differences.

Table 1. *Activity of glutamic acid decarboxylase in control and Huntington's choreic postmortem brain*

	$\mu mol\ ^{14}CO_2$ evolved/h/g tissue (mean ± S.E.M.)		
Area	Control	Huntington's chorea	$P<$
Globus pallidus	7.1 ± 1.3 (24)	1.9 ± 0.5 (14)	0.001
Substantia nigra	6.5 ± 1.0 (38)	2.2 ± 0.3 (40)	0.001
Griseum septum	6.4 ± 4.3 (4)	2.5 ± 1.7 (2)	N.S.
Caudate nucleus	5.1 ± 0.4 (68)	1.5 ± 0.2 (66)	0.001
Putamen	4.4 ± 0.4 (45)	0.9 ± 0.2 (41)	0.001
Dentate nucleus	4.3 ± 0.6 (21)	2.6 ± 0.6 (11)	0.025
Olfactory tubercle	3.1 ± 0.8 (11)	1.7 ± 0.3 (13)	N.S.
Hypothalamus	3.2 ± 0.5 (19)	4.4 ± 0.7 (18)	N.S.
Frontal cortex	3.0 ± 0.3 (30)	2.7 ± 0.3 (23)	N.S.
Hippocampus	1.9 ± 0.4 (15)	2.0 ± 0.4 (12)	N.S.

The number of brains are shown in parentheses.

The GAD activities of the globus pallidus, subtantia nigra, griseum septum, caudate nucleus and putamen were decreased by a highly significant degree, whereas the dentate nucleus is decreased, but only barely significantly different from controls. The olfactory tubercle, hypothalamus, frontal cortex and hippocampus had GAD activities that were not significantly different from controls. It should be noted that in no case does the GAD activity decrease in any area by more than 80% and that the remaining GAD activity, i.e. approximately 2 μmol/g/h, in the areas that had the greatest decrease is about the same as the activity in those areas that did not have any decrease. These activities are similar to the 80% decrease in GAD activity that McGeer *et al.* (1971) found in the substantia nigra after producing lesions in the striatum.

These data can be considered in the light of what is known about the pathological changes that have been seen in choreic brain. Although the greatest atrophy of the choreic brain is seen in the basal ganglia, where there is a weight loss of 50%, it should be realized that the largest overall cell loss in the choreic brain must be in cortical areas, since the average choreic brain is 200 g lighter than the average control brain and the loss from the basal ganglia accounts for only about 10% of this total loss.

It should also be noted that the postmortem brain from the rigid choreic patients has a greater tissue loss from the basal ganglia than the choreic group as a whole. This has been confirmed in histopathologic studies, where cell degeneration in the basal ganglia is severe (Byers *et al.*, 1973).

Although the histopathological examination of the choreic brains from cases that had the disorder for many years usually shows a loss of both small cells and large cells in the caudate nucleus, cases that die somewhat prematurely appear to have a greater loss of small cells than large cells. These small cells are short interneurons that have for some time been considered to have an inhibitory function in the brain, and it is likely that these small cells are the interneurons that use GABA as their neurotransmitter. In addition, there appears to be a striatonigral pathway of GABA-containing neurons since lesions of this tract in the striatum result in an 80% decrease in GAD activity in the substantia nigra (McGeer *et al.*, 1971). In the postmortem choreic brain we have found that the GAD activity in the substantia nigra is decreased to the same degree as in the caudate, putamen and globus pallidus (Table 1). This decrease, however, does not correlate with the normal histopathologic appearance described for the substantia nigra (Stone & Falstein, 1938). It has been our impression, however, from dissection of the substantia nigra, that the zona reticulata appears to be more atrophic than the zona compacta. This would coincide with our present neurochemical data on these two zones within the substantia nigra. The zona compacta of the normal brain contains twice the dopamine concentration and tyrosine hydroxylase (T-OH) activity as that in the zona reticulata. The GAD activities of the two zones of the substantia nigra are the same when taken at a midcaudal–rostral section of the substantia nigra. It would be reasonable, therefore, to assume that since we have found GABA decreased in the choreic substantia nigra there would be atrophic changes in the zona reticulata.

We have previously noted that there did not appear to be a decrease in GAD activity in the frontal cortex in spite of the fact that histopathological examination of this area reveals degeneration of cells and astrocytic proliferation. This was in agreement with Perry *et al.* (1973) where the GABA concentration was found to be normal in frontal cortex of the postmortem choreic brain. Perry has now examined the occipital cortex as well, an area that has the highest cortical GAD activity and after examining further cases he finds that there is a slight decrease in GABA concentration in the occipital area of choreic brain (T. L. Perry, unpublished work).

A number of neuropathological reports have suggested that there appears to be a greater atrophy in the frontal cortical area (Stone & Falstein, 1938), but J. A. N. Corsellis (unpublished work) believes that the atrophy of the cortex is fairly uniform throughout.

The hippocampus of the choreic brain is of special interest since it is often singled out as having a normal histopathological appearance. Our normal data for GAD activity (Table 1) would appear to support the histopathology. The hippocampus is also of interest since it contains short interneurons that are considered to be GABA-containing neurons not originating from fibre tracts of other areas of the brain, since sectioning of the fornix does not alter GAD activity of this area of the hippocampus (Fonnum, 1970).

The olfactory tubercle is of some interest since it has been found in animals to contain a high density of dopamine terminals. We have isolated this area in the human postmortem brain by subsectioning the anterior perforating substance and taking a small rectangular area that included the very small tubercle along the olfactory striae where we have found the highest dopamine concentration. Examination of the GAD activity of this area in choreic brain reveals that there was no significant difference from control brain (Table 1).

The hypothalamus is also of special interest in Huntington's chorea since hypothalamic dopamine has now been shown to have an influence on hormone-releasing factors in the hypothalamus. Since the clinical neurological manifestations of Huntington's chorea appear to be secondary to the hyperresponsive dopamine receptors, it would be logical to assume that there may be hormonal imbalances that could be secondary to these hyperresponsive dopamine receptors in the hypothalamus, and this disturbance may be reflected by altered concentrations of various pituitary hormones measured in the plasma of the choreic patient.

Growth hormone has been reported to be increased in the plasma following the administration of L-dopa (Boyd *et al.*, 1970). We have found that the growth hormone concentration in the plasma of choreic patients is increased, and that there is an increased growth hormone response to insulin-induced hypoglycaemia (Phillipson *et al.*, unpublished work). Choreic patients also have an exaggerated growth hormone response to the administration of L-dopa (Podolsky & Leopold, 1974).

In the postmortem choreic hypothalamus we found that GAD activity was increased rather than decreased, but this increase was not significant in the number of brains that we have examined so far (Table 1). Dopamine and noradrenaline concentrations and T-OH and DBH activities were also normal in the hypothalamus of the choreic brain (Table 2).

Tyrosine Hydroxylase (T-OH)

We have now been able to examine a much larger series of cases for T-OH activity and now report on values in the substantia nigra (see Table 2). The T-OH activity was twofold greater in the caudate than the putamen in both the control and choreic cases. There was no difference in T-OH activity between the control and choreic caudate and putamen. In the substantia nigra, however, there was twofold increase in T-OH activity of the choreic tissue per unit wet weight when compared to controls.

It has been our impression that the substantia nigra appeared darker than usual in the choreic brain stem. This has also been noted by J. A. N. Corsellis (personal communication). Increased extraneuronal pigment in the choreic substantia nigra has been noted by Forno & Jose (1973). However, in addition to being darker, the zona reticulata of the substantia nigra appears smaller. Since there are few cholinergic neurons in the substantia nigra it would appear that the greater majority of cells in the substantia nigra are GABA- and dopamine-containing neurons. As we have already shown that the GAD concentration is decreased by at least 75% and unless such cells are replaced by glial cells, for which there is little evidence, this would increase the concentration of intact

Table 2. *Tyrosine hydroxylase, dopamine, noradrenaline and dopamine β-hydroxylase in control and Huntington's chorea postmortem brain tissue*

	T-OH		Dopamine		Noradrenaline		DBH	
	Control	Choreic	Control	Choreic	Control	Choreic	Control	Choreic
Caudate	4.9 ± 1.4 (43)	4.1 ± 1.0 (45)	2.5 ± 0.5 (30)	1.0 ± 0.2* (33)				
Putamen	1.9 ± 0.7 (10)	1.8 ± 0.7 (19)	2.3 ± 0.4 (23)	1.8 ± 0.2 (29)				
Substantia nigra	3.7 ± 0.8 (32)	8.6 ± 2.4 (29)	0.4 ± 0.1 (5)	0.9 ± 0.1* (5)				
Hypothalamus			0.13 ± 0.05 (19)	0.4 ± 0.3 (18)	0.9 ± 0.4 (19)	0.27 ± 0.07 (18)	146 ± 21 (7)	147 ± 21 (6)
Olfactory tubercle			0.47 ± 0.1 (11)	0.38 ± 0.1 (13)				
Locus coeruleus							1320 ± 107 (12)	1145 ± 102 (11)

* = $P < 0.01$.
T-OH = tyrosine hydroxylase (nmol/g/h), dopamine (μg/g), noradrenaline (μg/g).
DBH = dopamine β-hydroxylase (nmol/g/h).
All values are means ±S.E.M. for the number of brains shown in parentheses.

dopaminergic neurons by almost twofold and may explain the twofold increase in both T-OH activity and dopamine concentration.

Dopamine

Whereas we had previously demonstrated a slight decrease in dopamine in the caudate, this decrease now appears to be significantly different from controls. However, in the putamen and olfactory tubercle there did not appear to be any significant difference in dopamine concentrations between control and choreic tissue. It was of interest to note that in the substantia nigra the dopamine concentration was increased to a similar degree as T-OH activity, i.e. twofold in the choreic group, compared to the controls. The other area of interest is the hypothalamus where there is also an increase in dopamine concentration although not significant, it is similar to the trend for GAD activity, being the only area in the brain where there is an increase in GAD activity.

Noradrenaline

The noradrenaline concentration was found to be decreased in the hypothalamus of the choreic brain when compared to controls.

Dopamine β-Hydroxylase

DBH activity was measured in the two areas of the brain where it is most active, the locus coeruleus and the hypothalamus (Table 2). The activity in the locus coeruleus was about 35% less than the activity that was found by R. E. Zigmond & L. L. Iversen (unpublished work) in the rat locus coeruleus measured immediately after death. Wise & Stein (1973) found that in the mouse brain kept under conditions similar to human post mortem conditions the activity of DBH decreased by about 40%, so it would appear that the activity noted in Table 2 would be similar to that found in the rat.

The DBH activity in the locus coeruleus and hypothalamus of choreic brain was not significantly different from normal, or from three schizophrenic brains that were examined.

The knowledge that dopaminergic neurons appear to be intact in the choreic brain gives further support to the suggestion that in Huntington's chorea there is a specific degeneration of the GABA containing neurons. Much work will be needed to determine why these neurons are selectively damaged, but it would appear from the data presented that since we have found a number of areas where GAD activity appears normal there is not a genetic defect in the production of GAD. Further information will be needed to determine the characteristics of the GAD in the various areas of the brain to determine if there are different GAD enzymes involved in the various areas that we have examined.

Pharmacology

Although it would appear that continued efforts must be made to determine the biochemical defect in Huntington's chorea well before extensive neurological damage has occurred, efforts are being made to treat the choreic patient with drugs that might increase the GABA concentration of the brain. We have given Lioresal (baclofen), a GABA-like analogue, but did not see any benefit from this agent. Efforts to block GABA-glutamate transaminase with agents such as Epilim (*n*-dipropyl acetate) have been tried and do not appear to be hopeful. Perry has tried aminooxyacetic acid and had obtained some encouraging preliminary results (T. L. Perry, unpublished work). Since GABA does not pass the blood–brain barrier large doses have been given in an attempt to overcome this barrier, but such large amounts by mouth could be dangerous (Perry *et al.*, 1974).

Deanol (2-dimethylaminoethanol), an agent that is thought to increase the acetylcholine content of cells, had some encouraging preliminary reports (Walker *et al.,* 1973), but we have found that in ten patients on a double blind trial for 6 months with deanol we were unable to document any improvement with this drug.

It would seem reasonable that before one spends a great deal of time in developing methods to increase the GABA concentration of the brain, an effort should be made to determine whether or not the GABA receptors are intact in choreic basal ganglia and still able to respond to GABA.

With the co-operation of a large number of consulting psychiatrists and pathologists throughout England and Wales a successful scheme for the collection of choreic brain has been developed. Much of this tissue has been used in the assays to date and tissues have been sent to various laboratories investigating Huntington's chorea around the world. It is hoped that through such a collection scheme we will be able to provide more tissues to established worthwhile investigations so that help could soon come to families that have been afflicted with this terrible disorder for generations.

Acknowledgements

The author wishes to thank Dr. Leslie L. Iversen and his group in the Medical Research Council Neurochemical Pharmacology for their help and advice. Mrs. Elizabeth Palfreyman and Miss Whitney Macauley Gordon provided excellent technical assistance. The research was supported in part by a grant from the National Fund for Research into Crippling Diseases.

Discussion

P. V. Taberner: The tissue you obtained was all post mortem. Had all these patients died from Huntington's chorea, and, if not, were any changes in enzyme activities observed at earlier stages of disease? Is it feasible at present to obtain biopsy tissue from patients in the early stages of the disease?

E. D. Bird: I think it will eventually become possible. I think that now most people know that research is going on in Huntington's chorea, the families with this disease will alert somebody that an earlier case has had a car accident, or died in unusual circumstances. We have had a few cases that way, but, again, they were cases that had died four or five years before the expected death. The changes were the same, with decreased GAD activity, and we would like to get them even earlier, in order to look at that and many other things. I think we will eventually get such material. Biopsy tissue would be the thing we would like to get, but at the moment we don't have any evidence of any abnormality in the frontal cortex.

M. G. Palfreyman: Is there any evidence for a decrease in serum prolactin in Huntington's chorea? There is some evidence for an increase of prolactin in Parkinsonism, which might be associated with increased sebaceous secretions, and decreased potency or fertility. It has been suggested that this increase in prolactin is the consequence of the reduced dopamine in this disease. If serum prolactin were reduced in Huntington's chorea, this might offer a means of biochemical screening for the latter disease.

E. D. Bird: No. I'm quite anxious to examine this.

P. J. Roberts: Have you investigated any other putative neurotransmitters in the caudate, for examination, 5-HT or histamine, in view of reports of altered urinary excretions of methyl histamine in choreic patients?

E. D. Bird: No, we haven't. I don't know whether Dr. Curzon can say anything on that.

G. Curzon: We haven't looked at histamine. CSF 5-HIAA is absolutely normal, or even non-significantly high. There is no indication of a 5-HT abnormality.

L. T. Fowler: I don't believe Dr. Bird mentioned the decrease in GABA concentration in choreic material, which has led people to suggest that a possible approach to therapy might be to elevate brain GABA levels.

E. D. Bird: We have looked at this, and found that GABA is decreased by about 50%. I did want to talk about therapy. I would suspect that if we do get a GABA substance that enters the brain, as we have areas of perfectly normal GABA, we are going to see complications to such therapy.

C. C. Jordan: Has anyone tried baclofen (Lioresal) at all, which is a GABA derivative which crosses the blood–brain barrier?

E. D. Bird: Yes, I tried in a few patients, and found it not really useful.

R. F. Metcalfe: Has any attempt been made to detect a biochemical lesion in young people who are genetically predisposed to Huntington's chorea?

E. D. Bird: No, we have not been able to detect anything. There are some people who have suggested that if you give levodopa to a non-choreic member of a choreic family, some may be susceptible to develop movements. However, we will have to wait about 10 or 12 years before the results of this study can be verified, i.e. whether those who develop chorea on levodopa prove to have the disease. I think there are a lot of ethical problems involved with such predictive tests.

C. D. Marsden: A social study has been done by Mrs. Barette on a large number of people who have Huntington's chorea in their family. They were asked if, given a test which may show that they would develop chorea subsequently, 90% of them said they wanted the test.

J. Gumpert: I found exactly the same.

C. D. Marsden: Families by and large are desperate for some sort of test, so the medical profession has to decide on the ethical issue of carrying it out.

E. D. Bird: I personally think that we should not do it, even though they want it, until we have something reasonable to treat them with.

J. D. Parkes: Dr. Bird has raised the question about growth hormones. Podolsky, after the original paper suggesting elevated growth hormone levels in Huntington's chorea, suggested also that the patients did not show a normal rise of growth hormone in response to levodopa [Podolsky *et al.* (1974) *J. Clin. Endocrinol. Metab.* **39**, 36–39]. Looking at the pathology, such patients do indeed have a lot of change in the hypothalamus. It is possible that presymptomatic carriers of the disease also have abnormal levels of growth hormone. This may be another possible biochemical approach to that problem.

References

Bird, E. D., Caro, A. J. & Pilling, J. B. (1974) *Ann. Hum. Genet. (London)* **37**, 255–260

Bird, E. D., Mackay, A. V. P., Rayner, A. N. & Iversen, L. L. (1973) *Lancet* **i**, 1090–1092

Bird, E. D. & Iversen, L. L. (1974) *Brain* **97**, 457–472

Boyd, A. E., Lebovitz, H. E. & Pfeiffer, J. B. (1970) *New. Engl. J. Med.* **283**, 1425–1429

Byers, R. K., Gilles, F. H. & Fung, C. (1973) *Neurology (Minneapolis)* **23**, 561–569

Cuello, A. C., Horn, A. S., Mackay, A. V. P. & Iversen, L. L. (1973) *Nature (London)* **243**, 465–467

Ehringer, H. & Hornykiewicz, O. (1960) *Klin. Wschr.* **38**, 1236–1239

Fonnum, F. (1970) *J. Neorochem.* **17**, 1029–1037

Forno, L. S. & Jose, C. (1973) *Advan. Neurol.* **1**, 453–470

Hall, Z. W., Bownds, M. D., Kravitz, E. A. (1970) *J. Cell Biol.* **46**, 290–299

Klawans, H. L. (1970) *Eur. Neurol.* **4**, 148–163

McGeer, P. L., McGeer, E. C., Wada, J. A. & Jung, E. (1971) *Brain Res.* **32**, 425–431

Minard, F. N. & Mushahwar, I. K. (1966) *Life Sci.* **5**, 1409–1413

Molinoff, P. B., Weinshilboum, R. & Axelrod, J. (1971) *J. Pharmacol. Exp. Ther.* **178**, 425–431

Perry, T. L., Hansen, S. & Kloster, M. (1973) *New Engl. J. Med.* **288**, 337–342

Perry, T. L., Hansen, S. & Urquhart, N. (1974) *Lancet* **i**, 995–996

Podolsky, S., Leopold, N. A. & Sax, D. S., (1974) *J. Clin. Endocrinol. Metab.* **39**, 36–39

Stone, T. T. & Falstein, E. I., (1938) *J. Nerv. Ment. Dis.* **88**, 602–626

Walker, J. E., Hoehn, M., Sears, E. & Lewis, J., (1973) *Lancet* **i**, 1512–1513

Wise, C. D. & Stein, L., (1973) *Science* **181**, 344–347

Chapter 10

Effects of Blocking Nigrostriatal γ-Aminobutyric Acid Receptors

By C. J. PYCOCK

University Department of Neurology, Institute of Psychiatry and King's College Hospital, Denmark Hill, London SE5 8AF, U.K.

Introduction

Abnormalities of cerebral biogenic amines have been implicated in the genesis of many human dyskinesias, including the choreas (for current review, see Marsden, 1975). Klawans (1970), for example, has marshalled evidence to suggest that excessive striatal dopaminergic activity is the pharmacological cause of chorea, in particular in Huntington's chorea.

More recently, however, attention has been focused on the possible involvement of γ-aminobutyric acid (GABA) in chorea. Perry and his colleagues (1973) noted that GABA levels of the substantia nigra and striatum were considerably reduced in patients dying of Huntington's chorea; and others have found a large and selective reduction in basal ganglia glutamic acid decarboxylase (GAD) activity (Bird *et al.*, 1973; Bird & Iversen, 1974; McGeer *et al.*, 1973*a*, *b*).

In view of this reported evidence on the involvement of GABA in chorea, a study of the effect of manipulating basal ganglia GABA mechanisms on behaviour was undertaken in the rat. (The results presented in this review are those of work initiated in conjunction with D. Tarsy, B. Meldrum and C. D. Marsden.)

Manipulation of Striatal γ-Aminobutyric Acid

Reductions of basal ganglia γ-aminobutyric acid (GABA) levels have been recorded in Huntington's chorea (Perry *et al.*, 1973). Indeed, the pathology of Huntington's chorea characteristically shows loss of small interneurons in the corpus striatum, while the substantia nigra, unlike that of Parkinson's disease, shows only mild cell degeneration (Bruyn, 1968; Bernheimer *et al.*, 1973). There are high levels of GABA in the striatum, so it is possible that some of the lost neurons will be directly associated with the GABA system. We have therefore studied the effect of focal injections into the caudate nucleus of the rat of drugs known to affect GABA neurotransmission.

Stainless steel guide cannulae were implanted into burrholes in the skulls of Wistar rats. These cannulae were stereotaxically placed vertically over different regions of the brain and held in position with dental cement. All subsequent injections were made in conscious hand-held animals by a microsyringe and needle fitted with a nylon cuff acting as a stop in order to govern the depth of injections. The co-ordinates used for the sites of injection in these experiments were

Anterior caudate nucleus	A 8.0, L ± 2.5, VO
Posterior caudate nucleus	A 6.5, L ± 3.5–4.0, VO
Globus pallidus	A 6.3, L ± 2.5, V−1.0
Thalamus	A 2.4, L ± 2.0, V−0.4
Ventral hippocampus	A 2.4, L ± 4.0, V−2.4

All co-ordinates were obtained from the stereotaxic atlas of the rat brain (König & Klippel, 1963).

In other experiments EEG recordings were monitored from the caudate nucleus and cortex during the injection of drugs into the striatum. Such animals were prepared with screws set in the skull that ended in the surface of the cortex. Striatal activity was recorded from needle electrodes, varnished except for the tip, and inserted through the guide cannulae into the left and right caudate nuclei.

The location of all needle tracts was subsequently determined by both macroscopical and histological examination.

Picrotoxin, a GABA receptor blocking agent (Precht & Yoshida, 1971; Crossman *et al.*, 1973) and the enzyme glutamic acid decarboxylase (GAD) inhibitor DL-*c*-allylglycine (Horton & Meldrum, 1973) were employed as the chief agents to manipulate basal ganglia GABA function.

Unilateral injection of picrotoxin (0.5–1.5 μg in 2–3 μl saline) into the striatum of the rat resulted in a rhythmic jerking movement of the contralateral limbs. This jerking motion has been termed 'focal myoclonus' (Marsden *et al.*, 1975) being reminiscent of the human syndrome. The myoclonus observed commenced between 4–9 min following picrotoxin injection (mean 5.8 ± 0.7 min, 20 animals) and lasted approximately 45–80 min (mean 52.0 ± 3.6 min). The time of onset and duration of myoclonus was approximately the same irrespective of the dose of picrotoxin. The phenomenon usually initially involved a sharp contraction of the flexor muscles of one limb, lifting it off the ground, followed by a somewhat slower phase of relaxation. Such an event lasted a fraction of a second, but would occur repetitively and rhythmically every 5–12 seconds. Between jerks the digits often remained flexed and the whole limb might be kept elevated off the floor surface on occasions. However, the animal apparently suffered no other obvious disability and would resume normal activity such as grooming or washing between the jerks. The flexed posture of the digits and limbs sometimes preceded the jerks by some 2–3 min. Almost always the myoclonus was monophasic, but occasionally a double jerk pattern was observed, and rarely (during the later part of the picrotoxin effect) the jerks built up in frequency merging into a sustained clonic seizure lasting a matter of some minutes, followed again by repetitive, rhythmic jerks. However, myoclonus often spread during the course of the experiment to other regions of the body. Thus a forelimb myoclonus subsequently involved the neck and head, the latter being pushed up and away from the jerking limb, while a hindlimb myoclonus often involved the lower trunk region. Such a spread of myoclonus is probably associated with diffusion of the drug through the striatum, although no direct studies have been made to verify this. However, if picrotoxin's diffusion rate is comparable to that measured for labelled haloperidol by Costall *et al.* (1972) following striatal injection, the drug would be expected to diffuse some 2 mm from the site of injection in one hour, thus covering a large area of the caudate nucleus. On occasions a single picrotoxin injection would cause jerks of both the opposite arm and leg. On such occasions the animal usually took refuge at the side of the cage for additional support. Histological location of the needle tracts suggests that forelimb and head myoclonus usually resulted from injection into the anterior region of the caudate nucleus; hindlimb and trunk myoclonus in general resulted following injection into the posterior region of the caudate nucleus. Such findings infer some topographical organization of the caudate nucleus. Postural deviation was not produced following unilateral striatal picrotoxin injection, and sustained circling behaviour was never seen.

Bilateral injection of picrotoxin (1 μg) into the anterior region of both caudate nuclei resulted in a synchronized flexion and extension of both forelimbs together with an extension of the head, a syndrome reminiscent of retrocollis in the human. Such an effect often produced violent limb contractions, so much so that the animal sat constantly upright on its haunches, and was on occasions thrown over backwards by the force of the contractions. The onset after picrotoxin injection of this movement disorder was of the same order as observed for unilateral injections, although its duration of approximately 90 min was slightly longer.

DL-*c*-Allylglycine (100–150 μg in 2–3 μl saline), an inhibitor of glutamic acid decarboxylase, caused similar effects when injected into one striatum. Again myoclonic jerks of the contralateral limbs, neck and head were observed, but only after a delay of some 45 min after injection (range 20–55 min). Once established the jerks lasted between 65 and 200 min. On one occasion repetitive forelimb jerks built up into repeated focal seizures. The latent period observed from the time of injection until the onset of myoclonus presumably reflects the mechanism by which this drug is acting. Allylglycine causes inhibition of the major GABA synthesizing enzyme GAD, and a similar latency between enzyme inhibition and seizure onset following systemic administration has been demonstrated in mice (Horton & Meldrum, 1973). Further work is required to establish the time course of changes in striatal GABA levels in relation to the onset of myoclonus after focal allylglycine injection in this rat model; such data would provide additional evidence as to the involvement of GABA in this bizarre phenomenon.

Indeed, the question arises as to whether these myoclonic jerks observed following striatal injections of picrotoxin and DL-*c*-allylglycine are due to a specific effect upon GABA neurotransmission or are the result of non-specific actions. Control injections of physiological saline into the caudate nucleus of the rat did not result in myoclonic jerks. Further evidence that the production of focal myoclonus directly involves GABAminergic neurons was provided from experiments using GABA itself. Alone, GABA did not induce myoclonus or any other abnormal movement when injected into the striatum. However, when it was given intrastriatally to rats in which picrotoxin or allylglycine-induced myoclonus was well established, GABA (100–750 μg) injected into the same site caused an inhibition of the jerks. Myoclonus ceased within a few minutes after GABA injection, but returned some minutes later with similar frequency. Control injections of saline in similar circumstances had no effect in abolishing or reducing drug-induced myoclonus. In another series of experiments, using animals in whom contralateral myoclonus had been produced by picrotoxin injection into the anterior caudate nucleus, carnitine (250–500 μg in 2 μl saline), γ-amino-β-hydroxybutyric acid, was subsequently injected into the same site 15–18 min after picrotoxin. In half the animals ($N = 6$) carnitine injection abolished picrotoxin-induced myoclonus within 2–5 min. In two of these three animals the myoclonic jerks ceased for 23 and 45 min respectively and then recurred: in the third myoclonus never reappeared. In two animals carnitine had no effect on picrotoxin-induced myoclonus, and in the remaining animal it decreased the amplitude and frequency of the jerks but did not abolish them.

To provide further support for the proposed GABAminergic involvement, initial experiments have shown the induction of myoclonus following intra-striatal bicuculline. Bicuculline, a GABA receptor blocking agent (Curtis *et al.*, 1970), in doses 1 to 3 μg induced a weak contralateral myoclonus with the time to onset directly comparable to that induced by picrotoxin, but with a much shorter period of duration of approximately 10–12 min. Control injections of vehicle caused no such movements. Subsequent injection of picrotoxin (1.5 μg) into the same site in these animals resulted in the production of the characteristic jerking in the same contralateral limb as had been observed much more weakly following bicuculline.

One must also consider the possibility that picrotoxin-induced myoclonus was due not to an effect within the striatum, but was the result of the diffusion of the drug into neighbouring regions, particularly back along the needle tract into the cerebral cortex. However, it has been established that the direct injection of picrotoxin into the cerebral cortex at the same site as the needle tract rarely resulted in the onset of myoclonus (in less than 10% of animals) whereas striatal location of the drug predictably caused jerking in over 85% of animals tested. In another series of experiments, picrotoxin (0.5–2 μg) was injected into other control sites in the brain. Injection of picrotoxin into thalamus, substantia nigra and ventral hippocampus did not result in the production of a contralateral myoclonus. The higher doses of drug often caused generalized seizures when injected into thalamus or hippocampus.

Similarly the EEG recordings lend some support to the striatal origin. Spikes were recorded from needle electrodes in the striatum before they could be identified in the electrocorticogram (Fig. 1). Striatal spikes were observed before the onset of myoclonus, continued regularly throughout the jerking phase and were abolished by a GABA injection which subsequently inhibited the jerks. When the GABA effect had worn off, both striatal spikes and limb jerks returned.

If such a phenomenon is indeed striatal in origin one might expect that lesions placed in the region of the globus pallidus would abolish drug-induced myoclonus as most of the recognized efferent neurons from the striatum pass into

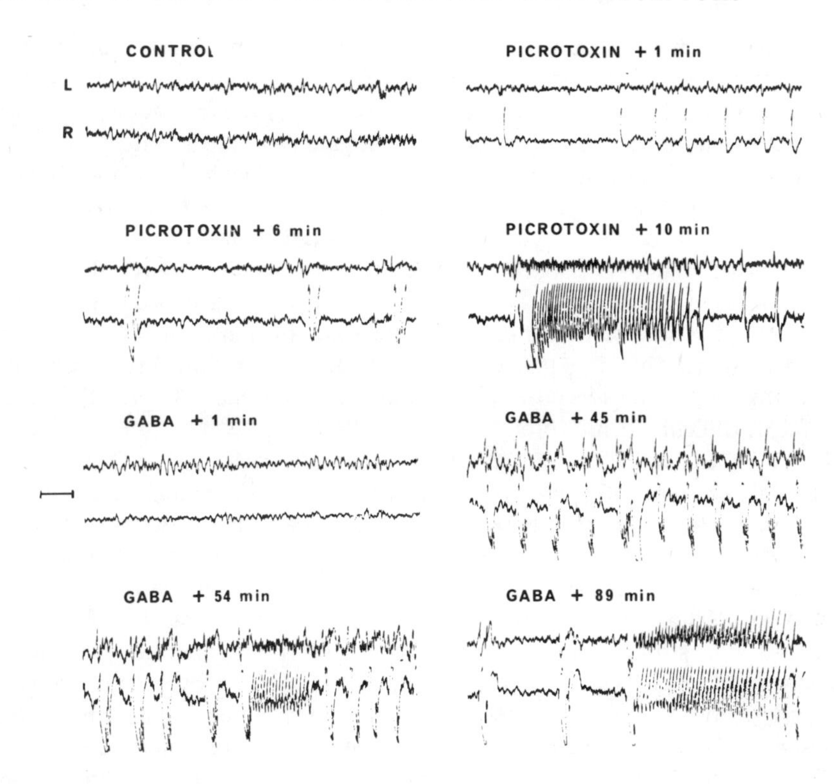

Fig. 1. *EEG records from a rat following focal injections of picrotoxin and GABA into the right striatum*

Bipolar recordings between left (L) and right (R) striatum and the ear. One minute after picrotoxin (0.75 μg in 2 μl) was injected into the right striatum, focal spikes were recorded in the absence of any motor activity. After 6 min triple spikes occurred in association with left fore- and hindlimb jerks. Subsequently bursts of spikes were accompanied by clenching and twitching of the left forelimb. GABA (500 μg in 2 μl) was injected into the right striatum 12 min after the picrotoxin injection. Focal spikes and limb jerks were absent for 40 min, but then reappeared and after 45 min were even more sustained than before GABA injection. Subsequently bursts of spikes associated with a raised left forelimb were frequent. Calibrations 50 μV and 1 s.

the pallidum. The globus pallidus, like other anatomically defined regions of the basal ganglia, is also rich in GABA (Bird *et al.*, 1973; McGeer *et al.*, 1973; Perry *et al.*, 1973). However, the injection of picrotoxin (1–2 μg) into one globus pallidus of the rat did not result in the production of the characteristic contralateral myoclonus: the drug caused a generalized hyperactivity with the animals running around their cage. Circling behaviour with accompanying body asymmetry was not observed. Injection of picrotoxin (1–2 μg) into the anterior region of the caudate nucleus after destruction of the globus pallidus by electrolytic coagulation did not result in the production of contralateral limb myoclonus in six animals. Thus it seems likely that destroying the efferent neuron pathways of the basal ganglia also destroys striatally drug-induced myoclonus. When the lesioned animals were subsequently tested with the dopamine agonist, apomorphine (0.5 mg/kg s.c.) all exhibited marked body asymmetry towards the lesioned side and rotated tightly in this direction at rates varying from 5–14 complete turns/min. Such rotational behaviour, following amphetamine, after unilateral electrolytic lesions of the globus pallidus has already been reported (Naylor & Olley, 1972), and, together with histology, provides a good indication as to the extent of the disruption of the basal ganglia efferent system.

Injection of picrotoxin into the striata of rats in which the nigrostriatal dopaminergic pathway had been previously cut by hemisection provided less convincing evidence since myoclonus was produced. However, it is difficult to ascertain the extent of damage incurred in such an animal model as although the dopaminergic afferent pathways to the striatum may have been completely destroyed, little information is known on the functional integrity of the efferent neuron systems. This, together with the induction of presumed supersensitive denervated dopamine receptors (Arbuthnott, this volume, Chapter 4), makes any result difficult to interpret.

In the light of current observations reported here, one may tentatively conclude that the myoclonus observed is caused by inhibition of a GABAminergic pathway. Further experiments were designed to see if the myoclonus initiated by picrotoxin was dependent upon other striatal neuron systems. Attention has already been drawn to the possible involvement of dopamine in the genesis of chorea (Klawans, 1970), but neither inhibition of cerebral catecholamine function, nor stimulation of catecholamine receptors affected picrotoxin-induced myoclonus. Neither depleting brain catecholamine levels with reserpine (10 mg/kg) or α-methyl-*p*-tyrosine (250 mg/kg), nor increasing catecholamine levels with L-dopa (100 mg/kg) together with the peripheral decarboxylase inhibitor, α-methyldopa hydrazine (25 mg/kg) prevented intrastriatal picrotoxin-induced myoclonus. Similarly blockade of central catecholamine receptors with chlorpromazine (10 mg/kg) did not abolish the phenomenon.

5-Hydroxytryptamine (5-HT) has not been implicated in chorea (see Marsden, 1975), although much interest has been aroused recently in its possible role in the bizarre and equally disabling disease, action myoclonus. This syndrome has been successfully treated in a few cases by the administration of the 5-HT precursor, 5-hydroxytryptophan (Lhermitte *et al.*, 1971; Chadwick *et al.*, 1974). However, manipulation of cerebral 5-HT function was without effect in our animal model. Elevation of neuronal 5-HT with 5-hydroxytryptophan (200 mg/kg) or presumed 5-HT receptor blockade with methysergide (4 mg/kg) had no effect on picrotoxin-induced myoclonus.

A possible abnormality of cholinergic neurons also has been suggested in Huntington's chorea (Aquilonius & Sjöström, 1971), a proposal supported by more recent findings that basal ganglia choline acetylase activity is often lower than normal at autopsy in patients with chorea (McGeer *et al.*, 1973*a*, *b*; Bird & Iversen, 1974). Pretreatment of rats with the anticholinergic scopolamine (10 mg/kg), however, did not modify intrastriatal picrotoxin-induced myoclonus.

In general, our results show a close correlation with those reported independently by McKenzie and colleagues (1972, 1975). These authors (McKenzie *et al.*, 1972) used the term chorea to describe movements produced by the injection of D-tubocurarine into the striatum of rats and cats. In both that paper and in their subsequent publication (McKenzie & Viik, 1975) they noted that picrotoxin would cause a similar phenomenon. However, from their description it seems likely that these movements were identical to those we have described as focal myoclonus. Both McKenzie's studies and our own agree that pretreatment with anticholinergics had no effect on these movements once established, and that a subsequent injection of GABA could temporarily abolish them. However, McKenzie *et al.* (1972) reported that haloperidol (5 mg/kg) and chlorpromazine (15 mg/kg) could reduce or block D-tubocurarine-induced myoclonus, whereas in our studies the latter agent was without effect on picrotoxin-induced myoclonus.

McKenzie & Viir (1975) later suggest that chorea induced by intrastriatal injection of D-tubocurarine, bicuculline or picrotoxin was due to GABA receptor blockade. Indeed, a recent paper by Curtis & Johnston (1974) has shown that D-tubocurarine is a GABA antagonist, although a very non-specific one. In this respect this relatively simple but reproducible behavioural model may prove useful for the investigation of new agents that may either directly affect the basal ganglia GABA mechanisms or function to prevent action myoclonus or focal epilepsies.

However, a certain caution is required at this stage as to the specificity of the model with regard to GABA activity in basal ganglia. The exact neuronal pathways involved are certainly difficult to ascertain. A similar phenomenon has been observed in this laboratory following intrastriatal injections of apomorphine and carbachol, while Dr. R. J. Naylor (personal communication) has reported similar movements using other dopamine agonists. Also, intrastriatal glycine showed a weak transient antagonism of picrotoxin-induced myoclonus. These effects may be due to indirect pharmacological action via the striatal GABA system, but alternatively may indicate a relatively non-specific phenomenon. Further experiments are obviously required to clarify this important point; the direct measurement of basal ganglia GABA levels and GAD activity at various stages during drug-induced myoclonus may provide an important clue as to the involvement of the GABAminergic system.

Manipulation of Substantia Nigra GABA

A number of studies have provided evidence for the existence of GABAminergic inhibitory pathway between striatum and substantia nigra in the rat brain. Uptake studies combined with electron microscopy suggest the presence of GABA nerve terminals in rat substantia nigra, and both GABA cell bodies and nerve terminals in caudate nucleus putamen and globus pallidus (Kim *et al.*, 1971; Hattori *et al.*, 1973; Okata & Hassler, 1973; Storm-Mathisen, 1975). These findings, together with the observation that a hemisection at the level of the ventromedial hypothalamus reduced GAD activity in the substantia nigra but not in the basal ganglia, point to the possible existence of a striatonigral GABAminergic pathway. Additional support for such a neuron system has been provided by electrophysiological evidence (Precht & Yoshida, 1971; Yoshida & Precht, 1971).

The following series of experiments were designed to seek behavioural evidence for the existence of such a GABA-mediated inhibitory synapse on the nigrostriatal dopaminergic system. In rodents a relative increase in activity of one nigrostriatal pathway is known to produce rotational behaviour towards the contralateral side (Ungerstedt, 1971). Unilateral manipulation of the proposed inhibitory GABA system would be expected to cause an imbalance of activity in the dopaminergic nigrostriatal system and thus circling behaviour.

Stainless steel guide cannulae were stereotaxically implanted at co-ordinates A 2.4 and L $\pm$ 2.0 (König & Klippel, 1963) in the rat skull above the substantia nigra. Injections were made using a needle fitted with a nylon cuff so that the needle tip was positioned in the substantia nigra zona compacta (V$-$2.4). Subsequently the site of injection was determined by histological examination.

Unilateral injection of picrotoxin (0.25–0.5 μg in 0.5 μl saline) into the substantia nigra resulted in contralateral rotation accompanied by a tight postural asymmetry (Tarsy *et al.*, 1975). Circling began between 1 and 6 min following drug injection and persisted continuously and regularly for between 10 and 26 min at frequencies of 3–6 turns/min. Lower doses of picrotoxin (0.05 and 0.1 μg) or control injections of saline into substantia nigra produced no such effect. Higher doses of picrotoxin (1 μg) resulted in hopping and running seizures frequently preceded by tonic contralateral head deviation. Injection of picrotoxin into other sites in the brain, as described in the previous section, never resulted in circling behaviour. That picrotoxin-induced rotational behaviour was directly dependent upon an intact nigrostriatal dopaminergic pathway was suggested by additional experiments in which this neuronal pathway was destroyed by prior injection of 6-hydroxydopamine into the substantia nigra. Such animals displayed rotational behaviour ipsilateral to the lesion following methylamphetamine (5 mg/kg i.p.) and contralateral to the lesion following apomorphine (1 mg kg i.p.) but failed to exhibit turning or postural deviation following an intranigral injection of picrotoxin.

The results of this study are consistent wlth an increase in activity of the ipsilateral nigrostriatal pathway resulting from the blockade of a GABA-mediated inhibition of nigral dopaminergic neurons. By itself, GABA (500 μg) injected directly into the substantia nigra caused no circling behaviour or marked postural deviation in the normal animal as may be predicted from this hypothesis. However, initial experiments have indicated that intranigral GABA may induce weak circling behaviour in normal animals stimulated with amphetamine or it may modify amphetamine-induced rotation in animals with previous destruction of one nigrostriatal dopaminergic system.

Conclusions

There is considerable anatomical, biochemical and physiological evidence for the existence of an inhibitory GABA descending pathway in the mammalian brain between the caudate nucleus putamen (neostriatum) and substantia nigra. Presumed manipulation of this neuron system with the GABA receptor blocking agent picrotoxin at these two sites in the rat results in two different behaviours. The injection of picrotoxin into the caudate nucleus induced a contralateral limb myoclonus but no turning behaviour; injection of picrotoxin into the substantia nigra resulted only in rotational behaviour (or generalized seizures) and did not cause myoclonus. Further work is required to completely define the specificity of each system, but nonetheless, these observations provide two animal models on which other pharmacological agents, both of well-defined or uncertain actions, can be tested. Because of the accumulating interest in the role of GABA in certain human diseases characterized by abnormal involuntary movements, such as Huntington's chorea, it is hoped such animal models will provide stimulation for further experiments.

Acknowledgement

This work was supported by the Parkinson's Disease Society.

Discussion

J. C. Watkins: On the question of specificity of the system affected by picrotoxin, and the antagonism of the picrotoxin effects by GABA, it might be useful to try other convulsants which do not work on the GABA system (e.g. strychnine, excitatory amino acids) and inhibitory amino acids other than GABA (e.g. glycine and taurine) and try to reverse the effects.

C. J. Pycock: GABA by itself has no effect. I haven't tried glutamate or glycine.

J. C. Watkins: What about taurine?

C. J. Pycock: No, I haven't tried that yet.

J. C. Watkins: Because I felt that in the antagonism of the picrotoxin effect with GABA, it might have been a non-specific effect of damping down excitation by GABA acting on other synapses perhaps. You might decide whether it was non-specific or not by injecting taurine.

C. J. Pycock: Yes.

L. L. Iversen: Is it possible to construct a dose response for this phenomenon?

C. J. Pycock: No, it seems an all or none action, especially as regards turning behaviour.

References

Aquilonius, S.-M. & Sjöström, R. (1971) *Life Sci.* **10**, 405–414

Bernheimer, H., Birkmayer, W., Hornykiewicz, O., Jellinger, K. & Seitelberger, F. (1973) *J. Neurol. Sci.* **20**, 415–455

Bird, E. D. & Iversen, L. L. (1974) *Brain* **97**, 457–472

Bird, E. D., Mackay, A. V. P., Rayner, C. N. & Iversen, L. L. (1973) *Lancet* **i**, 1090–1092

Bruyn, G. W. (1968) *Handb. Clin. Neurol.* **6**, 298–378

Chadwick, D. W., Reynolds, E. H. & Marsden, C. D. (1974) *Lancet* **ii**, 111

Costall, B., Naylor, R. J. & Olley, J. E. (1972) *Neuropharmacol.* **11**, 645–663

Crossman, A. R., Walker, R. J. & Woodruff, G. N. (1973) *Brit. J. Pharmacol.* **49**, 696–698

Curtis, D. R., Duggan, A. W., Felix, D. & Johnston, G. A. R. (1970) *Nature (London)*, **226**, 1222–1224

Curtis, D. R. & Johnston, G. A. R. (1974) *Ergebn. Physiol.* **69**, 98–188

Hattori, T., McGeer, P. L., Fibiger, H. C. & McGeer, E. C. (1973) *Brain Res.* **54**, 103–114

Horton, R. W. & Meldrum, B. S. (1973) *Brit. J. Pharmacol.* **49**, 52–63

Kim, J. S., Bak, I. J., Hassler, R. & Okata, Y. (1971) *Exp. Brain Res.* **14**, 95–104

Klawans, H. L., Jr. (1970) *Eur. Neurol.* **4**, 148–163

König, J. F. R. & Klippel, R. A. (1963) *The Rat Brain*, Williams and Wilkins, Baltimore

Lhermitte, F., Peterfalvi, M., Marteau, R., Gazengel, J. & Serdaru, M. (1971) *Rev. Neurol.* **124**, 21–31
Marsden, C. D. (1975) *Modern Trends in Neurology* (Williams, D., ed.), vol. 6, Butterworths, London
Marsden, C. D., Meldrum, B. S., Pycock, C. & Tarsy, D. (1975) *J. Physiol.* **246**, 96P
McGeer, P. L., McGeer, E. C. & Fibiger, H. C. (1973*a*) *Lancet* **ii**, 623–624
McGeer, P. L., McGeer, E. C. & Fibiger, H. C. (1973*b*) *Neurology* (*Minneapolis*) **23**, 912–917
McKenzie, G. M. & Viik, K. (1975) *Exp. Neurol.* **46**, 229–243
McKenzie, G. M., Gordon, R. J. & Viik, K. (1972) *Brain Res.* **47**, 439–456
Naylor, R. J. & Olley, J. E. (1972) *Neuropharmacologia* **11**, 91–99
Okata, Y. & Hassler, R. (1973) *Brain Res.* **49**, 214–217
Perry, T. L., Hansen, S. & Kloster, M. (1973) *New Engl. J. Med.* **288**, 337–342
Precht, W. & Yoshida, M. (1971) *Brain Res.* **32**, 229–233
Storm-Mathisen, J. (1975) *Brain Res.* **84**, 409–427
Tarsy, D., Pycock, C., Meldrum, B. S. & Marsden, C. D. (1975) *Brain Res.* **89**, 160–165
Ungerstedt, U. (1971) *Acta Physiol. Scand. Suppl.* **367**, 69–93
Yoshida, M. & Precht, W. (1971) *Brain Res.* **32**, 225–228

Chapter 11

The Muscarinic Receptor for Acetylcholine in Huntington's Chorea

By C. ROBIN HILEY

Department of Pharmacology and Therapeutics, University of Liverpool, Liverpool L69 38X, U.K.

Introduction

Since Hornykiewicz commenced his study of the postmortem catecholamine content of the brains of patients who died while suffering from Parkinson's disease and other disorders of the extrapyramidal motor system (Ehringer & Hornykiewicz, 1960) there has been growing interest in the neurochemical analysis of diseases of the central nervous system. As a result of Hornykiewicz's studies, levodopa was successfully introduced as a therapeutic agent in Parkinson's disease and it is possible that similar advances may be made in the treatment of other diseases of the central nervous system if it is possible to define the biochemical changes which occur.

Although there are many disorders which might profitably be subjected to neurochemical analysis, Huntington's chorea is perhaps a particularly suitable candidate since it is fairly easy to identify and there is no effective treatment for either the movement disorder or the dementia. This disease has recently been investigated in order to determine which transmitter systems are affected (Bird *et al.,* 1973; McGeer *et al.,* 1973*a*, *b*; Bird & Iversen, 1974; Bird, this volume, Chapter 9). The most extensive studies are those which have been performed in Cambridge by Dr. Bird and his colleagues. They have measured the postmortem concentrations of dopamine and γ-aminobutyric acid (GABA), the activity of their synthetic enzymes, tyrosine hydroxylase and glutamic acid decarboxylase, and the activity of choline acetyltransferase. Unfortunately all these parameters only indicate whether there has been a change of transmitter function at the presynaptic cell. The only component of a transmitter system that is associated with the postsynaptic cell is the receptor, and the means to study these macromolecules are not routinely available. However, some workers have developed methods which allow determination of the concentration and properties of some neurotransmitter receptors, including those for glycine (Young & Snyder, 1973), γ-aminobutyric acid (Zukin *et al.,* 1974) and acetylcholine (Burgen *et al.,* 1974*a*; Yamamura & Snyder, 1974).

We have used an affinity label for the muscarinic acetylcholine receptor, [^{3}H]propylbenzilylcholine mustard (N-2′-chloroethyl-N-[2″,3″-$^3H_2$]propyl-2-aminoethyl benzilate; [^{3}H]PrBCM), to determine the concentration and properties of this receptor in the central nervous system (Burgen *et al.,* 1974*a*; Hiley & Burgen, 1974). The highest concentration of receptor occurred in the caudate nucleus, which is reported to be rich in cells which respond to the microiontophoretic application of acetylcholine (Bloom *et al.*, 1965) and which appears to contain only muscarinic receptors for acetylcholine (McLennan & York, 1966).

The postmortem caudate nucleus was studied by Bird & Iversen (1974), who found that choline acetyltransferase activity was reduced in only 50–60% of the cases of Huntington's chorea which they studied. The other choreic brains in their sample had an enzyme activity which was the same as that found in the control caudate nuclei. Similarly, McGeer *et al.* (1973*b*) found that decreases in choline acetyltransferase were unevenly distributed amongst the regions of this nucleus. A study of the muscarinic receptor in this disease is therefore of particular interest since it should reveal whether the postsynaptic component of the cholinergic system shows a similar pattern of degeneration. This chapter describes the results of such a study, together with other experiments on the nature and concentration of the muscarinic receptor in human brain.

Materials and Methods

[^{3}H]PrBCM was used to determine the concentration of muscarinic receptor in homogenates of human brain obtained post mortem. The procedures used in the assay and for the collection of tissue have been described previously (Hiley & Bird, 1974; Hiley & Burgen, 1974).

Binding studies with the reversibly acting homologue of [^{3}H]PrBCM, *N*,*N*-dimethyl-*N*-[2′,3′-$^3H_2$]propyl-2-aminoethyl benzilate ([^{3}H]PrBCh: 0.8 Ci/mmol) were carried out using the microcentrifugation technique of Terenius (1974). Homogenates were diluted with Krebs–Henseleit solution to give a concentration of 0.6 mg–protein/ml and were preincubated for 15 min at 30 °C before the addition of [^{3}H]PrBCh to give the required concentration of ligand. The incubation was continued for a further 15 min and was terminated by centrifugation at 14000 *g* for 30 s in a microcentrifuge (Jobling, Stone, Staffordshire). The surface of the pellet was rinsed twice with Krebs–Henseleit solution, and the radioactivity in the pellet was determined by liquid scintillation counting in a Nuclear-Chicago Unilux II spectrometer at an efficiency of 31% determined by internal standardization with [^{3}H]toluene (Packard). Correction for non-specific binding of [^{3}H]PrBCh was by counting pellets from a parallel series of samples in which 3×10^{-5} M atropine was present throughout.

Protein concentration was measured by the method of Lowry *et al.* (1951) using bovine serum albumin as the standard.

Results and Discussion

Binding of [^{3}H]PrBCh to homogenates of caudate nucleus

Figure 1 is the double reciprocal plot obtained when a homogenate of a normal caudate nucleus (brain C42) was incubated with several concentrations of [^{3}H]PrBCh. It can be seen from the intercept on the abscissa that the equilibrium dissociation constant describing the binding of [^{3}H]PrBCh to the human muscarinic receptor is 1.0×10^{-8} M. In a similar experiment using a homo-

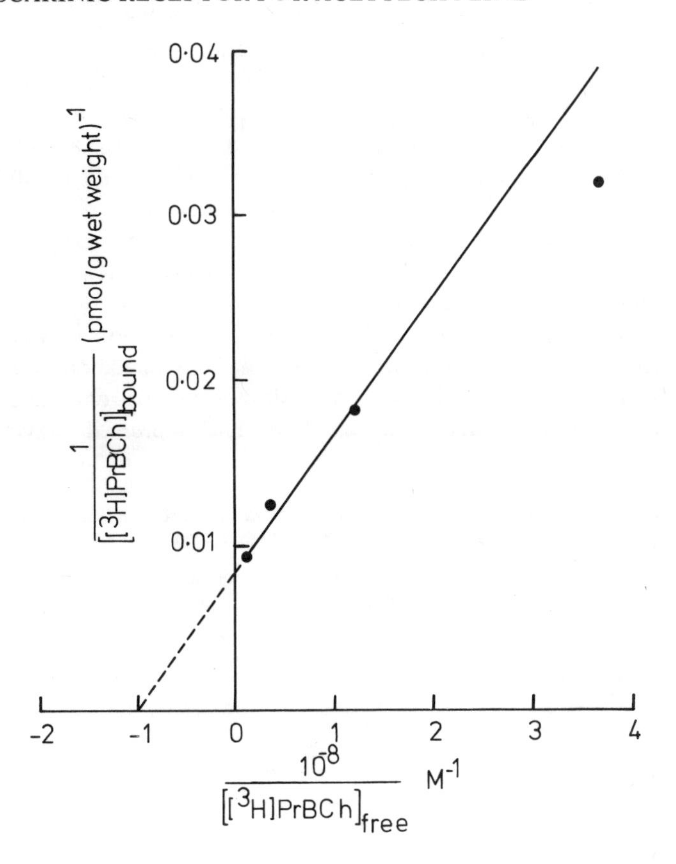

Fig. 1. *Double reciprocal plot of* $1/[^3H]$*PrBCh bound to homogenates of normal human caudate nucleus versus* 1/*concentration of* $[^3H]$*PrBCH free in solution*

The experiment was performed using the microcentrifugation technique of Terenius (1974) as described in the text.

The total receptor concentration is the reciprocal of the intercept on the ordinate and is therefore 122 pmol/g wet weight of tissue.

genate of rat cerebral cortex as the source of muscarinic receptor the equilibrium dissociation constant was found to be 1.8×10^{-8} M. In a previous study we found that the atropine-binding constants in the cerebral cortex of five species were within a threefold range (Burgen *et al.*, 1974*a*) and thus there would seem to be little variation in the antagonist binding site between mammalian species.

The concentrations of receptor present in homogenates of brain areas from choreic brains are too low to allow determination of the binding constant of $[^3H]$PrBCh. Therefore it has not been possible to determine whether there has been a change in the molecular morphology of the muscarinic receptor in Huntington's chorea.

Receptor Concentration Studies

In a previous study we showed that the muscarinic receptor concentration was greatly reduced in the caudate nucleus and putamen of choreic brains relative to control samples (Hiley & Bird, 1974). The mean values are shown in Table 1, together with the activities of choline acetyltransferase and the protein concentrations for the nuclei included in the sample. Two features of these data are particularly interesting. Firstly, there is no significant difference in protein concentration between control and diseased brains. Since tyrosine hydroxylase activity is normal in these nuclei in choreic brains (Bird & Iversen, 1974) it is unlikely that the decrease in receptor concentration is a consequence of a non-specific loss of protein in Huntington's chorea.

The second feature of interest is the lack of correlation between receptor concentration and choline acetyltransferases activity in the choreic brains. In our sample there were four brains which showed normal choline acetyltransferase activity in the caudate nucleus, seven had depressed enzyme activity.

Table 1. *Muscarinic receptor concentration, choline acetyltransferase activity and protein concentration in human postmortem basal ganglia*

Caudate nucleus	(*n*)	Receptor (pmol/g wet weight)	Choline acetyltransferase (μmol acetylcholine/g wet weight/h)	Protein (mg/g wet weight)
Control	(14)	13.9 ± 0.5	11.9 ± 1.5	69 ± 3
Huntington's chorea				
—Normal ChAc activity	(4)	8.9 ± 1.6	15.4 ± 1.9	74 ± 4 (both groups)
—Low ChAc activity	(7)	8.8 ± 1.0	3.7 ± 0.9	
Putamen				
Control	(7)	18.9 ± 2.1	39.6 ± 3.0	79 ± 4
Huntington's chorea	(6)	8.7 ± 0.8	11.0 ± 6.2	79 ± 6

Data from Hiley & Bird (1974).
(*n*) Represents the number of brains in the sample and values given are mean ± standard error.

Although the difference in activity of the enzyme in the two groups is highly significant when subjected to Student's *t* test ($P < 0.001$), there is no difference in the mean receptor concentrations. A similar trend appeared in the putamen, but the sample groups were too small to allow statistical analysis.

There are two likely explanations for this lack of correlation between receptor concentration and enzyme activity. Two biochemical subtypes may exist within the disease which might affect cholinergic systems to different degrees. Such a situation may exist in the dopaminergic system where Bird & Iversen (1974) found that dopamine concentration was significantly increased in the putamen of patients who suffered from the Westphal, or rigid, form of the disease. An alternative explanation is that those brains in which both muscarinic receptor concentration and choline acetyltransferase activity were lowered were from patients who had progressed to a later stage of the disease than those which displayed only a depressed receptor concentration. That is, there may be a primary loss of postsynaptic cholinoceptive cells which is followed some time later by degeneration of the presynaptic cholinergic neurons. An extension of these studies combined with examination of clinical records may allow differentiation between these alternatives.

More recently, the receptor concentrations in several brain areas have been compared in one normal and one choreic brain. The results are shown in Table 2. Although care must be taken not to assign too much significance to limited data, there are several features of these results which deserve comment. In all the areas where direct comparison may be made between the two brains there is a decrease in the muscarinic receptor concentration in the choreic brain whilst the protein concentration is generally unchanged. The observed decreases are quite large and range from a 35% reduction in the hippocampus to a complete loss of receptor in the globus pallidus. A direct comparison of neocortical areas was not obtained, but the concentration in the cingulate gyrus of the choreic brain is very similar to that found in the normal frontal cortex.

Table 2. *Comparison of muscarinic receptor and protein concentration in several areas of postmortem human brain in Huntington's chorea with a control*

	Receptor concentration (pmol/g wet weight)		Ratio	Protein concentration (mg/g wet weight)	
	C78	H67	H67/C78	C78	H67
Frontal cortex	32.4	—	—	73	—
Cingulate gyrus	—	29.7	—	—	58
Hippocampus	28.8	18.6	0.65	76	77
Amygdala	37.8	13.8	0.37	86	91
Putamen	99.0	14.1	0.14	88	70
Globus pallidus	13.8	0	0.00	93	101
Cerebellar cortex	14.7	3.0	0.20	78	77

Receptor concentration in homogenates of postmortem human brain samples was determined by the method of Hiley & Burgen (1974) using [^{3}H]PrBCM with a specific activity of 0.6 Cl/mmol. Protein concentration was measured by the method of Lowry *et al.* (1951).

C78 was a control brain and H67 was a brain from a patient who died while suffering from Huntington's chorea.

Previous studies have found little variation between areas of the neocortex in other species (Hiley & Burgen, 1974; Yamamura *et al.*, 1974) and if the human is similar then it is possible that this result indicates that Huntington's chorea does not affect the postsynaptic component of the cholinergic system in the cerebral cortex. Bird & Iversen (1974) found unchanged activities of choline acetyltransferase in the frontal cortex in Huntington's chorea and so it is possible that the cholinergic system in the neocortex remains intact.

Two other points should be noted. Firstly, Table 3 shows that there is excellent agreement between the receptor concentrations of the several brain areas in the human brain and those of the monkey brain as determined by Yamamura *et al.* (1974) when the values are expressed as a percentage of the concentration in the putamen. This observation lends support to the assumption made above concerning the neocortical receptor distribution.

Secondly, it can be seen that the receptor concentrations found in this study are higher than those reported by Hiley & Bird (1974). We have found that our present batch of [^{3}H]PrBCM had a radiochemical purity of only 33% as originally prepared and the purified material used in these studies had a specific activity of 0.6Ci/mmol. It is likely that the batch used in our previous

Table 3. *Comparison of muscarinic receptor concentrations in different areas of normal human and monkey (Macaca mulatta) brains*

Area	Normalized receptor concentration	
	Human (C78)*	Monkey†
Frontal cortex	0.33	0.39
Cingulate gyrus	—	0.48
Hippocampus	0.29	0.45
Amygdala	0.38	0.44
Putamen	1.00	1.00
Globus pallidus	0.14	0.15
Cerebellar cortex	0.15	0.11

* Receptor concentrations from Table 2 normalized to putamen = 1.00.
† Receptor concentrations from Yamamura *et al.* (1974) normalized to putamen = 1.00.

study also had a lower radiochemical purity than we reported despite the rigorous procedures used in the preparation of the compound (Burgen *et al.*, 1974*b*). Since none of that batch is available for analysis it is not possible to investigate the discrepancy further. However, we have found complete agreement between receptor concentrations determined in rat cerebral cortex with the present batch of [^{3}H]PrBCM and by binding of [^{3}H]PrBCh and [^{3}H]atropine (O. O. Birdsall, C. R. Hiley & O. O. Hulme, unpublished work). Although the absolute receptor concentrations reported before may be underestimates, the comparison between the control and choreic brains remains valid since the same batch of [^{3}H]PrBCM was used throughout that study.

In conclusion, it seems that Huntington's chorea is accompanied by degeneration of cholinoceptive neurons in several brain areas and that the neocortex may be unaffected. If these conclusions are supported by studies on larger numbers of brains, it would suggest that there may be a severe limitation of the effectiveness of drugs which increase acetylcholine levels or which prolong its activity after release from the presynaptic cell. If there are not sufficient cells present to respond to the increased acetylcholine activity, or if there are not enough receptors to detect it, the normal function cannot be restored to the affected neuronal tracts. The number of cells or receptors may not be a limiting factor in all cases of the disease since physostigmine has been reported to reduce choreiform movements in some patients (Aquilonius & Sjostrom, 1971). Since γ-aminobutryric acid-releasing cells are also affected in Huntington's chorea, effective treatment may require the activity of both transmitters to be increased.

Summary

Two radio-labelled antagonists have been used to study the muscarinic receptor for acetylcholine in Huntington's chorea. The receptor concentration was found to be decreased in several brain areas relative to control brains but it was not possible to determine whether the molecular morphology had changed in the disease. It is concluded that drugs which increase acetylcholine levels in the brain, or which prolong its activity after release from neurons, may be of limited value in treatment of the disease.

References

Aquilonius, S. M. & Sjostrom, R. (1971) *Life Sci.* **10**, 405–414
Bird, E. D. & Iversen, L. L. (1974) *Brain* **97**, 457–472
Bird, E. D., Mackay, A. V. P., Rayner, C. N. & Iversen, L. L. (1973) *Lancet* **i**, 1090–1092
Bloom, F. E., Costa, E. & Salmoiraghi, G. C. (1965) *J. Pharmacol. Exp. Ther.* **150**, 244–252
Burgen, A. S. V., Hiley, C. R. & Young, J. M. (1974*a*) *Brit. J. Pharmacol.* **51**, 279–285
Burgen, A. S. V., Hiley, C. R. & Young, J. M. (1974*b*) *Brit. J. Pharmacol.* **50**, 145–151
Ehringer, H. & Hornykiewicz, O. (1960) *Klin. Waschr.* **38**, 1236–1239
Hiley, C. R. & Bird, E. D. (1974) *Brain Res.* **80**, 355–358
Hiley, C. R. & Burgen, A. S. V. (1974) *J. Neurochem.* **22**, 159–162
Lowry, O. H., Rosebrough, N. J., Farr, A. L. & Randall, R. J. (1951) *J. Biol. Chem.* **193**, 265–275
McGeer, P. L., McGeer, E. C. & Fibiger, H. C. (1973*a*) *Neurology* (*Minneapolis*) **23**, 912–917
McGeer, P. L., McGeer, E. C. & Fibiger, H. C. (1973*b*) *Lancet* **ii**, 623–624
McLennan, H. & York, D. H. (1966) *J. Physiol.* (*London*) **187**, 163–175
Terenius, L. (1974) *Acta Pharmacol. Toxicol.* **34**, 88–91
Yamamura, H. I. & Snyder, S. H. (1974) *Proc. Nat. Acad. Sci. U.S.A.* **71**, 1725–1729
Yamamura, H. I., Kuhar, M. J., Greenberg, D. & Snyder, S. H. (1974) *Brain Res.* **66**, 541–546
Young, A. B. & Snyder, S. H. (1973) *Proc. Nat. Acad. Sci. U.S.A.* **70**, 2832–2836
Zukin, S. R., Young, A. B. & Snyder, S. H. (1974) *Proc. Nat. Acad. Sci. U.S.A.* **71**, 4802–4807

General Discussion on Dyskinesias

C. D. Marsden: As a framework for this discussion, perhaps I might try to summarize the very complex and often contradictory data that we have discussed concerning the dyskinesias.

We think that these abnormal involuntary movements are due usually to disordered function of the basal ganglia, for when pathology exists that is where it is most commonly found. The dyskinesias, in general, are probably generated by abnormal neuronal activity in globus pallidus and thalamus. Surgeons have employed stereotactic lesions at both sites to control a wide variety of dyskinesias. The abnormal neuronal signals are transmitted via descending pathways to spinal motor neurons to produce the distorted patterns of muscle contraction.

With regard to the commonest site of origin in the basal ganglia, greatest attention has been focused on the nigrostriatal dopaminergic pathway and the striatal dopamine receptor. This emphasis was decided historically by the discovery of the importance of this dopaminergic system in Parkinson's disease. When it was found that levodopa caused a vast array of dyskinesias in Parkinsonian patients, naturally we all looked at the striatal dopamine receptor for the source. However, it is clear, not least from the work on Huntington's chorea, that the other chemical substances in the striatum, such as acetylcholine, 5-HT and GABA, are probably vitally concerned with the overall function of this area in the brain. Unfortunately, we do not yet have a clear understanding of the complex neuronal machinery of the striatum and pallidum, either in physiological or neurotransmitter terms. It is not certain yet whether the dopamine receptor in striatum lies on cholinergic or GABAminergic neurons, or where the axons of these cells pass to. Eventually striatal outflow modulates pallidal activity, which in turn affects activity in thalamic and other brain regions. However, at present we only know that if you give drugs such as neuroleptics or levodopa to humans, dyskinesias may appear. What is happening in the basal ganglia in response to these drugs to cause dyskinesias is the theme of the meeting.

J. C. Watkins: How watertight is the evidence that levodopa exerts its therapeutic effects by action on dopaminergic systems? Dopa has structural similarities to glutamate, in that it has a system of phenolic groups, which are acidic, separated by a similar distance from an α-amino carboxylic acid grouping, as in the γ-carboxyl group of glutamate. Moreover, Biscoe, Evans, Headley, Martin and Watkins (unpublished work) have shown that L-dopa has a weak glutamate-like excitatory action on neurons of the rat and frog spinal cord. 6-Hydroxydopa (so far tested only on the frog) is, in fact, considerably stronger than L-glutamate, possibly because it ionizes in a way which enables an even stronger interaction with glutamate receptors than the putative transmitter itself. Perhaps these actions of dopa, and of its possible metabolites, if not involved in the therapeutic action of the drug may, nevertheless, contribute to its side effects.

C. D. Marsden: Could I ask you if you have any similar information on the dopamine agonists presently being evaluated clinically in Parkinson's disease?

J. C. Watkins: No we haven't, but we could look into that. I think their structure is sufficiently dissimilar to rule out a glutamate-like action. That would, of course, be very good evidence that the glutamate system is not involved in the therapeutic action of levodopa, although it may well be involved in the side effects.

N. G. Bower: Does HA 966 (the glutamate antagonist) affect the excitation by L-dopa?

J. C. Watkins: We haven't tried it.

G. Curzon: The ring trihydroxylated derivative detected in urine after levodopa treatment is a trihydroxyphenyl acetic acid [G. H. Wada & J. H. Fellman (1972) *Biochemistry* **12**, 5212–5217]. Although this has an identical ring structure to 6-hydroxydopamine, it does not have the side chain amino group required for uptake by dopamine neurons. Neither is it derived from 6-hydroxydopamine, but from 3,4-dihydroxyphenyl pyruvic acid. Therefore, formation of trihydroxyphenyl acetic acid does not provide good evidence that a 6-hydroxydopamine-like substance is involved in the effects of dopa.

B. E. Leonard: With reference to the treatment of Huntington's chorea, it has recently been found that apomorphine has a beneficial effect on many of the dyskinetic symptoms of the disease [E. S. Tolosa & S. B. Sparber (1974) *Life Sci.* **15**, 1371–1380]. There is now evidence that there are at least two types of dopamine receptor in the mammalian striatum, one inhibitory and one excitatory. It is possible that the beneficial effects of apomorphine could be due to its ability to block the excitatory and stimulate the inhibitory receptors. From studies of the invertebrate brain, Van Rossum and colleagues have shown that apomorphine acts on dopamine receptors in this way [H. A. J. Struycker-Boudiere *et al.* (1973) *Biochem. Pharmacol. Suppl. part 2*, p. 549]. Do you have any comment on the therapeutic effect of apomorphine in Huntington's chorea and its mechanism of action? Perhaps Dr. Woodruff would like to comment on the nature of the different types of dopamine receptors in the invertebrate, and possibly the vertebrate brain.

E. D. Bird: I think one has to be careful when evaluating drugs in Huntington's chorea, and I would stress the need to evaluate in a double blind fashion, as we are dealing with a group of patients that seem to have the ability to improve simply by frequent visits to the physician.

C. D. Marsden: Apomorphine has also been found to reduce the intensity of dopa-induced dyskinesias in Parkinson's disease [S. E. Düby *et al.* (1972) *Arch. Neurol.* **24**, 474–480]. But I would hesitate to attribute these beneficial effects of apomorphine on dyskinesias to dopamine receptor activation. The drug probably does much more than just this; for example, there is the question as to whether apomorphine is a 'partial agonist'.

G. N. Woodruff: I would like to reply to Professor Leonard's question about the two types of dopamine receptor. Several years ago we looked at the structural requirements for dopamine inhibitory activity on dopamine receptors on snail neurons. These receptors are highly specific and in terms of agonist activity closely resemble the dopamine receptors in dog kidney and in the striatal adenyl cyclase system. More recently Van Rossum's group have shown that other neurons in snail brain are excited by dopamine and these receptors are much less specific, for example, they are activated by isoprenaline [H. A. J. Struycker-Boudiere *et al.* (1974) *Arch. Int. Pharmacodyn.* **209**, 324–331]. As a result of our invertebrate work we have developed the dopamine analogue ADTN. We have shown that ADTN is active on both excitory and inhibitory invertebrate dopamine receptors. We have also shown that ADTN is equipotent with dopamine when iontophoretically applied onto rat striatal neurons. In other studies we find that ADTN is similarly equipotent with dopamine in increasing striatal cyclic AMP levels. When we inject ADTN into the nucleus accumbens of conscious rats we find a long lasting stimulation of locomotor activity. Injections of ADTN into the caudate nucleus cause stereotyped behaviour.

R. J. Naylor: I should like to offer some explanation for the apparent paradox of the similarities between the acute dyskinesias observed after L-dopa therapy and the tardive dyskinesias induced by the neuroleptic drugs. In recent intrastriatal dopamine injection studies in the rat and guinea-pig we have observed the development of abnormal involuntary movements, in particular the development of a marked hyperactivity and gnawing/biting or licking reactions. It was considered that the hyperactivity may be related to the antiakinetic effect of L-dopa and the gnawing reactions similar to the oro-bucco-lingual dyskinesias which contribute to the L-dopa and neuroleptic-induced dyskinesias in man. At the present time both the antiakinetic and dyskinetic effects are generally considered to be mediated via the same dopaminergic mechanism. However, using intrastriatal dopamine-induced dyskinesias as an experimental model of abnormal involuntary movements, we have obtained a differential inhibition of the two effects and this strongly suggests the presence of at least two types of dopaminergic receptor mechanisms. We hypothesize that a DA_1 mechanism mediates the hyperactivity response, whilst a DA_2 mechanism mediates the dyskinesias. Thus, whereas the hyperactivity response (DA_1 system) is susceptible to inhibition by a wide range of phenothiazine, butyrophenone, dibenzazepine and other neuroleptic agents, the oro-bucco-lingual effects are very resistant. In contrast, pimozide will completely inhibit the oro-bucco-lingual effects (DA_2 receptor blockade), albeit in high doses, with *relatively* less effect upon the DA_1 system. A novel neuroleptic drug oxiperomide has a similar effect to pimozide.

It is possible that with a moderate dosage of neuroleptic agent in the clinic, DA_1 blockade is achieved with the development of hypokinesia. The increased synthesis and release of dopamine known to occur after neuroleptic treatment may then activate the second DA_2 mechanism, which we have shown to be *relatively resistant to neuroleptic blockade,* and result in oro-bucco-lingual effects. This could explain the very high doses of neuroleptic agents required to inhibit tardive dyskinesias and the unusual activity of pimozide to reduce dyskinesias. The studies may have important implications for the design of neuroleptic agents to specifically inhibit the DA_2 system (and thus L-dopa and neuroleptic tardive dyskinesias) *without* an accompanying more generalized motor depression. More important, it may allow the development of new neuroleptic agents which may not induce dyskinesias in the first instance.

C. D. Marsden: These very elegant pharmacological experiments add considerable weight to the suggestion of the presence of two populations of dopamine receptors in the striatum [H. Klawans (1973) *The Pharmacology of Extrapyramidal Movement Disorders,* Karger, Basel; C. D. Marsden (1975) *Advan. Neurol* 6, 141–166]. As you point out, Dr. Naylor, the critical outcome of this concept is that it offers the possibility of finding drugs that selectively stimulate the population of receptors concerned with akinesia, thereby avoiding dyskinesias; or drugs which selectively block the population concerned with dyskinesias, thereby avoiding the problem of drug-induced Parkinsonism. Obviously, if this could be achieved it would offer considerable advantages for the therapy of both Parkinson's disease and the dyskinesias. On the other hand, if both the anti-Parkinsonian effect of, and the dyskinetic action of dopamine stimulation are due to activation of the same striatal dopamine receptors, then it would be impossible to dissociate the two therapeutic consequences.

D. F. Sharman: Do there have to be only two types of striatal dopamine receptors?

R. J. Naylor: No!

A. R. Green: I was gratified to see the results of pimozide. Locomotor activity following monoamine oxidase inhibition and L-dopa administration to rats is blocked by haloperidol, chlorpromazine, and α-fluphenthixol (but not by β-fluphenthixol). Pimozide even at doses of 10 mg/kg has no effect on this locomotion.

H. S. Bachelard: We have heard much about dopaminergic systems, but, with the exception of Dr. Bird and Dr. Iversen's studies in Huntington's chorea, almost nothing has been mentioned about other transmitters. Professor Marsden suggested the historical reasons for this, but I'd like to ask if the lack of knowledge on other transmitters is because they have been studied with negative results, or because they have not been studied. If so, can the clinicians give the biochemists some guidelines.

C. D. Marsden: One of the major problems everybody faces is the difficulty of getting postmortem material. Dr. Bird has achieved almost a miracle in laying his hands on a sufficient body of human material of Huntington's chorea. There are 20 or 30 other diseases with names like Huntington's chorea. Most of such patients die in a general hospital or at home, and never come to post mortem. Unless you have an army of people chasing up patients like this to get the brain, you'll never get an answer to the questions you ask. I'm afraid nobody has ever looked at the brains of patients with dystonia musculorum deformans for transmitter function, or in spasmodic torticollis. One could trot out the 20 or 30 different names of diseases which have not been examined yet. It's a question of getting hold of the material.

L. L. Iversen: It's very difficult to do, and also impossible to do because we don't know what many of the other transmitters are yet.

D. A. Bender: We have heard some very elegant experiments concerning the dopaminergic effects of the neuroleptic drugs; they are all apparent either *in vitro* or acutely *in vivo.* The time course of the clinical antischizophrenic effects of neuroleptic drugs is slow, with gradual onset of behavioural improvement on drug administration over several days, and gradual behavioural deterioration over several days on drug withdrawal. Should we not be looking for neurotransmitter effects which show the same sort of time course as the neuroleptic effect, rather than more immediate neurotransmitter effect?

We have shown serum tryptophan changes in schizophrenics (indicative of serotonin changes) on chlorpromazine withdrawal and reinstatement apparently matching the time course of behavioural changes.

L. L. Iversen: That's a very apposite question, and one which arises over and over again whenever clinicians start talking to biochemists. We have trotted out the old adages that psychotropic drugs are slow in onset of action clinically and slow in offset; really, the pharmacologist doesn't have any answer to that. But we would query whether systematic studies have really been made on these drugs to know when you obtain steady state concentrations in tissues. We know there are pharmacokinetic questions here. In the case of chlorpromazine, for example, this is a highly lipid soluble material; the drug enters in huge quantities into the lipid stores of the body, and when the drug is withdrawn one can still measure chlorpromazine metabolites in urine and plasma for weeks, or even months after the drug is withdrawn.

D. B. Calne: I would like to support what Dr. Iversen has said. In some respects it would be surprising if giving a drug to man produced clinical effects as quickly as you can obtain them in a test tube; in the latter you have your preparation immediately accessible. There was a lot of argument about, for instance, the slow onset of levodopa action in Parkinson's disease. But when it was combined with a decarboxylase inhibitor the therapeutic action could occur almost immediately.

A. V. P. Mackay: I think pharmacokinetic variables must be the first candidates for explanation of time courses of response to drugs in man. Only when one knows the precise extracellular concentration of a drug in brain tissue can one discard a pharmacokinetic explanation of delayed response—it may bear a very complicated relationship even to plasma drug concentration.

G. Curzon: As the time course of the clinical effects of levodopa has been brought up by Dr. Calne, I wonder whether the argument of Dr. Naylor, that dopa dyskinesia without anti-Parkinsonian action occurs in some patients, necessarily means two kinds of receptors. Could it not also be interpreted in terms of a more rapid expression of the dyskinetic effect than of the beneficial response?

B. S. Meldrum: During the course of today's presentations, we have all become uncomfortably aware of the gap between the clinical descriptions given by Dr. Parkes and Professor Marsden and the various rodent models of movement disorders that have been described. It is possible, however, to reproduce certain syndromes very closely indeed in subhuman primates. We have recently observed an idiosyncratic response to neuroleptics in baboons (*Papio papio*) that is very similar to acute dystonic reactions in man [B. S. Meldrum, G. Anlezark & M. Trimble (1975) *Eur. J. Pharmacol.* **52**, 203–213]. Out of 16 baboons so far tested, two have responded to both haloperidol (0.6–1.0 mg/kg i.v.) and pimozide (0.5–2.5 mg/kg i.v.) with an orofacial dyskinesia (principally forced mouth opening with tongue protrusion, plus 'compulsive gnawing') and a severe dystonia of the trunk and limbs. This response begins within 5–30 min of drug administration and lasts for 6–7 hours. The intravenous injection of benztropine (0.2 mg/kg) promptly arrests all aspects of the dystonia, for a period of 90–180 min. This effect of benztropine appears to be related to its anticholinergic properties rather than to its effect on dopamine uptake, as the administration of eserine (0.1 mg/kg) a few minutes after benztropine leads to the reappearance of 'compulsive gnawing' and trunk dystonia. This syndrome provides a model of acute dystonic reactions to neuroleptics that could serve to test various pharmacologic and biochemical hypotheses about the nature of this syndrome.

D. B. Calne: What happens if you push up the dose in the animals who do not develop dyskinesias?

B. S. Meldrum: Only low or moderate doses were given.

D. A. Brown: Am I correct that some of the neuroleptic drugs produce a long-term irreversible dystonia?

C. D. Marsden: We must distinguish acute dystonic reactions, which occur after a single dose of a neuroleptic drug (and to which most of the biochemistry presently undertaken as acute experiments may be relevant), from chronic tardive dyskinesia, which appears after many months of neuroleptic treatment. The latter, which is clinically very different from acute dystonic reactions, can be irreversible, although how often is by no means certain [see C. D. Marsden, D. Tarsy & R. J. Baldessarini (1975) in *Psychiatric Aspects of Neurologic Disease*, pp. 219–265, Grune and Stratton, New York].

D. A. Brown: It seems to me that the chronic form of tardive dyskinesias cannot be explained by any of the material discussed so far. Has this been imitated in primates?

B. S. Meldrum: A symptom complex resembling Parkinsonism (including tremors, masque facies, postural rigidity and salivation) has been observed in baboons receiving fluphenazine (5–15 mg/kg daily) [J. Dreyfuss, B. Beer, D. D. Devine, B. F. Roberts & E. C. Schreiber (1972) *Neuropharmacology* **11**, 223–230]. A dyskinetic syndrome that may be comparable to tardive dyskinesias has been described in rhesus monkeys receiving chlorpromazine (30 mg/kg daily) for 3–9 months [G. W. Paulson (1973) *Advan. Neurol.* **1**, 647].

D. A. Brown: What mechanism is operating there?

D. F. Sharman: It is known that the effects of some neuroleptics (e.g. thioridazine) on cerebral dopamine metabolism are attenuated following prolonged administration. It has been suggested that this effect is due to other pharmacological actions of the drugs, in particular their antiacetylcholine action. If such a mechanism applied to the action of neuroleptic drugs on some other aspects of the dopamine-containing neuron systems in the brain and, after long term administration of neuroleptic drugs, this attenuating mechanism was more persistent than the other actions of the drugs, then on drug withdrawal the system might behave as if it were more sensitive to dopamine.

D. A. Brown: What happens when you withdraw the drug? These tardive dyskinesias go on for years after the drugs are withdrawn.

C. D. Marsden: Yes, for years.

D. A. Brown: Even if the tardive dyskinesias are due to altered receptor sensitivity or something like that, unless there is some physical lesion of nerve cells, you would expect them to be reversed by restoration of normal receptor turnover at membranes.

C. D. Marsden: No physical lesion has been identified with certainty as yet.

D. A. Brown: I feel certain that there must be one

C. D. Marsden: Your point is well taken, for nearly all biochemical work on neuroleptic drugs has been undertaken as acute studies. People are only just beginning to look at what happens after years of administration of these neuroleptic drugs.

A. V. P. Mackay: I wonder if I can bring this discussion back to Huntington's chorea. I think it must be a major responsibility of neurochemists working in this area to look as energetically as possible for the primary inborn error of metabolism, which is clearly not any of the transmitter synthesizing enzymes we have been discussing. I would like to mention some work which I have been performing in Edinburgh with Dr. Piers Emson and Janet Dickinson. It seemed to us that by looking at a range of biochemical parameters in an easily accessible peripheral tissue such as blood, there was a chance of getting a little nearer to whatever inborn error of metabolism represents the source of the demonstrated neurochemical pathology in Huntington's chorea. We obtained blood platelets from six patients suffering from Huntington's chorea and compared several biochemical phenomena present in these platelets with those obtained from six closely matched controls. The activities of monoamine oxidase, choline acetyltransferase and acetylcholinesterase, and the uptake of tritiated dopamine and GABA were similar in Huntington's chorea and control groups. However, the metabolism of L-[l-C^{14}]glutamic acid by the platelets of Huntington's chorea was consistently lower than controls. The exact nature of this enzyme defect is not yet clear—enzyme activity was assessed by the evolution of $^{14}CO_2$, which may not necessarily reflect glutamic acid decarboxylase activity with formation of GABA. An interesting possibility is that the enzyme deficit found in the Huntington's chorea platelet may, in fact, lie in the tricarboxylic acid cycle. We are at present working to define the precise nature of this metabolic abnormality and plan to investigate the occurrence of the abnormality in the families of Huntington's chorea sufferers.

B. S. Meldrum: Were the GAD assays conducted with or without added pyridoxal phosphate?

A. V. P. Mackay: We have not left out pyridoxal phosphate from the assays we have done.

G. W. Arbuthnott: Could I ask one of the clinicians whether pyridoxine injections have been tried in Huntington's chorea?

E. D. Bird: No double blind trial has been done. I have given pyridoxine to one patient, for other reasons, but with no improvement over 2 or 3 months treatment.

W. E. Davies: Is the GAD activity of platelets inhibited by chloride or by allylglycine?

A. V. P. Mackay: I cannot make any other statements about the characteristics of this enzyme, until we have taken the work further.

H. S. Bachelard: Could I ask the clinicians if there is any genetic component to the susceptibility to developing chronic tardive dyskinesias? Also, is there any effect related to age?

C. D. Marsden: Although it has been suggested that patients with drug-induced Parkinsonism have an increased incidence of Parkinson's disease in family members [N. C. Myrianthopolous, A. A. Kurland & L. T. Kurland (1962) *Arch. Neurol.* **6**, 5], this has been denied by others [H. J. Hasse & P. A. H. Janssen (1965) *The Action of Neuroleptic Drugs,* North Holland, Amsterdam]. I am unaware of any similar data for chronic tardive dyskinesias.

With regard to age, there is debate as to whether old age predisposes to drug-induced tardive dyskinesias [S. Brandon, H. A. McClelland & C. Protheroe (1971) *Brit. J. Psychiat.* **118**, 171–184; P. F. Kennedy, H. I. Hershon & R. J. McGuire (1971) *Brit. J. Psychiat.* **118**, 509–518], but certainly children develop these dyskinesias [F. J. Ayd (1974) *Int. Drug Ther. Newslett.* **9**, 25–28; P. Polizos, P. M. Engelhardt & S. P. Hoffman (1973) *Psychopharmacol Bull.* **9**, 34–35]. Interestingly, while the typical drug-induced tardive dyskinesia occurring in adults over the age of 50 is orofacial dyskinesia, perhaps with some distal chorea, that in young adults may include severe axial dystonia, while children develop more obvious limb dystonia and chorea. This pattern almost exactly matches the manifestations of idiopathic torsion dystonia in relation to age of onset.

I wonder if we could bring this discussion round to the question as to how best to avoid these chronic tardive dyskinesias. It seems crucial to decide whether a high inherent anticholinergic activity in a neuroleptic protects not only against drug-induced Parkinsonism and possibly acute dystonias and akathisia [see R. J. Miller & C. R. Hiley (1974) *Nature* (*London*) **248**, 596], but also against tardive dyskinesias. The evidence on this point is conflicting.

D. B. Calne: What about clozapine, which has a very high inherent anticholinergic activity yet doesn't seem to cause extrapyramidal side effects?

C. D. Marsden: This is a fascinating drug. Certainly it doesn't appear to cause much Parkinsonism, acute dystonias or akathisia, possibly due to its anticholinergic activity. Nor have tardive dyskinesias been described. But most of the evidence has been based on short-term trials in which these unwanted actions have been lumped together as 'extrapyramidal side effects'. If clozapine really does not produce tardive dyskinesias in long-term use, then this crucial question may be answered.

D. F. Sharman: Which of the neuroleptics has the highest incidence of tardive dyskinesias?

C. D. Marsden: Piperazine phenothiazines are reputed to produce the highest incidence of tardive dyskinesia [A. Faurbye, P. J. Rasch, P. B. Peterson, G. Brandborg & H. Pakkenberg (1964) *Acta Psychiat. Scand.* **49**, 10–27], but this evidence is not certain and most neuroleptics have been reported to cause this late complication, even thioridazine [G. E. Crane & T. A. Smeets (1974) *Arch. Gen. Psychiat.* **30**, 341–343]. Careful comparative trials to establish this point would take years of treatment with a single drug, and this has just not been done.

H. S. Bachelard: This question of long-term side effects doesn't necessarily have to be thought of in structural terms. One could conceive perhaps of such actions in terms of changes in protein synthesis, or a long-term effect on genetic expression of enzyme synthesis.

L. L. Iversen: Are neuroleptics effective in controlling drug-induced tardive dyskinesia? There are some reports of clozapine being especially effective.

C. D. Marsden: A variety of neuroleptics have been shown to control tardive dyskinesias, including a variety of phenothiazines [H. Kazamatsuri, C. Chien & J. O. Cole (1972) *Arch. Gen. Psychiat.* **27**, 491–499; K. Singer & M. N. Cheng (1971) *Brit. Med. J.* **iv**, 22–25; G. E. Crane (1973) *Brit. J. Psychiat.* **122**, 395–405], butyrophenones such as haloperidol [H. Kazamatsuri, C. Chien & J. O. Cole (1972) *Arch. Gen. Psychiat.* **27**, 100–103; H. Kazamatsuri, C. Chien & J. O. Cole (1973) *Am. J. Psychiat.* **130**, 479–483]. pimozide [D. B. Calne, L. E. Claveria, P. F. Teychenne, L. Haskayne & I. C. Lodge-Patch (1974) *Arch. Neurol.* **30,** 423], and, as you say, clozapine [F. J. Ayd (1974) *Int. Drug. Ther. Newslett.* **9**, 5–12; G. M. Simpson & E. Varga (1974) *Curr. Ther. Res.* **16**, 679–686]. I have used chlorpromazine, perphenazine, thiopropazate, trifluperazine, thioridazine and pimozide, as well as tetrabenazine [E. Bandrup (1961) *Am. J. Psychiat.* **118**, 551–552; R. B. Godwin-Austin & T. Clark (1971) *Brit. med. J.* **iv**, 25–26]. All will control tardive dyskinesias to some degree, but the extent of benefit depends on how high you push the dose. Often one can only adequately stop the dyskinesia at the expense of unacceptable drug-induced Parkinsonism, and obviously it is unwise to use drugs that cause this complication to treat it.

R. Balázs: I think that experimental studies on tardive dyskinesias using animal models may raise difficult problems. These conditions develop after central neurons, which are thought to be responsible for the symptoms, have been exposed to drug treatments for a very long period of time, for years. It is open to question whether in experimental animals, in comparison to man, the time needed to induce alterations in nerve cells, which lead ultimately to dyskinesias, is shortened in proportion to the briefer animal life span. Thus the relevance of drug trials in experimental animals, especially if the results are negative, is questionable.

The other question I would like to raise relates to Dr. Bird's presentation. Are there any indications of premature aging in Huntington's chorea patients? It seems that normal aging in experimental animals is associated with a decrease in the activities of various transmitter enzymes in the brain, without a corresponding reduction in nerve cell numbers. Since the onset of the symptoms of Huntington's chorea is in middle life, this may relate to the premature aging of certain non-replicating cell types in the central nervous system.

E. D. Bird: These patients do seem to have some signs of premature aging. I should point out that in our biochemical studies, the control brains were actually from somewhat older people than those from choreics. Also we couldn't show any real relation between GAD activity and age.

J. Lagnado: Is there any information as to whether some of the neuroleptics are also mutagenic?

J. D. Parkes: I know of none. Can I return to a point of Dr. Naylor's and Professor Marsden? It is possible to get reversal of orofacial dyskinesia with, say, pimozide, without apparent drug-induced Parkinsonism.

D. M. Bowen: Dr. Balázs's suggestion that the changes in cerebral enzyme activity in Huntington's chorea may be due to premature aging is supported by a demonstration of abnormally high lipofuscin content in cortical biopsy samples from choreic patients [Jellez-Nagel *et al.* (1974) *J. Neuropathol. Exp. Neurol.* **33**, 308–332]. There is also evidence that other abiotrophies, such as motor neuron disease and senile dementia, may also reflect an exacerbation of brain aging [D. M. Bowen, *et al.* (1974) *Lancet* **ii**, 1247–1249].

R. A. Webster: Evidence has been presented that in Huntington's chorea there is a reduction in GABA function in the basal ganglia, i.e. reduced GAD activity and lowered GABA levels. In contrast, in drug-induced dyskinesia, either from levodopa or the neuroleptics, there is increased dopamine activity. Perhaps we should not separate the two systems as much as we have. Is there any evidence that levodopa or dopamine can depress GABA function either by reducing GAD activity, or the release of GABA or even its postsynaptic action?

E. D. Bird: Chronic levodopa therapy for 8 months or less has been reported to depress brain GAD activity, which anyway is reduced in that illness [K. G. Lloyd, L. Davidson & O. Hornykiewicz (1973) *Advan. Neurol.* **3**, 173–188]. However, after chronic treatment for a year or more, striatal GAD levels had returned to untreated control values. There was, though, an initial drop in striatal GAD activity.

L. L. Iversen: These authors also went on to do experiments in rats, in which chronic oral levodopa for 109 days increased striatal GAD levels.

C. D. Marsden: We must also consider the reverse relationship between GABA and dopamine. There is evidence from many disciplines, biochemical, pharmacological and physiological, for the existence of a strionigral GABA-mediated pathway that inhibits the nigrostriatal dopaminergic system [see D. Tarsy, C. Pycock, B. Meldrum & C. D. Marsden (1975) *Brain Res.* **89,** 160–165], as discussed by Dr. Pycock. The interrelationship between GABA and dopamine is reciprocal, as seems to be the case for all other putative neurotransmitters in the basal ganglia.

A. V. P. Mackay: Can I ask a general question? What is the evidence for a direct proportional relationship between the activity of a transmitter synthesizing enzyme, or even the steady state concentration of a neurotransmitter, and synaptic performance? It may be misleading to consider that there is any such relation.

L. L. Iversen: I agree that one cannot transpose biochemical concentrations of enzyme activity or transmitter substance to the level of synaptic activation. One cannot put too much functional implication into small biochemical changes. But the changes we are looking for in human material are all or none. As we have seen in Hornykiewicz's work in Parkinson's disease, there is essentially no dopamine left in the striatum, and in Huntington's chorea the average loss of GAD activity is 75–80%. We are talking about massive changes, not subtle nuances.

H. S. Bachelard: I think this is an important point. A change in an enzyme level certainly need not imply any change in the level of a product or substrate. Even if it did, this need not relate to synaptic events. But, having said that, there are a number of important genetic disorders involving the brain in which the only change found has been a slight deficit in enzyme kinetic properties, not an all or none change. So a slight change in enzyme properties can have diabolical effects.

G. W. Arbuthnott: We have examined the relationship between transmitter synthetic enzyme activity and the output of transmitter in recent experiments on intracranial self-stimulation behaviour. Rats with chronic indwelling electrodes near the noradrenaline containing locus coeruleus can be trained to stimulate themselves through these electrodes.

Such stimulation is associated with an increase in the concentration of the final metabolite of noradrenaline 4-hydroxy-3-methoxyphenylglycol in cerebral cortex, indicating that the stimulation is associated with increased noradrenaline output. Animals allowed to stimulate themselves through electrodes in the region of locus coeruleus for 5 days also have significantly higher tyrosine hydroxylase activity in the stimulated locus.

These experiments, carried out in collaboration with Dr. Zigmond in Cambridge, go some way towards answering Dr. Mackay's query, since they suggest that regular increases in nerve/cell firing are associated with raised tyrosine hydroxylase activity.

L. J. Fowler: With reference to Dr. Bachelard's point concerning enzyme levels, one approach might be to design specific irreversible enzyme inhibitors to depress the enzyme level. This does not, of course, eliminate the problem of whether the increased transmitter concentration inside the cell will be reflected by increased synaptic activity.

H. H. Hillman: Has Dr. Bird considered the possibility that the changes in GABA metabolism in the brains of his patients with Huntington's chorea are a consequence of the abnormal motor activity rather than the cause of it?

E. D. Bird: I would have thought that if this were the case, we would see a similar change in other diseases with hyperactivity. But then we don't have that material yet.

D. B. Calne: By the time Huntington's chorea patients die they are usually chairbound or bedridden, and often rigid; in terms of calories they probably burn rather less than normal by this stage.

E. D. Bird: Although Huntington's chorea patients do have tremendous appetites earlier in the disease while they have choreiform movements, this hyperphagia seems to persist even in the late stages of the illness, when the chorea usually decreases.

D. A. Howell: There is good morbid anatomical evidence that the striatum is directly involved in Huntington's chorea. But could the dementia be related to changes in the thalamus, which we have noted to be very small. Have you analysed the thalamus biochemically?

E. D. Bird: No, we haven't. Dr. Dom in Belgium (unpublished work) has done cell counts on the thalamus, and has found a decrease in the small cell population.

R. G. Hill: Could I refer back to the drug-induced dyskinesias again? It occurs to me that there may be a parallel between acute and chronic tardive dyskinesias, and the acute and chronic actions of convulsants. An animal will recover from an acute convulsion, or an acute drug-induced dyskinesia, and subsequently show no abnormality. However, convulsions continuing for an extended period can produce chronic focal changes in the brain that sustains convulsive activity after the initial stimulus is removed. Neuroleptics produce a change in the pattern of neuronal activity in the basal ganglia. If this continues for a sufficient period of time, a self-sustaining focus may again be created, thereby giving the tardive effect. I would be interested to know whether anticonvulsant drugs have been tried against the dyskinesias.

C. D. Marsden: Barbiturates [J. H. Evans (1965) *Lancet* **i**, 458–460] and benzodiazepines [R. B. Goodwin-Austin & T. Clark (1971) *Brit. med. J.* **iv**, 25–26] do not help tardive dyskinesias much. But barbiturates [J. Gailitis, R. R. Knowles & A. Longobardi (1960) *Ann. Inter. Med.* **52**, 538] and diazepam [A. D. Korczyn & G. J. Goldberg (1972) *Brit. J. Psychiat.* **121**, 75] have been reported to stop acute dystonic reactions.

Diphenylhydantoin has been shown to antagonize the therapeutic effects of levodopa in Parkinson's disease, and levodopa-induced dyskinesias, but to enhance chorea in Huntington's disease [J. S. Mendez, G. C. Cotzias, I. Mena & P. S. Papavasiliou (1975) *Arch. Neurol.* **32**, 44–46].

E. H. Reynolds: There is another parallel to be drawn with the epileptic situation mentioned by the previous speaker. Many of our anticonvulsant drugs have been in use for much longer than phenothiazines, for example, phenobarbitone for approximately 50 years, and diphenylhydantoin for over 30 years. And yet it is only in the last 10 years, and especially the last five years, that we have recognized so many of the chronic metabolic and structural implications of such therapy. For example, peripheral neuropathy or cerebellar changes. We are, I think, as clinicians extremely naïve if we imagine we can give drugs for so many years without producing permanent structural changes, and I hope no one will think of adding the chronic complications of anticonvulsant therapy to the chronic complications of phenothiazine therapy!

A. P. Green: Metoclopramide was reported this morning to cause rutting in pigs without being a neuroleptic. It has also been reported to cause catalepsy without being a neuroleptic [B. Costall & R. J. Naylor (1973) *Psychopharmacologia* **32**, 161–170]. However, in animal tests for neuroleptic activity carried out at our laboratories the drug has behaved as a neuroleptic. We have also shown the drug to increase both striatal and mesolimbic homovanillic acid. Perhaps the lack of clinical evidence for the neuroleptic action of the drug is due to its short duration of action.

R. J. Naylor: Metoclopramide is of interest since this agent apparently lacks antipsychotic activity in the clinic although it exhibits a spectrum of activity in animal models virtually indistinguishable from that of the antipsychotic neuroleptic agents. Both extrapyramidal and mesolimbic dopamine mechanisms have been implicated with metoclopramide and neuroleptic effects. However, we have recently shown that whereas the neuroleptic agents are very potent and specific inhibitors of the hyperactive response induced by injecting dopamine into the nucleus accumbens septi, metoclopramide is inactive even in supramaximal cataleptic doses. It is suggested that the metoclopramide blockage of dopamine receptors in the mesolimbic system is readily overcome by an increased dopaminergic activity. The postulated importance of an enhanced mesolimbic dopaminergic activity to the development of psychoses and antipsychotic drug action makes this result of particular interest.

C. D. Marsden: Metoclopramide is a very interesting drug, for on other pharmacological evidence [A. Dolphin, P. Jenner, C. D. Marsden, C. Pycock & D. Tarsy (1975) *Psychopharmacologia* **41**, 133–138] it appears to be a striatal dopamine receptor antagonist, and it certainly does mimic other neuroleptics in causing a rise in both striatal and mesolimbic HVA [L. Ahtee & G. Buncombe (1974) *Acta Pharmacol. Toxicol.* **35**, 429–432; E. Perringer, P. Jenner & C. D. Marsden (1975) *J. Pharm. Pharmacol.* **27**, 442–444]. We have also found that metoclopramide enhances the rate of dopamine disappearance after α-methyl-*p*-tyrosine and the rate of dopa accumulation after NSD 1015, indicating increased dopamine synthesis and release. So on all counts it seems to block at least striatal dopamine receptors. Yet we could find no effect of the drug on Parkinsonism or levo-dopa-induced dyskinesias, although it effectively counteracts levodopa-included emesis [D. Tarsy, J. D. Parkes & C. D. Marsden (1975) *J. Neurol. Neurosurg. Psychiat.* **38**, 331–335]. The clinical evidence that metoclopramide is not a neuroleptic is not entirely convincing, and perhaps its short duration of action is relevant here.

The time has come to close the discussion. Clearly this subject of the dyskinesias is of considerable interest to both clinicians and research workers. The subject is of relevance not only in relation to the neurological problems posed by patients with dyskinesias, but also to psychiatry in view of the design of the best neuroleptic agents. I am sure we would all wish to thank the organizers of this session, our hosts and the speakers.

PART II

EPILEPSY

Chapter 12

Clinical and Biochemical Aspects of Epilepsy

By PHILLIP HARRIS

Department of Surgical Neurology, Western General Hospital and the Royal Infirmary, Edinburgh; and Medical Research Council Brain Metabolism Unit, University Department of Pharmacology, 1 *George Square, Edinburgh EH*8 9*JZ, U.K.*

Epilepsy is a clinical paroxysmal neurological disorder due to sudden synchronous high voltage discharges arising from a number of hyperexcitable neurons. There is a tendency to recurrent attacks of the abnormality. Neurochemical and electrical changes occur with disturbance of the inhibitory mechanisms and spread of electrical discharges beyond the locus.

Epilepsy is a 'symptom complex' and therefore we must attempt to discover the aetiology and, if discovered, decide on (*a*) elective and (*b*) symptomatic treatment—but usually require both.

It may be very difficult to define the primary area of abnormality and difficult to demonstrate the spread and indeed at times decide which of two separate foci, for example in each temporal lobe of the brain, is the primary one.

Pathology, Pathophysiology and Pathogenesis

Three factors exist regarding epilepsy:

(1) an individual predisposition, which is constitutional or hereditary;

(2) occurrence of an epileptogenic focus in the brain;

(3) precipitating factors which give rise to electrochemical changes.

The 'epileptic focus' may result from many different pathological conditions, including trauma, vascular lesions, neoplasms, infections and atrophy of the brain. Metabolic and nutritional factors may induce epilepsy, for example hypoglycaemia, various electrolyte imbalances, pyridoxine (vitamin B_6) deficiency and disturbance of fat metabolism such as that which occurs in lipid storage diseases. Various toxic states including lead and carbon monoxide poisoning and uraemia and certain allergies may also be important factors causing epilepsy. Certain endocrine aspects may also be important including the changes that occur at puberty, during pregnancy and premenstrually. Age is very important, thus in young children febrile convulsions may develop and these may cause temporal lobe epilepsy later in life; also in children petit mal (Fig. 1) is a common form of epilepsy but rarely occurs in adults although a child with petit mal may later develop grand mal seizures. Precipitating causes include sleep, noise and the flickering of the television screen (two examples of reflex epilepsy), also hunger, emotional disturbances etc. Activating factors include drugs such as pentylenetetrazol, photic stimulation, sleep and hyperventilation.

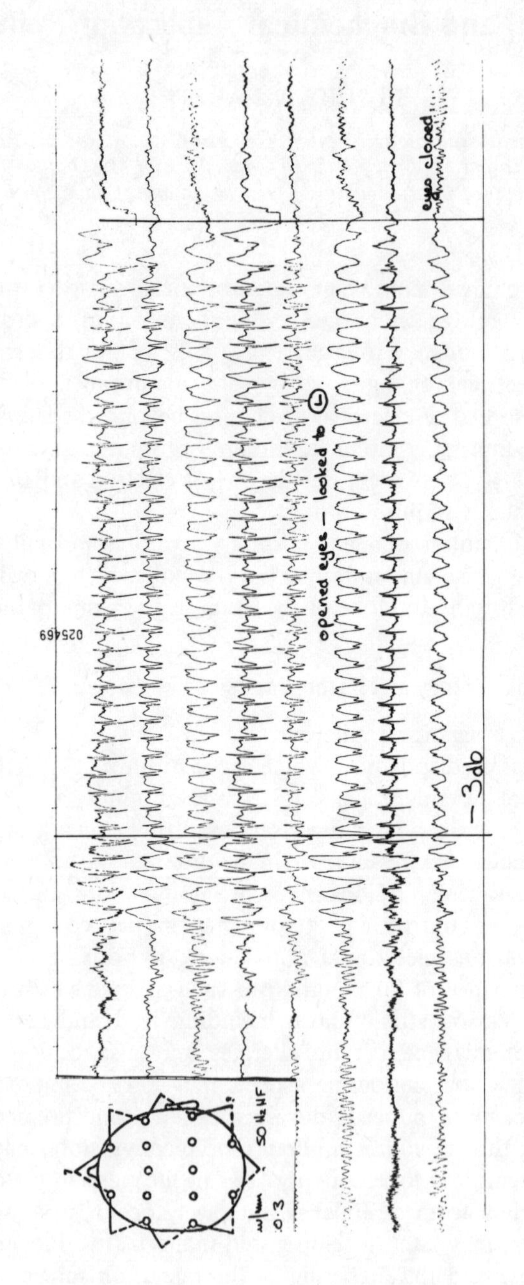

Fig. 1. *EEG showing a typical petit mal attack*

Much has been learned from studies of post-traumatic epilepsy, and in about one-third of such patients the epilepsy first appears within three months of the injury. The degree of destruction of brain tissue bears a definite relationship to the likelihood and severity of epilepsy occurring especially if the brain lesion is in the region of the Rolandic sulcus. Trauma initially destroys some cerebral neurons and disturbs the functions of others, and secondary effects ensue with metabolic changes including release of epileptogenic agents such as acetylcholine, and there appear to be disturbances of inhibitory transmitters including γ-aminobutyric acid (GABA). Electrolyte changes, anoxia and brain swelling etc. may also tend to occur. In time, there may be hyperexcitability of recovering neurons, with neuronal instability (Phillips, 1954). Much work is progressing in this field of neurochemistry, and much of it is only possible in experimental animal situations. The exact mechanism whereby anticonvulsant drugs may modify biochemical changes and thus the seizure activity of the central nervous system is not yet fully understood. Is the action directly on the 'epileptogenic focus', or do the drugs act by limiting or preventing propagation of the spread of impulses through the brain? Or, indeed, do some drugs act in both of these ways; and some by some other mechanisms? Another useful experimental tool is the electroencephalogram (EEG), and very sophisticated electrical studies have been used by Schmidt *et al.* (1959), who carried out microelectrode studies of electrical discharges from single cells in the epileptic focus; and the work of A. Earl Walker in Baltimore (1959) on preferential subcortical pathways in experimental epileptic monkeys. These and other studies have added greatly to our understanding of the physiology of the normal brain and the pathophysiology which may occur in epilepsy. There are, of course, many ill-understood and indeed unknown factors, thus a similar type of pathological lesion may develop in man or be made to occur in experimental animal models without the appearance of epilepsy; also the type of epilepsy that may occur, the number of attacks and the response to drug and other therapy varies greatly. In addition, for example in post-traumatic epilepsy, about 50% of patients will cease to have epilepsy spontaneously as the years go on whatever treatment may have been used. EEG abnormalities may occur at some point distant to the 'epileptogenic lesion'.

In temporal lobe epilepsy, sometimes called mesial temporal sclerosis (Falconer, 1972), a single pathological lesion may be found in the mesial temporal structures, including the hippocampus (Ammon's Horn), the amygdala, and hippocampal gyrus and the uncus. This type of epilepsy is considered to be a later result of a severe febrile convulsion which has occurred at about the age of 6 months to 4 years, causing hypoxia in the particularly sensitive mesial temporal structures resulting in a degree of gliosis and an epileptogenic lesion, but in addition there are genetic or predisposing factors in these patients.

Thus much is now known about the establishment of an epileptogenic focus and the spread of the discharge. In certain types of epilepsy, consciousness may be disturbed, possibly due to a synchronization of cerebral cortical rhythms from thalamic–reticular impulses. According to Gastaut & Fischer-Williams (1960) grand mal fits depend on a 'thalamic discharge which involves the non-specific reticular structures and is projected to the cortex and what may be considered a generalized recruiting response transmitted along the diffuse cortical projection pathways'. These authors considered that petit mal was a result of thalamic discharge occurring in a person with a very effective inhibitory mechanism. In about 20% of patients with epilepsy, no obvious cause can be identified. Some 'focal cerebral lesions' may cause epilepsy, but many do not. In any event, the location and nature of the focal brain abnormality does not bear a direct relationship to the frequency of the epileptic seizures.

Incidence and Frequency

One person in 200 of the population suffers from epilepsy. The condition may start at any age although it is much commoner in adults than in young children, and may present as a single type of fit in an individual patient or such a patient may have more than one type of fit during a period of time, and as the years go on, especially if the fits begin in childhood, they may change their character. In about one-third or more of patients there is no obvious pathological lesion or metabolic or toxic factor to be found, even so these patients do not necessarily have 'idiopathic' or 'cryptogenic' epilepsy.

About 5% of patients admitted to hospital because of a blunt (closed) head injury develop epilepsy, usually within three months following their trauma. Indeed this is a similar interval to that which occurs in patients who have had open head injuries, although the incidence of epilepsy in these patients is in the region of 45 to 50% (Russell & Whitty, 1952; Phillips, 1954).

The commonest cause of epilepsy beginning in later life is usually progressive disease of the brain and in particular brain neoplasms. Epilepsy occurs in 50% of patients with a supratentorial neoplasm; it occurs in over 90% of patients with oligodendrogliomas. In 40% of patients with intracranial neoplasms epilepsy is the leading symptom.

Classification

Individual patients may have more than one type of fit, giving rise in man to nosological, diagnostic and therapeutic problems. These features may be considered in the classification of epilepsy:

Generalized or partial seizures—the latter with elementary symptomatology (e.g. Jacksonian epilepsy) or with complex symptomatology (e.g. temporal lobe epilepsy with disorders of behaviour and abnormalities of perception);

'Symptomatic' or 'functional'—depending on whether a demonstrable lesion in the brain is present or not;

Genetic factors; the type of pathology; EEG abnormalities—localization and spread; the age of the patient when the seizures begin; drug susceptibility; metabolic state; multiple lesions and multiple factors.

Penfield & Jasper (1954) mainly took into account the location and spread of the abnormal electrical rhythm, correlating this with the clinical features and the anatomical location of the apparent origin of the seizure. Thus focal epilepsy (cerebral cortex), centrencephalic (grand mal and petit mal) and cerebral seizures (unlocalized). Symonds (1955) felt that complete evaluation of each patient was necessary (see Table 1). In 1964 the International League Against Epilepsy along with other international scientific bodies proposed a classification of epilepsy, which was revised by Gastaut in 1969. The nature of the onset of the fit was

Table 1. *Symond's Classification (1955)*

	Clinical	Anatomical	Physiological (EEG)	Pathological	Therapeutic
Central epilepsy	Major—generalized minimal lapses and jerks	Central	Bilateral synchronous symmetrical discharge	Idiopathic (genetic)	Dione responsive
Partial epilepsy	Variable, focal onset depending on location	Variable focal	Focal abnormality	Anatomical lesion present	Phenobarbitone and diphenylhydantoin

stressed. Masland (1974) gave a slightly modified form of the WHO glossary of terms relating to different types of epilepsy, which provides a single uniform terminology:

Classification by aetiology, by seizure pattern and the EEG, by anatomy, and by the age or circumstances of occurrence.

Indeed 'classification of epilepsy' is a study in itself, there is as yet no ideal, accurate, and yet all-embracing classification available.

Diagnosis of Epilepsy and Clinical Problems Encountered

In man, the diagnosis of epilepsy mainly depends on a full study of the clinical features, along with special investigations and in particular EEG, which has become much more sophisticated and more useful in recent years, and various radiological studies including plain radiographs of the head and various contrast radiographs and in very recent times the use of computerized transverse axial tomography of the brain (Ambrose, 1973); this remarkable British invention is said to be the greatest advance in radiodiagnosis since the discovery of Roentgen rays. The apparatus permits distinction between objects with minute differences in density, is quite safe and is 'non-invasive'. Using it, very small abnormalities in the head and brain may be distinguished, thus it would appear to have a very important place in the early diagnosis of brain lesions causing epilepsy. Up to 20% of patients with epilepsy have a normal EEG, especially in adults with grand mal seizures. The diagnosis of epilepsy can be difficult, thus Jeavons (1975) found that 20% of 420 patients seen in two epileptic clinics did did not have epilepsy, but some had syncope, psychological abnormalities, migraine, breath-holding attacks, night terrors or narcolepsy. It is not intended in this chapter either to discuss special techniques of electroencephalography or to discuss the symptomatology of various types of epilepsy.

There are many clinical problems concerned with epilepsy, both medical and social, including problems of aetiology, localization of an epileptogenic focus, the control of the fits and the many social problems that these patients and their families present. The development of epilepsy clinics and the appearance of 'epileptologists' appear to be inevitable. There are difficulties in studying groups of epileptic patients for example in relation to prognosis, because of the wide

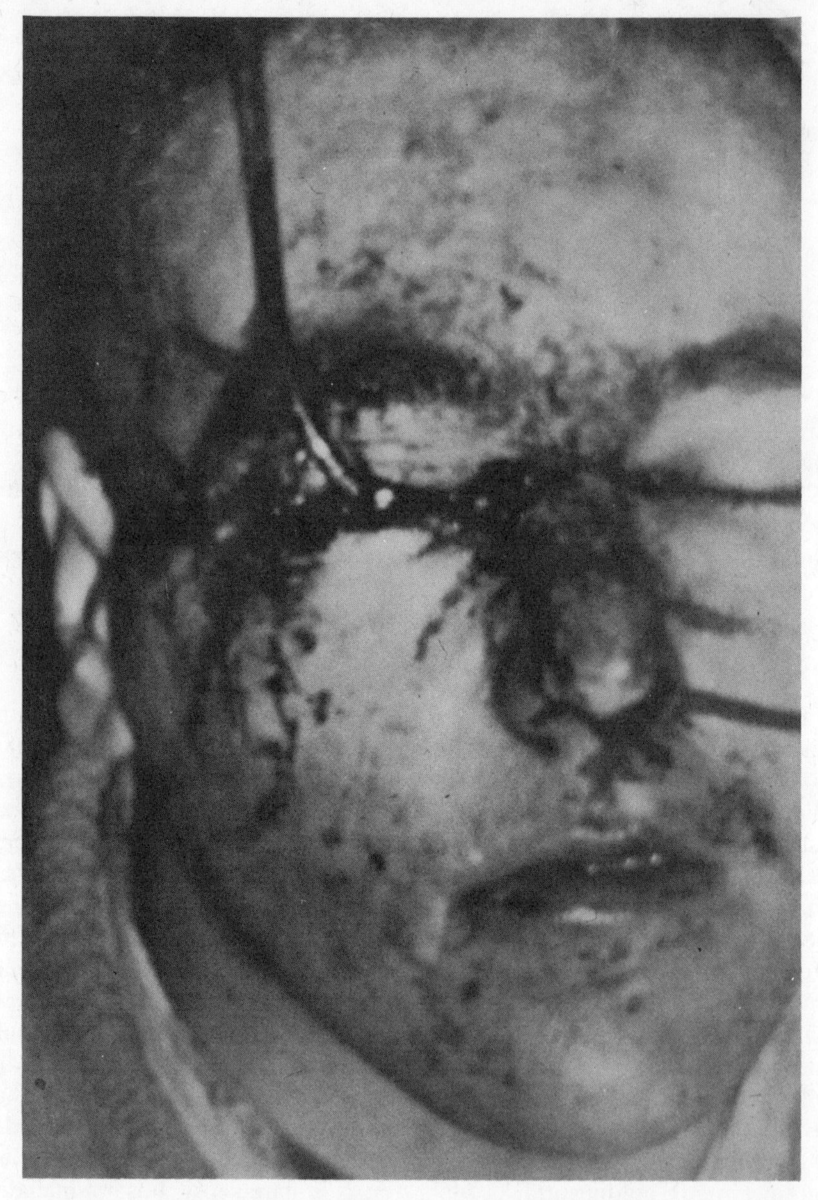

Fig. 2*a*. *Fishing gaff penetrating the right eye and orbit, and the right frontal lobe of the brain*

variety of variables present in any series and because of the different criteria used, and the difficulty in comparing one series with another. In many human clinical situations the prognosis of the cause or causes of epilepsy is of paramount importance and that of the epilepsy itself of secondary importance. The prognosis may be poor in epilepsy resulting from brain neoplasms, and in temporal lobe epilepsy. The prognosis is better regarding grand mal seizures, thus Williamsen & Dahl (1970) showed that 60% of such patients became fit-free. But Jüül-Jensen (1967) noted that when anticonvulsant drugs were stopped 40% of patients may relapse during the next five years. The prognosis is on the whole poor if a patient has a combination of different type of epilepsy; or if the patient has frequent fits and a significant degree of mental deterioration. Kiørboe (1974) showed that even a single EEG study may be helpful regarding prognosis, and if the EEG improves so usually will the prognosis.

Epilepsy may develop some years following a head injury. This is shown for example in a 25-year-old man who sustained an orbital and frontal lobe injury when a fishing gaff accidently entered his right frontal lobe producing a haematoma there and causing drowsiness and progressive left-sided hemiparesis (Fig. 2*a*).

A radiograph of the skull demonstrates the location of the barbed end of the fishing gaff (Fig. 2*b*). At operation the fishing gaff was removed and a large

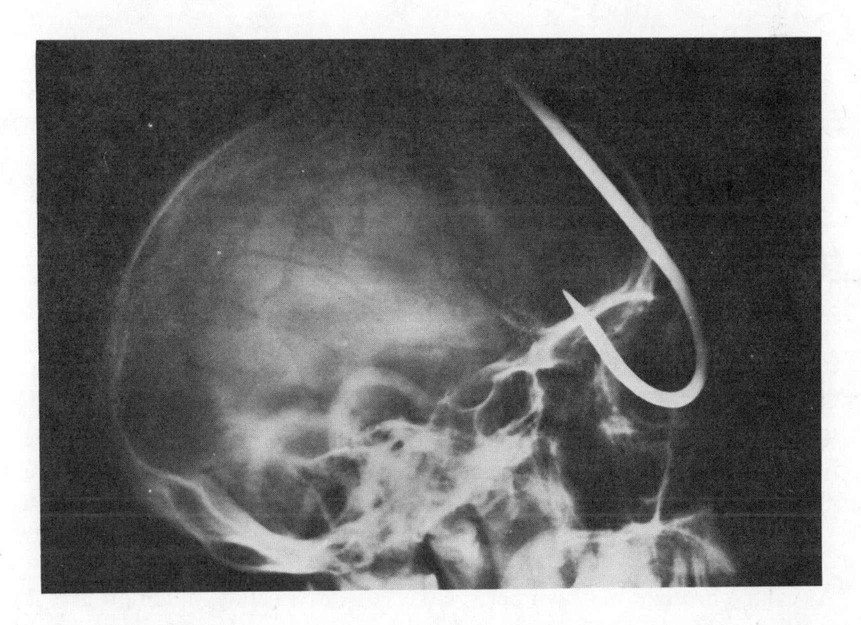

Fig. 2*b*. *Lateral radiograph showing the fishing gaff entering the head and the right frontal lobe of the brain*

haematoma was evacuated from the right frontal lobe. The patient made an excellent postoperative recovery without any neurological abnormalities, but some five years following the accident he began to develop epilespy (Fig. 2*c*). The EEG showed a right medial frontal spike epileptogenic area. This patient has responded well to anticonvulsant drug therapy, but it was two years before the EEG became normal.

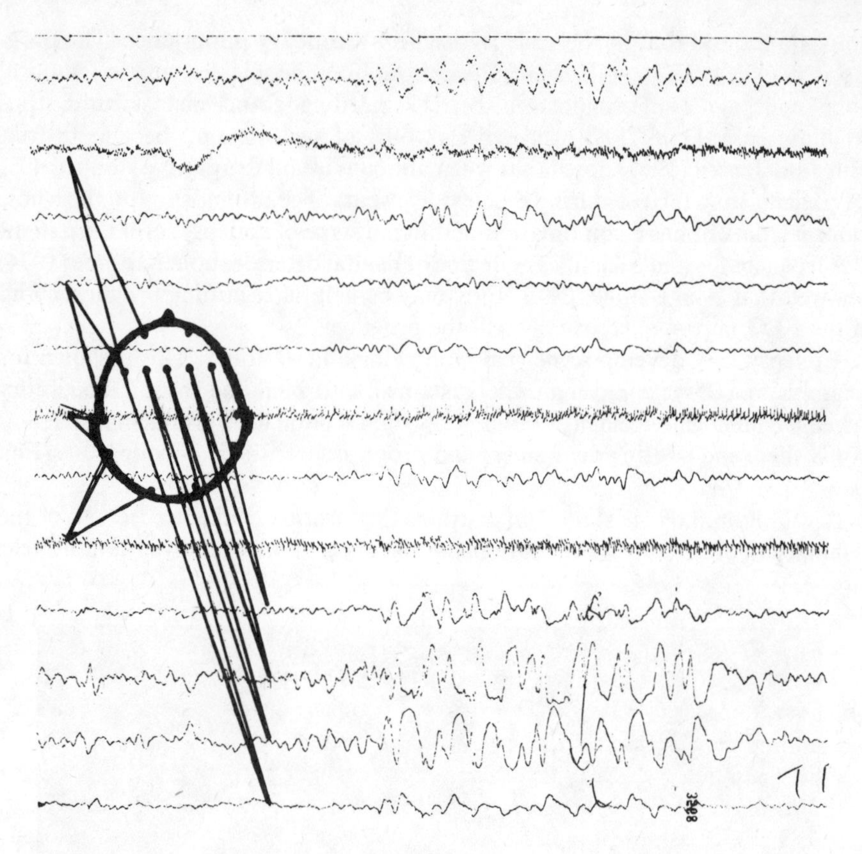

Fig. 2*c*. *EEG five years after the injury, showing evidence of epilepsy, with secondary bilateral synchrony, right more than left*

Drug Therapy

The anticonvulsant drug therapy of epilepsy has become more scientific and rational due to better experimental and clinical studies, and knowing that the aim of treatment is to allow patients to live a reasonably normal life. It must be said however that the ideal anticonvulsant drug does not yet exist, even for one form of epileptic seizure.

As will be brought out by other speakers at this meeting, much progress has been made in the biochemical study of epilepsy, and it is hoped that the results of these studies will permit pharmacologists to develop better anticonvulsant drugs. Blood level estimations of drugs would appear to be very important, but unfortunately at present not many clinicians have the laboratory facilities necessary to carry out these studies.

Understandably, experimental animal studies are carried out on all anticonvulsant drugs before they are used in the clinical human situation, but there would not appear to be a clear relationship between fits experimentally induced in animals and those occurring in man regarding the response to drugs. Different anticonvulsant drugs in some patients certainly appear to control different types of epileptic seizure, but about 20% of patients with epilepsy do not respond at effects, and their mechanism of action. How are anticonvulsant drugs taken up peutic use of anticonvulsant drugs, and Woodbury (1974) mentions that knowledge is required of the absorption, distribution, biotransformation and excretion in the body of these drugs as well as pharmacological and toxic effects, and their mechanism of action. How are anticonvulsant drugs taken up and distributed in the central nervous system, and what is their site of action? To be successful in use an anticonvulsant drug should give maximal control of epilepsy, with minimal side effects and there should be low toxicity—both short and long term. The drugs should have a sufficiently long half-life, the dose should be easy to decide on and adjust and the drugs should not require to be administered too frequently. Also there should be minimal or absence of drug interaction.

There are problems in carrying out anticonvulsant drug trials in patients. It is difficult to get a homogeneous group of patients, either in-patients or out-patients, and each of these groups present particular problems concerning the type of epilepsy that they most likely have. There are problems regarding the reliability of a patient taking the drug or drugs regularly; thus some patients with epilepsy have quite significant behaviour and personality disorders. Some anticonvulsant drugs may activate certain types of epilepsy, for example phenytoin is good for grand mal epilepsy but may activate petit mal seizures. The EEG is of limited use in evaluating the response to anticonvulsant drug therapy. The aim in treatment is to increase the threshold to epilepsy and therefore to control and confine the electrochemical abnormality.

In the past few years certain new anticonvulsants have become available, including carbamazepine (Tegretol) and sodium dipropylacetate (Epilim). This latter drug has been assessed clinically by Jeavons & Clark (1974), who found that it will control fits in patients whose epilepsy has previously been intractable, and who state that it is the drug of choice in patients with absences and three cycles per second spike and wave discharges. It appears to potentiate other anticonvulsants.

Some clinicians, including myself, feel that there is a place for the use of anticonvulsant drugs prophylactically, for example in certain patients who have sustained head injuries, and in patients who have had brain operations for certain conditions such as brain neoplasms and intracranial vascular lesions. Little is known, to date, regarding the use of anticonvulsants in this way and there do not appear to be any controlled studies.

Anticonvulsant drugs *per se* rarely cure epilepsy, but treatment of any metabolic, nutritional or toxic condition causing epilepsy may cure the disorder—however, if it returns, so may the epilepsy reappear. On the other hand, as will be mentioned below, certain surgical operations can in fact cure some forms of epilepsy.

Surgical Treatment of Epilepsy

The surgical treatment of epilepsy represents one of the earliest types of surgical operations (in the form of trephining) carried out in man, going back almost to prehistoric times.

In selected patients, surgical excision or ablation of an 'epileptogenic focus' or interruption of pathways of spread of the impulses in the brain from this focus may be highly beneficial to the patient (Rasmussen, 1974).

The pioneering work was carried out in man, but animal experimental work is not only providing much more information concerning the nature of epilepsy, but is providing the neurosurgeon with vital new information so that the indications for operation and new types of operation can be better understood and planned.

In the clinical human situation there is often a clear indication for operative treatment for obvious pathological lesions such as brain neoplasms and arteriovenous malformations of the brain, and the excision of such a lesion, if possible with the adjacent 'epileptogenic cerebral tissue', may abolish, and certainly should greatly reduce, the liability of the patient to have the symptom complex of epilepsy. About 40% of patients operated on for slow growing gliomas of the cerebrum lose the tendency to epilepsy, and this percentage rises to 50% with cerebral meningiomas; and after the excision of an arteriovenous malformation of the brain, where epilepsy has been a main feature, some 44% of patients will no longer be subject to epilepsy.

The main medication regarding the surgical treatment of epilepsy *per se* is for those patients who do not have an obvious pathological lesion of the brain, and certainly do not obviously have a lesion which is steadily expanding such as an intracranial neoplasm. Thus, initially, the various special investigations pre-

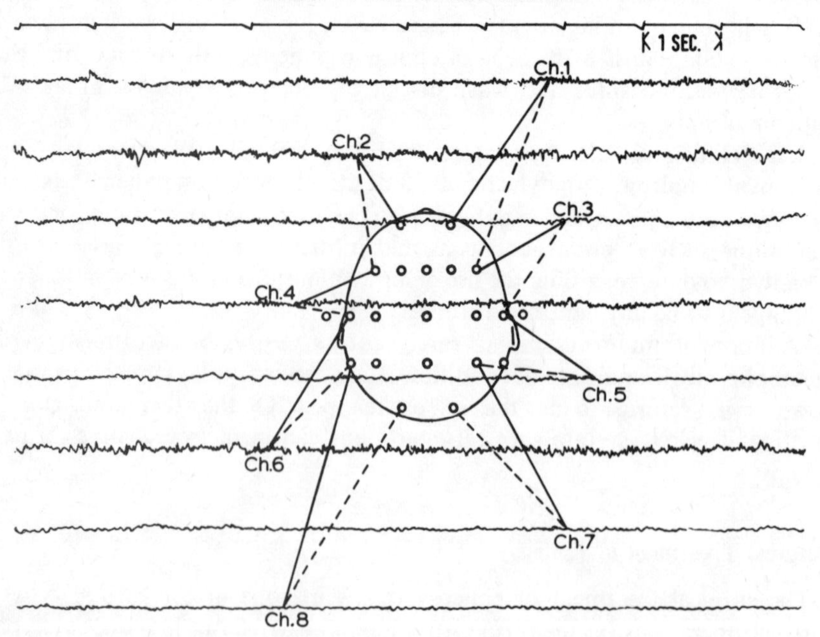

Fig. 3*a*. *Patient with left-sided temporal lobe epilepsy; scalp EEG preoperatively*

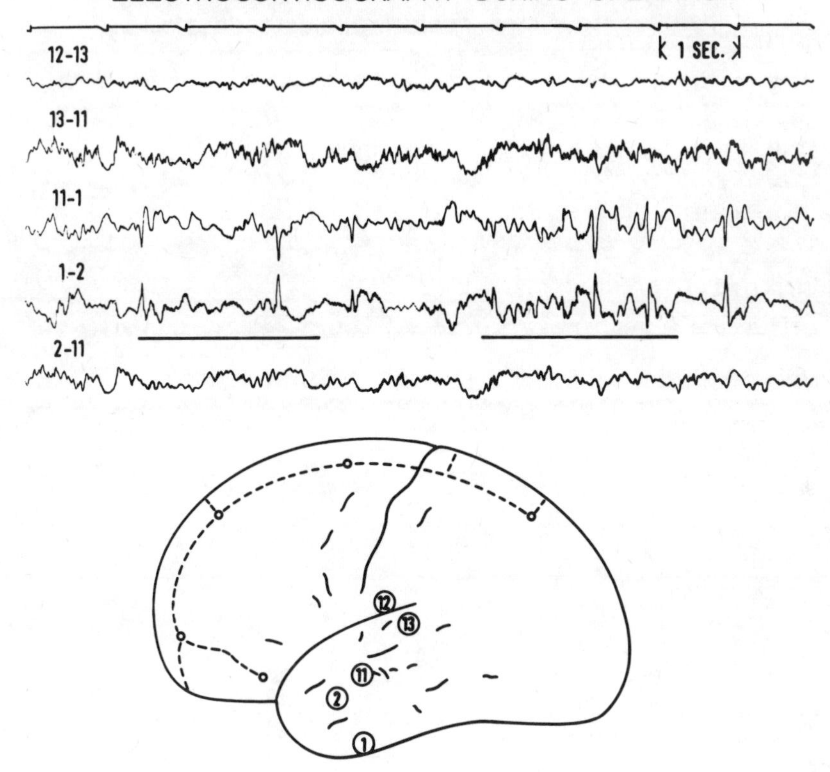

Fig. 3*b*. *EcoG during operation, showing spike focus*

viously mentioned must be carried out. Criteria for selection for operative treatment for such patients include these features:

The demonstration of an 'epileptogenic focus' which appears to be causing the epilepsy, knowledge that the patient's epilepsy is not responding satisfactorily to full drug treatment, and that the seizures are frequent and are seriously affecting the patient's normal life and activities, also that the patient has had fits for at least three or more years. It is necessary to take into account the natural history of the patient's epilepsy; thus post-traumatic epilepsy has a tendency to abate some three or four years following the head injury, and in young people the type and frequency of fits often changes with time. In addition it is essential to realize that the part of the brain being operated upon can be safely reached and dealt with without causing neurological symptoms and signs, and that the surgical procedure used is most likely to produce a neurological and social success.

Certain cortical cerebral scars may be excised along with neighbouring epileptogenic cortex with a cure rate that reaches about 50% (Penfield, 1952).

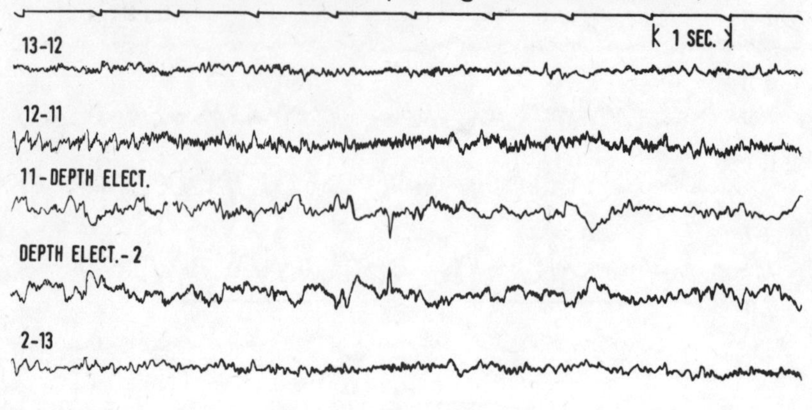

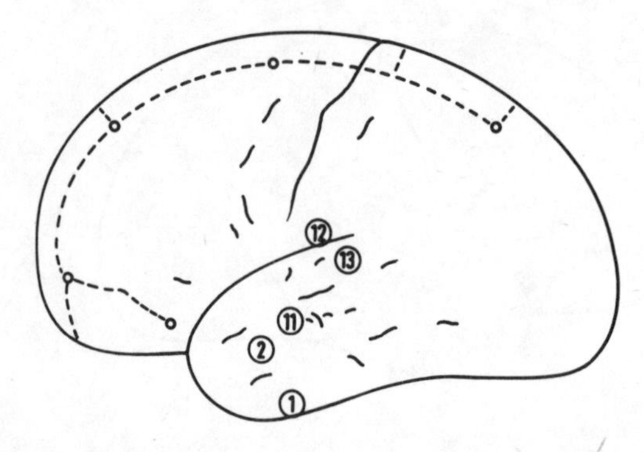

Fig. 3*c*. *Depth ECoG during operation, showing spike focus*

In some of these patients a partial temporal, frontal or occipital lobectomy may be the best form of treatment. Regarding temporal lobe epilepsy, it is now well known that the commonest lesion is 'mesial temporal sclerosis' (Falconer & Taylor, 1968), although other pathological abnormalities may be seen on histological examination of the temporal lobe. The temporal lobectomy should include the amygdala and the anterior part of the hippocampus (Figs. 3*a*, *b*, *c* and *d*). When the criteria for selection of patients for temporal lobectomy have been adhered to, and the appropriate operation has been carried out, the cure rate concerning the patient's epilepsy reaches 40%, and in various series an impressive additional percentage of patients are greatly benefited by this operation. Certain associated psychological and psychiatric abnormalities may also be helped by this procedure.

Stereotaxic EEG investigations and stereotaxic surgery for epilepsy are relatively new, although in 1950 Spiegel & Wycis reported on thalamic recordings in man with special reference to seizure discharges, and described coagulation of the dorsomedial nucleus of the thalamus for 'centrencephalic' seizures. Stereo-

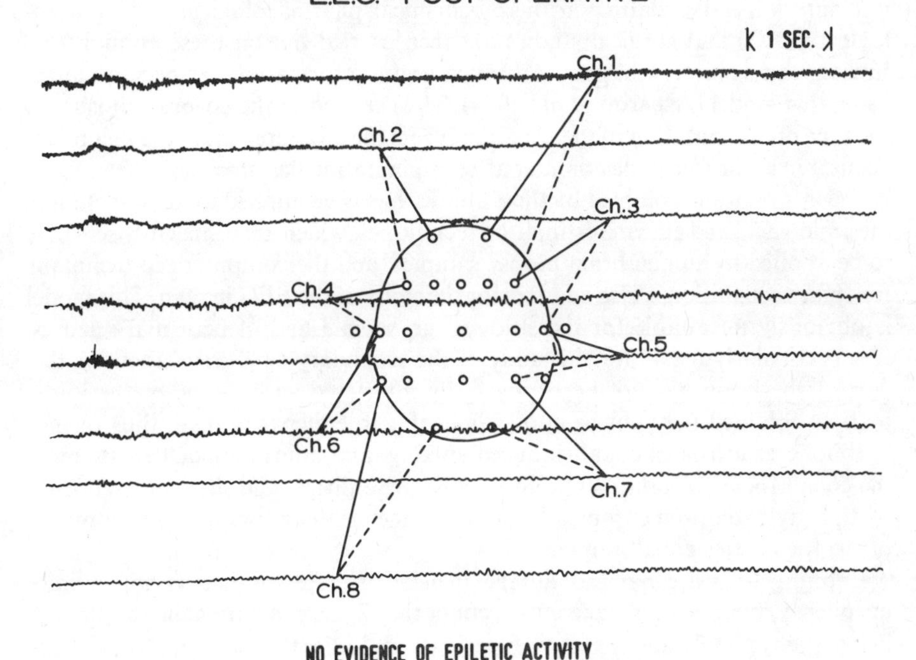

Fig. 3*d*. *Scalp EEG three years after temporal lobectomy: no evidence of epileptic activity*

taxis depth electrocorticography and electrical stimulation have a place in the experimental study of epilepsy in animals, and also in the clinical situation in man, to help to locate 'epileptogenic foci'. Various stereotaxic operations are now being carried out in selected patients with intractable epilepsy: thus stereotaxic ablative lesions are made in the central, dorsomedial, or the anteroventral nuclei of the thalamus, and commissurotomy and Forel-H-Tomies are carried out by some neurosurgeons for certain patients with intractable grand mal or petit mal seizures (Ramamurthi *et al.*, 1974; Sano, 1974). Stereotaxic amygdalotomy may be carried out by a stereotaxic technique for certain patients with temporal lobe epilepsy (Vaernet, 1974). Several workers in this field report very good results, but the follow-up in most of the reported series is still very short.

Since the results of the surgical treatment of certain types of intractable epilepsy appear to be so good that many of the patients treated surgically no longer require antiepileptic drugs, one wonders why so few neurosurgeons have taken up this type of treatment for what is a very common and serious condition.

Research

In animals it is possible to prepare a reproducible 'epileptic animal model'. Thus in Edinburgh, in the Medical Research Council Brain Metabolism Unit, the model that we are using is the rat with implantation of cobalt gelatin into the frontal cortex which gives rise to an early primary epileptic focus and a secondary focus on the opposite side of the brain which appears after some 7 days of the implantation of the gelatin. Various biochemical, pharmacological, neurophysiological (EEG) and surgical studies are then carried out on these animals (see Emson, this volume, Chapter 14). Much useful information has been obtained using this model (Ashcroft *et al.,* 1974). Thus section of the corpus callosum at the time of implantation of the cobalt prevents the development of the secondary focus, but at the fourth day or later after implantation has much less effect. Some of the information obtained by these studies has been applied to man, including the amino acid and enzyme estimation techniques, which are sufficiently sensitive to be applied to human brain biopsy samples, and the computerized technique for spike recognition is also suitable for the study of the EEG in man. This model is obviously unsuitable for the study of grand mal and of petit mal seizures. It is particularly useful for the study of the secondary or 'mirror' focus as this part of the brain has not been physically interfered with. Some of the information obtained may not be at all applicable to epilepsy in man, thus phenobarbitone exacerbates cobalt-induced spiking, and ethosuximide can suppress the cobalt focus for up to 30 minutes, also these drugs, and the administration of diphenylhydantoin chronically, do not influence the subsequent development of the focus after cobalt implantation.

Certainly there are many advantages in using primates for these various studies in epilepsy, but one must take into account the expense, maintenance and handling of such animals, and for many experimental studies the use of small mammals would appear to be quite satisfactory. Walker & Udvarhelyi (1965) carried out studies on the cortical and subcortical spread of focal discharges in monkeys, and found that there was a stage where a critical mass of cerebral neurons discharge to produce a generalized fit. This was contrary to the 'centrencephalic origin' view of generalized epilepsy. In Chapter 13 of this volume, Dr. Meldrum discusses the relationship of experimental epilepsy to clinical problems, and provides up-to-date information on his experimental studies in the baboon *Papio papio*, which can be made to develop a type of epilepsy with some of the characteristics of human epilepsy. Long-term studies of such animals, over a period of years, would be of great value in helping our understanding of epilepsy.

Purpura *et al.* (1972) discuss experimental animal models used in the study of epilepsy.

There is certainly a more rapid development of epilepsy in animals low in the evolutionary scale and in animals with an unconvoluted forebrain, such as the rat, but not the cat.

There have been attempts to produce experimental animal models of petit mal epilepsy, including acute experiments in the cat (Marcus, 1972).

Focal electrical stimulation is not a good experimental model for interictal studies, but may be used to study epileptic susceptibility, and is used by some neurosurgeons for the investigation and treatment of certain patients with tractable epilepsy, to study the location of epileptic areas and the spread of abnormal electrical discharges.

Walker & Mayanagi (1974) produced and have studied temporal lobe epilepsy in monkeys by inserting alumina cream into the temporal lobes and then studying these animals clinically and by the EEG and activating them by giving intravenous Metrazol.

In recent times special interest has been focused by some workers on the place of the cerebellum in epilepsy (Cooper *et al.*, 1973). It has been found that electrical stimulation of the cerebellar cortex or nuclei may alter or stop seizure discharges. The cerebellum appears, via the Purkinje cells, to have inhibitory functions over cortical excitability. Indeed, there has now been reported some early clinical experience in the use of chronic cerebellar stimulation to control certain types of intractable epilepsy in man.

There are a number of limitations in using animal experimental models to study epilepsy, and these include species difference, the (usual) lack of a predisposition to epilepsy in these animals, the difficulty in causing delayed epilepsy in animals and in producing an animal model where spontaneous recurrent attacks of epilepsy with a local epileptic focus occurs. Psychological aspects, and social and familial studies cannot be carried out, nor can studies on the prognosis of epilepsy be done. In addition, the individual response of these animals to various drugs is not shown as is so commonly seen in the human situation. There would not appear to be a clear relationship between fits produced in experimental animals and those occurring in man regarding the response to drugs, certainly over a prolonged period of time. When discussing the condition of 'epilepsy', whether in animals or in man, it is necessary to be certain what one means precisely by this term, and some experimental animal models have conditions which are quite different from clinical epilepsy seen in man.

On the other hand, there are certain advantages in using experimental animal models, and several of these points have been brought out above. Thus biochemical, pharmacological and EEG studies can be carried out in a controlled way in a large number of 'subjects', and pathological changes may be readily studied. It appears to be possible to produce febrile convulsions and fits due to cerebral hypoxia, and to various metabolic disturbances in animals, and the development of a secondary or mirror focus is proving to be a useful lesion for biochemical research. In animals it is possible to make precise predetermined lesions in various parts of the brain using stereotaxic methods, and this is very different from, for example, post-traumatic epilepsy in man, where the damage may be localized to one small area but where there is every likelihood that wider areas and more distant areas of the brain have also been disturbed when the injury occurred. In addition in man, various extrinsic factors such as metabolic and electrolyte disturbances may influence the development of epileptogenic foci in the human brain following trauma.

We must not be complacent concerning our present medical and surgical treatment of epilepsy. We must forever be searching for better drugs and for better operations, utilizing the vital basic information provided by neurophysiological and biochemical studies, and work very closely with pharmacologists. However good some animal models may be in studying some aspects of epilepsy, man will remain the ultimate 'model'. His symptoms and signs and response to drugs and operations must be recorded and studied most assiduously, but at all times the ethics of new investigations and treatments require to be fully considered and kept to the forefront (Declaration of Helsinki, 1964; Smith, 1973; Strauss, 1973).

The Future

It is hoped that more vigorous attempts will be made to prevent the development of epilepsy in man, certainly many head injuries could be prevented and other causes of 'organic' or 'symptomatic' epilepsy could be recognized at an earlier stage and treated accordingly to prevent epilepsy occurring. Not enough is yet known about the hereditary aspects of epilepsy for genetic counselling to be applied in the clinical situation.

It is hoped that some biochemical means might be introduced to prevent or limit the development of an epileptic focus in the brain, for example after a head injury has occurred or before a major brain operation is being undertaken.

The development of better drugs, and a better understanding of their correct use is hoped for, including the possibility of developing 'depot' drugs which might require to be taken only once a day or even less frequently but which will still produce a steady therapeutic blood and tissue level of the drug.

Almost one third of patients with significant epilepsy have temporal lobe epilepsy, and in this respect the importance of febrile convulsions is now appreciated. The challenge of preventing these convulsions is discussed by Lennox-Buchtal (1974), by preventing and vigorously treating febrile illnesses especially in children between the ages of 6 months and 4 years and stopping any febrile convulsion that may occur immediately with a suppository or an injection of diazepam.

However, for very many years to come the appropriate treatment of patients with epilepsy will remain a combination of science and art. Devoted clinical attention to every aspect of the epileptic person's disabilities is essential if the patient's epilepsy is to be kept under proper control.

References

Ambrose, J. A. (1973) *Brit J. Radiol.* **46**, 549, 736

Ashcroft, G. W., Dow, R. C., Emson, P. C., Harris, P., Ingleby, J., Joseph, M. H. & McQueen, J. K. (1974) in *Epilepsy. Proceedings of the Hans Berger Centenary Symposium* (Harris, P. & Mawdsley, C. eds.), pp. 115–124, Churchill Livingstone, Edinburgh

Cooper, I. S., Ricklan, M. & Snider, R. S. (eds.) (1973) *The Cerebellum, Epilepsy and Behaviour*, Plenum Press, London

Declaration of Helsinki (1964), World Medical Association

Falconer, M. A. (1972) *J. Med. Sci.*, **141**, 12, 147–161

Falconer, M. A. (1974) *Lancet* **ii** 767–770

Falconer, M. A. & Taylor, D. C. (1968) *Arch. Neurol. Psychiat. (Chicago)* **19**, 353–361

Gastaut, H. 1969. *Epilepsia Suppl.* **10** s2–s13

Gastaut, H. & Fischer-Williams, M. (1960) *Handbook of Physiology* (Field, J. ed.), sect. 1, vol. 1, p. 329, Washington, D.C.

Jeavons, P. M. (1975) *Hosp. Update* January, 11–22

Jeavons, P. M. & Clark, J. E. (1974) *Brit. Med. J.* **i**, 584

Jüül-Jensen, P. (1967) *Acta Neurol. Scand. Suppl.* **43**, 31, 166

Kiørboe, E. (1974) in *Handbook of Clinical Neurology*, vol. 15 *The Epilepsies* (Magnus, D. & De-Gaasr, A.M.L., eds.), pp. 783–799, North Holland Publ. Co., Amsterdam

Lennox-Buchtal, M. (1974) in *Handbook of Clinical Neurology*, vol. 15 *The Epilepsies* (Magnus, O. & De-Haast, A.M.L., eds.), p. 260, North Holland Publ. Co., Amsterdam

Marcus, E. M. (1972) in *Experimental Models of Epilepsy* (Purpura, D. P., Penry, J. K., Tower, D. M. & Walker, R., eds.), pp. 113–146, Raven Press, New York

Masland, R. L. (1974) in *Handbook of Clinical Neurology,* vol. 15 *The Epilepsies* (Magnus, O. & De-Haast, A. M.L., eds.), pp. 24–28, North Holland Publ. Co., Amsterdam

Penfield, W. (1952) *J. Neurol. Neurosurg. Psychiat.* **15**, 73

Penfield, W. & Jasper, H. (eds.) (1954) *Epilepsy and the Functional Anatomy of the Human Brain,* Little, Brown, Boston

Phillips, G. (1954) *J. Neurol. Neurosurg. Psychiat.* **17**, 1–10

Purpura, D. P., Penry, J. K., Tower, D. M. & Walker, R. (eds.) (1972) *Experimental Models of Epilepsy,* Raven Press, New York

Ramamurthi, B., Kalyanarman, S., Sayeed, Z. A. & Dharmapal, N. (1974) in *Epilepsy. Proceedings of the Hans Berger Centenary Symposium* (Harris, P. & Mawdsley, C. eds.), pp. 203–214, Churchill Livingstone, Edinburgh

Rasmussen, T. (1974) in *Epilepsy. Proceedings of the Hans Berger Centenary Symposium* (Harris, P. & Mawdsley, C., eds.), pp. 227–239, Churchill Livingstone, Edinburgh

Russell, W. R. & Whitty, C. W. M. (1952). *J. Neurol. Neurosurg. Psychiat.* **15**, 93–98

Sano, K. (1974) in *Epilepsy. Proceedings of the Hans Berger Centenary Symposium* (Harris, P. & Mawdsley, C., eds.), pp. 215–221, Churchill Livingstone, Edinburgh

Schmidt, R. P., Thomas, L. B. & Ward, A. A. Jr. (1959) *J. Neurophysiol.* **22**, 285–296

Smith, H. C. (1973) in *The Cerebellum, Epilepsy and Behaviour* (Cooper, I. S., Ricklan, M. & Snider, R. S., eds.), pp. 343–365, Plenum Press, London

Spiegel, E. A. & Wycis, H. T. (1950) *Electroencephalogr. Clin. Neurophysiol.* **2**, 23–29

Strauss, M. B. (1973) *New Eng. J. Med.* **288**, 1183–1184

Symonds, C. P. (1955) *Brit. Med. J.* **i**, 1235–1238

Vaernet, K. (1974) in *Epilepsy. Proceedings of the Hans Berger Centenary Symposium* (Harris, P. & Mawdsley, C., eds.), pp. 222–226, Churchill Livingstone, Edinburgh

Walker, A. E. (1959) in *Clinical Neurosurgery. Proc. Congr. Neurol. Surg.* (R. G. Fisher, ed.), pp. 69–103, Williams and Wilkins, Baltimore

Walker, A. E. & Mayanagi, Y. (1974) in *Epilepsy. Proceedings of the Hans Berger Centenary Symposium* (Harris, P. & Mawdsley, C., eds.), pp. 48–54, Churchill Livingstone, Edinburgh

Walker, A. E. & Udvarhelyi, G. B. (1965) *Arch. Neurol.* **12**, 357–380

Williamsen, R. & Dahl, E. (1970) *Nord. Med.* **84**, 1038–1041

Woodbury, D. M. (1974) in *Epilepsy. Proceedings of the Hans Berger Centenary Symposium* (Harris, P. & Mawdsley, C., eds.), pp. 78–95, Churchill Livingstone, Edinburgh

Chapter 13

Experimental Epilepsy: Its Relation to Clinical Problems

By B. S. MELDRUM

University Department of Neurology,
Institute of Psychiatry and King's College Hospital,
Denmark Hill, London SE5 8AF, U.K.

The term 'epilepsy' refers to a great diversity of phenomena with very varied aetiologies and patterns of clinical evolution. Thus although it is convenient to speak collectively of 'experimental—or animal—models of epilepsy', it is absurd to speak of an animal model of epilepsy. Any single phenomenon in animals, however clearly 'epileptic', cannot possibly be a model for all the forms of epilepsy occurring in man. Further, an admirably detailed scientific study of the aetiology and nature of an epileptic syndrome in one species may have no relevance to phenomena observed in man. The research strategy that has seemed to me most appropriate is to define the clinical problems, and then look for experimental situations that can be used to solve these problems.

This chapter is therefore divided into sections describing experimental models appropriate to the search for drugs or procedures that

(1) diminish the probability of a seizure occurring in an epileptic patient;
(2) terminate prolonged seizures (*a*) in non-epileptics, (*b*) in epileptics;
(3) prevent brain damage secondary to status epilepticus;
(4) prevent the development of fits or epilepsy.

The term 'antiepileptic' will not be used here: it is commonly employed to describe drugs in category 1 but ought logically to refer to drugs in category 4.

(1) Drugs or Therapies that Diminish the Probability of a Seizure Occurring in an Epileptic Patient

There are over 20 drugs currently marketed as anticonvulsants in the United Kingdom and some of them undoubtedly diminish the incidence of seizures in a significant proportion of epileptic patients. However, when the long-term control of seizures is assessed it is found to be inadequate in about 50% of patients. An additional problem is that of toxic side effects. These can occur acutely in relation to a high blood level of the anticonvulsant drug or its metabolites, or can result from chronic poisoning. Nystagmus, ataxia, dizziness and dysarthria are common acute effects of high doses of diphenylhydantoin or carbamazepine. Slowing and drowsiness may accompany high doses of phenobarbitone. Long-term side effects include gum dysplasia or hypertrophy (diphenylhydantoin), anaemia (oxazolidinediones, diphenylhydantoin) and calcium depletion from bone (diphenylhydantoin, primidone).

The search for new anticonvulsant drugs has a clear justification, but has it a clear enough goal? The operationally-defined goal [(1) above] is meaningful as long as controlled trials of anticonvulsants can be conducted. However, the research scientist in the pharmaceutical company needs a definition of the goal expressed in biochemical or pharmacological terms. It is not yet possible to give such a definition. The object of some basic research is to provide a biochemical description of the fault in the epileptic brain. There is no good evidence that such a primary fault exists in most epileptics. The best definition of the pharmaceutical scientist's goal might refer to some normal physiological function, such as inhibitory transmission, which, if chemically modified, could reduce the probability of seizures regardless of their primary aetiology. An alternative to studying the nature of epilepsy is to study the nature of anticonvulsant drugs. Thus chemists identify structural features common to known anticonvulsant drugs in order to synthesize new compounds with different combinations of these 'anticonvulsant' features. Likewise some pharmacologists look at simple *in vitro* or *in vivo* effects of known anticonvulsants in the hope that drugs found to be active in these systems may prove to be more potent as anticonvulsants. Clearly these approaches, and those described below, require an animal test system that allows the anticonvulsant potency of the drug to be assessed, and can indicate which class of patient is likely to be benefited by the drug.

(i) A Pharmacological Approach

Experiments employing drugs with known actions may indicate mechanisms involved in various types of epilepsy, and thus suggest appropriate targets for the synthetic chemist. The excessively synchronous neuronal discharges that underlie all epileptic phenomena are presumably initiated or maintained by some abnormality of function in synapses or in neuronal or glial membranes. Thus studying whether the tendency to seizures in appropriate animal models is enhanced or diminished by drugs with known actions on membranes or synapses can identify critical functions.

Membrane functions

A high extracellular potassium concentration can initiate seizure activity, especially in the hippocampus (Glaser, 1972). One of the functions of the glia in the brain is to regulate the ion content of the extracellular fluid (Kuffler & Nicholls, 1966; Henn *et al.*, 1972). Possibly the most consistent abnormal cytological feature in epileptogenic lesions in the brain is the presence of large numbers of reactive or fibrous astrocytes (see Meldrum, 1975*a*), hence it was natural to propose that these cells are defective in regulating extracellular potassium concentration (Pollen & Trachtenberg, 1970). However, direct measurements of membrane properties of such reactive glia have tended to show the opposite, i.e. that they may be more effective (Grossman & Rosman, 1971; Glötzner, 1973).

The main pharmacological approach to this topic has been the focal application of compounds known to inhibit (Na, K)-ATPase activity in membranes.

Ouabain perfused through the lateral ventricle or injected focally at various sites (e.g. the septal nuclei) induces sustained focal seizure activity (Baldy-Moulinier *et al.*, 1973). Several authors have sought evidence that (Na, K)-

ATPase activity is abnormal in various experimental forms of epilepsy (see Bachelard, this volume, Chapter 18). The clinical syndromes in which a reduction of cerebral (Na, K)-ATPase might be profitably looked for tend to be those that are non-fatal and do not require neurosurgery (e.g. febrile convulsions and petit mal). If further evidence from animal studies confirms a role for changes in ATPase activity in epileptogenesis, then ways will be found for investigating this enzyme activity in man.

Synaptic transmission

Acetylcholine. An excitatory neurotransmitter function is now well established for acetylcholine at several sites within the nervous system including the neocortex, hippocampus, striatum, hypothalamus and spinal cord (see reviews by Krnjević, 1974; Tebecis, 1974). Inhibition of cerebral acetylcholinesterase activity by the systemic administration of compounds such as the organophosphorus insecticides (Grob, 1963) leads to generalized convulsions, and the local application of acetylcholine to the eserinized cortex induces focal discharges. Thus it is natural to ask if an abnormality in cerebral acetylcholinesterase activity could be responsible for some forms of epilepsy (see reviews by Tower, 1960, 1969). Anticholinergics that modify cerebral activity can arrest or prevent seizures due to anticholinesterase activity but do not prevent seizures in primate models or those human syndromes in which they have been tried (Meldrum *et al.*, 1970).

Glycine. That glycine acts as an inhibitory transmitter within the spinal cord has been demonstrated by a variety of biochemical and physiological procedures including microiontophoresis (see Curtis & Johnston, 1974). Seizures induced by systemically administered strychnine or related alkaloids (e.g. bruceine) have long been known to be primarily spinal rather than cerebral, and it is now apparent that they act by competing with glycine for inhibitory receptor sites (see Straughan, this volume, Chapter 17).

There is no naturally occurring syndrome in man that shows the same seizure pattern. Although there is little evidence to implicate a primary disorder of glycine-mediated inhibition in human epilepsy, the possibility that some myoclonic syndromes involve a failure of spinal inhibitory mechanisms requires further study.

γ-Aminobutyric acid(GABA). The evidence that GABA mediates postsynaptic inhibition in many brain areas, including the cerebral cortex, cerebellum, hippocampus, thalamus and striatum, is presented in two recent reviews (Curtis & Johnston, 1974; Krnjević, 1974). It is probable that it also mediates presynaptic inhibition on primary afferent pathways.

Drugs that block the action of GABA at postsynaptic sites or that inhibit its synthesis can be shown to induce seizures in animals (for review see Meldrum, 1975*b*).

Most compounds that inhibit glutamate decarboxylase (GAD, L-glutamate 1-carboxylase; EC.4.1.1.15) do so by interfering with the synthesis or coenzymic function of pyridoxal phosphate, and are therefore likely to produce inhibition of other cerebral decarboxylases and transaminases, including GABA-transaminase (GABA-T, 4-aminobutyrate: 2-oxoglutamate aminotransferase,

Table 1. *Inhibition of glutamate decarboxylase activity in mouse brain by convulsant drugs* Horton and Meldrum (1973), Meldrum *et al.* (1975*c*)

Drugs	Dose	Survival	% Inhibition	
	(mmol/kg i.p.)	(min)	−PLP	+PLP
Thiosemicarbazide	0.27	55	12	0
	2.2	27	24	9
Thiocarbohydrazide	0.07	40	30	11
	0.75	10	15	18
Methyldithiocarbazinate	0.12	30	14	1
	0.98	4	64	9
4-Deoxypyridoxine	1.18	90	58	7
Allylglycine	1.74	90	37	39
3-Mercaptopropionic acid	1·13	2	42	53

Mice were decapitated at the time of the first convulsion (TSC or TCH) or directly before the time at which convulsions started in similarly treated mice (other four drugs). GAD activity was assayed in brain homogenates with no added co-factor (−PLP) or with exogenous pyridoxal phosphate 0.1 mM (for 4DP) or 0.5 mM (other drugs) (+PLP).

EC.2.6.1.19). Glutamate decarboxylase appears to be the enzyme most vulnerable to a reduction in the availability of pyridoxal phosphate. Compounds that inhibit GAD activity by mechanisms other than pyridoxal phosphate antagonism, e.g. allyglycine, and the competitive inhibitor 3-mercaptopropionic acid, are also convulsants. Measurement of the percentage inhibition of glutamate decarboxylase in mouse brain homogenates prepared before seizure onset in mice receiving convulsant doses of different types of GAD inhibitor (4-deoxypyridoxine, allylglycine and 3-mercaptopropionic acid) indicates that a comparable degree of inhibition (40–60%) precedes convulsions (Horton & Meldrum, 1973—see Table 1). However, studies with hydrazine derivatives (Balzer *et al.*, 1960; Medina, 1963; Meldrum *et al.* 1975*c*) have repeatedly failed to show a comparably consistent correlation between whole brain GAD inhibition and the onset of convulsions. Some of the apparent discrepancies can probably be explained in terms of a concomitant inhibition of GABA-T activity (see data tabulated in Meldrum 1975*b*). Thus *N*,*N*′-dimethylhydrazine (1·6 mmol/kg) produces 39% inhibition of GAD but is not convulsant in rats, probably because of the 76% inhibition of GABA-T activity (Medina, 1963). This is clearly not the explanation for the fact that convulsions seen after a relatively long latency following thiosemicarbazide, thiocarbohydrazide and methyldithiocarbazinate are preceded by only moderate GAD inhibition (12–30% see Table 1), as the associated inhibition of GABA-T is very slight (Wood & Abrahams, 1971; Meldrum *et al.*, 1975*c*). There is evidence for a selective accumulation of hydrazides by certain cell types within the brain (Knyihár *et al.* 1971). Thus it is possible that these hydrazides are producing a cytologically selective inhibition of GAD. However, hydrazides may also possess a convulsant action additional to their capacity to inhibit GAD.

Initial studies of the convulsant action of hydrazides and pyridoxal phosphate antagonists reported that reflex or sensory seizures were very readily induced in animals or man after such poisoning (Reilly *et al.*, 1953; Pfeiffer *et al.*, 1956; Balzer *et al.*, 1960). Subsequent studies in the two major animal models of sensory epilepsy—audiogenic seizures in rodents (Lehmann, 1964) and photically-induced epilepsy in the baboon *Papio papio* (Meldrum *et al.*, 1970; Meldrum & Horton, 1971; Horton & Meldrum, 1973; Meldrum *et al.*, 1975*a*) have shown that subconvulsant doses of all types of GAD inhibitor potently enhance these natural syndromes of sensory epilepsy (Table 2). Similar syndromes of reflex

Table 2. *Glutamate decarboxylase inhibitors, GABA-receptor blockers, pentylenetetrazol and seizures in photosensitive baboons (Papio papio)*
Meldrum *et al.* (1970), Meldrum & Horton (1971), Horton & Meldrum (1973), Meldrum & Horton (1974)

Drug	Dose-enhancing syndrome (μmol/kg i.v.)	Convulsant dose (μmol/kg i.v.)	EEG seizure onset	Seizure latency (min)
Isoniazid	500–700	700–1100	Focal	7–40
Thiosemicarbazide	40–80	40–110	Focal	60–240
4-Deoxypyridoxine	200–500	530–870	Focal	15–45
Allylglycine	900–3100	4000–5000	Focal	60–120
3-Mercaptopropionic acid	90–280	280–750	Diffuse	4–6
Picrotoxin	0.8–1.3	1.7–2.5	Diffuse	2–10
Bicuculline	—	0.6–1.8	Generalized	0.1–0.2
Pentylenetetrazol	36–90	90–180	Diffuse/ generalized	0.5–1.5

epilepsy can be induced in normal rats, cats or monkeys by administering hydrazides or 4-deoxypyridoxine (Wada & Asakura, 1969; Meldrum & Horton, 1971), or in man with isoniazid, thiosemicarbazide or 4-deoxypyridoxine (Reilly *et al.*, 1953; Pfeiffer *et al.*, 1956). Semicarbazide in the cat reduces dorsal root potentials and presynaptic inhibition (Bell & Anderson, 1972) and there is other evidence that interfering with GABA synthesis can diminish presynaptic inhibition (see Straughan, this volume, Chapter 17). It is possible therefore that it is impairment of inhibition at the level of the primary afferent neuron that is responsible for sensory seizures after GAD inhibitors. However, some interneurons at subsequent levels of the afferent pathway are also probably GABA-ergic (e.g. in the specific thalamic relay nuclei) so that an action here is also possible. The GABA receptor blocker, picrotoxin, also potently enhances the sensory induction of seizures. However, the compound that is best known for its capacity to enhance reflex epilepsy, namely pentylenetetrazol, has not yet been shown to block the inhibitory action of GABA in microiontophoretic experiments. This is possibly because of the lower potency of pentylenetetrazol compared with picrotoxin and because of the difficulty of applying pentylenetetrazol by iontophoresis.

In baboons, convulsant doses of GAD inhibitors or of compounds known to block the inhibitory action of GABA at postsynaptic inhibitory receptor sites (such as picrotoxin, and bicuculline) give rise to seizures that have three different patterns of onset. Either, with a relatively long latency, a seizure discharge originates in one hemisphere in the posterior part of the cerebral cortex (hydrazines, pyridoxal phosphate antagonists and allylglycine), or seizure activity begins after a few minutes, spasmodically and diffusely throughout the cortex (3-mercaptopropionic acid and picrotoxin), or it begins with an extremely short latency throughout the cortex (bicuculline). Seizures following bicuculline appear to originate subcortically and comparison of close arterial injection of bicuculline to the cortex with systemic injections suggests that the cortex is relatively insensitive to bicuculline (B. S. Meldrum & L. Symon, unpublished work).

Within the cortex inhibitory interneurons are activated by recurrent collaterals and other pathways, and their normal functioning probably prevents the excessively synchronous or sustained discharge of neurons that underlies all manifestations of epilepsy. Thus it is not surprising that drugs which block the synthesis of GABA or prevent its postsynaptic action cause seizures. This raises the question whether seizures could be prevented by drugs which augment or prolong the action of GABA at inhibitory synapses. GABA is largely inactivated by reuptake into glia and into nerve terminals. Many drugs have been shown to inhibit the high affinity uptake of brain slices or synaptosomes for GABA, but in most cases this is a non-specific inhibition and a comparison of different drugs provides no evidence that blocking GABA reuptake has an antiepileptic action. The further metabolism of GABA is to succinate and involves two mitochondrial enzymes. Drugs which inhibit the further metabolism of GABA and produce an increase in brain GABA content, such as aminooxyacetic acid, hydrazinopropionic acid, and di-*n*-propylacetate can be shown to possess an anticonvulsant action, especially in reflex epilepsy (see Meldrum, 1975).

Various anticonvulsant drugs in relatively high concentrations *in vitro* inhibit GABA transaminase or succinic semialdehyde dehydrogenase (B. S. Meldrum, R. W. Horton & M. C. B. Sawaya, unpublished work). This is perhaps evidence that the drug molecules contain a group not dissimilar to GABA or succinic semialdehyde, rather than proof of the mode of action of the drugs. Valium and phenobarbitone have long been known to enhance or prolong presynaptic inhibition in the spinal cord (review Schmidt, 1971). It is not yet clear how this effect is brought about, but it is likely to be of importance in their anticonvulsant action (see Straughan, this volume, Chapter 17).

Monoamines. Pharmacological experiments in rodent test systems have frequently shown that treatments which deplete the brain of monoamines (e.g. reserpine) lower seizure thresholds, whereas treatments which increase brain amine content (e.g. monoamine oxidase inhibitors) tend to raise seizure thresholds (Maynert, 1969). In some test systems these changes appear to be dependent primarily on catecholamines, and in others on serotonin. Thus dopamine and noradrenaline appear to protect against electroshock in rats, whereas 5-hydroxytryptamine (5-HT) protects against pentylenetrazol (Rudzik & Johnson, 1970; Jobe *et al.*, 1974).

In baboons with photosensitive epilepsy, parenteral administration of L-5-hydroxytryptophan, the immediate precursor of 5-HT, completely prevents photically induced myoclonus or EEG paroxymal responses (see Table 3). L-Tryptophan alone does not have a protective action but it blocks myoclonic responses in animals pretreated with moderate doses of a monoamine oxidase inhibitor (Meldrum *et al.*, 1972). L-3,4-Dihydroxyphenylalanine (L-dopa) given intraperitoneally also reduces photically induced epileptic responses only in animals pretreated with a monoamine oxidase inhibitor.

Administration of enzyme inhibitors blocking the synthesis of amines in the brain has not provided conclusive data. An initial enhancement of photically induced responses is followed by a prolonged reduction in responsiveness when tryptophan hydroxylase is inhibited with *p*-chlorophenylalanine (Wada *et al.*, 1972).

Fairly consistent results have been obtained with drugs acting on amine

Table 3. *Effects of amino acid precursors and drugs modifying the metabolism or action of cerebral monoamines on the EEG and on epileptic responses to intermittent photic stimulation (ILS) in photosensitive baboons (Papio papio)*

Meldrum, *et al.* (1972), Wada *et al.* (1972), Meldrum & Balzamo (1971), Meldrum *et al.* (1975b)

Compound	Dose (mg/kg)	EEG changes: Background rhythms	EEG changes: Paroxysmal activity	Response to ILS: EEG	Response to ILS: Myoclonus
L-3,4-Dihydroxyphenylalanine	20–500	++	−	0	0
L-Tryptophan	200–600	−	+	0	0
5-Hydroxytryptophan	10–35	−−	++	−−	−−
Tranylcypromine	15–30	++	−	−−	−−
4-Methyl-*p*-tyrosine	100–600	−	+	0	0
p-Chlorophenylalanine	200–600	−−	++	+−	−
Apomorphine	0.5–1	++	−−	−−	−−
Piribedil	2.5–10	−		−	−
Galoperidol	0.6–1.2	−	++	++	(±)
Pimozide	0.5–2.5	−(+)	+	+(−)	

0 = No change.

+ or ++ = Mild or marked increase in frequency of background rhythms
or incidence of spontaneous spikes and waves,
or incidence of ILS-induced spikes and waves
or ILS-induced myoclonus.

− or −− = Mild or marked—slowing of background rhythms,
—decrease in incidence of spontaneous spikes and waves
—decrease in response to ILS.

receptor sites. Thus drugs believed to act as dopamine agonists, such as apomorphine, piribedil, and ergocornine, diminish or abolish photically induced myoclonus (Meldrum *et al.*, 1975*b*; G. Anlezark & B. S. Meldrum, unpublished work). Neuroleptics that are believed to act as dopamine antagonists show variable effects on myoclonic responses. Galoperidol dramatically modifies the EEG, increasing the incidence of both spontaneous and photically induced spikes and waves.

Among compounds believed to act on serotoninergic transmission, the hallucinogenic indoles, dimethyltryptamine and psilocybin, slow the background EEG rhythms, increase spontaneous spikes and waves and diminish photically induced epileptic responses (Meldrum & Naquet, 1971). As this effect is similar to that of L-5-hydroxytryptophan it is likely that a 5-HT agonist effect is involved. Similar actions of these indoles and 5-HT on single unit activity have been demonstrated at the lateral geniculate and at the nuclei of the median raphe (Curtis & Davis, 1962; Haigler & Aghajanian, 1974).

Numerous ergot alkaloids and related compounds can block photically induced myoclonic responses (see Table 4). Although most of these compounds are potent blockers of the smooth muscle contracting action of 5-HT, there is evidence from numerous biochemical and pharmacological studies that they act centrally as 5-HT agonists (Curtis & Davis, 1962; Haigler & Aghajanian, 1974; Ellaway & Trott, 1975) and as dopamine agonists (Fuxe *et al*., 1974). Although actions on noradrenergic and other systems are possible, it is reasonable to suggest that the effects of lysergic acid diethylamide, methylergometrine and ergocornine on the background EEG and on photically induced myoclonus arise from a combination of dopamine and 5-HT agonist actions.

Table 4. *Effects of hallucinogenic and related indole derivatives on the EEG and on epileptiform responses to intermittent photic stimulation (ILS) in photosensitive baboons (Papio papio), Meldrum & Naquet (1970), Vuillon-Cacciuttolo & Balzamo (1972), G. Anlezark A. B. S. Meldrum (unpublished work 1972).*

		EEG changes		Response to ILS	
Drug	Dose (mg/kg i.v.)	Background rhythms	Paroxysmal activity	EEG	Myoclonus
Dimethyltryptamine	2–4	— —	+	—	— —
Psilocybin	2–4	— —	+	—	— —
Lysergic acid diethylamide (LSD 25)	0.05–0.15	+ +	— —	— —	— —
2-Bromlysergic acid diethylamide (BOL-148)	2–4	—		—	—
Methysergide	2–5	—	+	—	— —
Methylergonovine (methylergometrine)	0.4–0.8	+ +	— —		
,, ,,	0.8–2	—	+	— —	— —
Dihydroergotoxine	0.1–2	0		0	0
Ergocornine	1–2	+ +	—	— —	— —
Methergoline	2–4	0		—	— —
Ergotamine	0.1–0.5	0	0	— —	— —

0 = No change.
+ or + + = Mild or marked increase in frequency of background rhythms
or incidence of spontaneous spikes and waves,
or incidence of ILS-induced spikes and waves
or ILS-induced myoclonus.
— or — — = Mild or marked —slowing of background rhythms
—decrease in incidence of spontaneous spikes and waves,
—decrease in response to ILS.

In the light of these observations it is natural to ask if any epileptic syndromes in man are related to deficient functioning in aminergic systems. Action myoclonus occurring after ischaemic or traumatic brain damage possibly has such a basis, as it can be dramatically relieved by the administration of L-5-hydroxytryptophan (Lhermitte *et al*., 1972). Other epileptic syndromes do not usually show such a therapeutic response but the possibility of a deficiency in aminergic systems is supported by several reports that the concentration in the cerebrospinal fluid of the amine metabolites, 5-hydroxyindoleacetic acid and homovanillic acid is reduced in adults and children with epilepsy (Shaywitz *et al*., 1975; Papeschi *et al*., 1972). Changes in the cerebral content and turnover of monoamines have frequently been reported in animals receiving very high doses of anticonvulsant drugs especially the benzodiazepines (Lidbrink *et al*., 1974). Recently an increase in the concentration of 5-hydroxyindoleacetic acid has been described in the cerebrospinal fluid of epileptics receiving therapeutic doses of phenobarbitone and diphenylhydantoin (Chadwick *et al*., 1975). Thus it is possible that a change in aminergic activity contributes to anticonvulsant drug action.

(ii) Classical Screening Tests for Anticonvulsant Drugs

Seizures induced by electroshock in rodents or cats were first used systematically as a screening test for anticonvulsant drugs by Putnam & Merritt (1937). Subsequently a variety of chemically (strychnine, picrotoxin and pentylenetetrazol) and electrically induced seizures in rats and mice have been used to screen potential anticonvulsants (reviews by Millichap, 1969; Swinyard, 1973). These tests have led to the introduction of several major classes of anticonvulsants, including the hydantoins (Merritt & Putnam, 1938*a*, *b*), the oxazolidinediones, such as trimethadione (Everett & Richards, 1944) and the succinimides, such as phensuximide and ethosuximide (Chen *et al.*, 1951). There are very marked differences in the comparative efficacy of anticonvulsant drugs according to the test system employed (see Table 5). Phenobarbitone is active against both electrically and chemically induced convulsions, but hydantoin is ineffective in most simple test systems, except that it abolishes the tonic-extension phase of maximal electroschock seizures. The oxazolidinediones and succinimides are active against pentylenetetrazol seizures. These and other observations have led to the generalization that drugs active against maximal electroshock seizures are likely to be effective in grand mal and focal epilepsy, and drugs active against pentylenetetrazol more effective in petit mal. It is obvious that as a preliminary screen both electroshock and chemically induced seizures should be employed. It cannot be assumed that any new class of anticonvulsant drugs will conform to the same pattern of correlation between laboratory tests and clinical usefulness as existing drugs. It is also possible that these simple acute screening procedures could reject a useful anticonvulsant. In fact, I think it likely that most research directors in pharmaceutical companies if presented today with screening test data on several current drugs, including sulthiame, trimethadione and possibly hydantoin, would reject them as unworthy of further investigation.

(iii) Testing Anticonvulsants in Subhuman Primates

Seizures used in rodent screening tests are clinically and neurophysiologically very different from seizures occurring spontaneously in man. Furthermore, drug actions and metabolisms are likely to be different in rodents and man. Hence numerous authors have sought subhuman primate models of epilepsy, in the hope that they might be neurophysiologically and pharmacologically closer to the human disease (reviews by Chusid & Kopeloff, 1969; Naquet & Lanoir, 1973; Meldrum *et al.*, 1975*a*). The local application of alumina cream to the sensorimotor cortex to produce focal motor seizures was described nearly 40 years ago (Hawk & Bergheim, 1937) and has since been extensively investigated in baboons and monkeys (Kopeloff *et al.*, 1954; Chusid & Kopeloff, 1969). Tremors, myoclonus, focal seizures and generalized seizures occur spontaneously in these animals, and can be enhanced by a variety of convulsant drugs (including antihistamines, but not anticholinesterases). Anticonvulsant drugs can be most

Table 5. *Comparative potencies of various anticonvulsant drugs in rodent and baboon test systems and in human epileptic syndromes*
Swinyard *et al.* (1963), Millichap (1969), Eadie & Tyrer (1974), Meldrum *et al.* (1975*a*), Meldrum *et al.* (1975*d*)

Drug	Mice (mg/kg, s.c. for ED_{50}) MES (ton. ext.)	Pentyl	Audiogenic runn.	Audiogenic ton. ext.	Baboons (mg/kg i.v.)	Man daily oral dose (mg/kg)	Syndrome
Phenobarbitone	21	25	20	5	15	2–4	g.m., focal
Primidone	13	74			100	9–15	g.m., t.l.e.
Diphenylhydantoin	14		35	5	50	6–7	g.m., t.l.e.
Trimethadione	630	488	550	375		55	p.m.
Ethosuximide		135			100	20	p.m.
Sulthiame					125	10–20	
Carbamazepine					40	15–20	t.l.e.
Diazepam					1	0.1–0.3	myoclonic st.ep.
Donazepam					0.15	0.05–0.1	g.m., p.m. myoclonic

Epileptic syndromes: g.m. = grand mal; t.l.e. temporal lobe epilepsy; p.m. = petit mal absence; st.ep. = status epilepticus.
ton. ext. = tonic extension; runn. = running fit.

conveniently tested by looking for protection against the enhancement of responses after a convulsant drug. Benzodiazepines (chlordiazepoxide, 1–5 mg/kg; diazepam 0.2 mg/kg) protect against pentylenetetrazol-induced enhancement in all animals. This model, although it clearly resembles some forms of human focal epilepsy, is too complex for use as a screening test and has not so far been widely applied as a preclinical trial assessment.

The discovery of a natural syndrome of photosensitive epilepsy in baboons, *Papio papio*, from the Casamance region of Senegal (see Naquet & Meldrum, 1972) has provided a potential animal test system that is very similar to some human syndromes. This syndrome has proved to be of great value for studying the influence on epileptic phenomena of drugs modifying synaptic transmission (see Meldrum & Balzamo, 1972; Meldrum *et al.*, 1975*a*). The natural syndrome offers some special advantages for the pretrial assessment of drugs but it is not a suitable screening procedure. One problem is the day to day variability in responsiveness shown by some of the animals. This is, of course, similar to the clinical situation but necessitates complex and expensive experimental designs. A prolonged trial of clonazepam in *Papio papio* (Killam *et al.*, 1973) revealed a remarkable similarity to the situation that has emerged in clinical trials in man. Initial control of seizure responsiveness was achieved with doses comparable to those used in man, but after 1–3 months a proportion of the animals showed an escape from control similar to that observed in some patients.

The natural syndrome of photosensitivity can be made more convenient for the rapid assessment of anticonvulsant drugs if animals are pretreated with a subconvulsant dose of 2-amino, 4-pentenoic acid. Between 2 and 8 hours after such pretreatment all animals show a consistently high level of sensitivity to

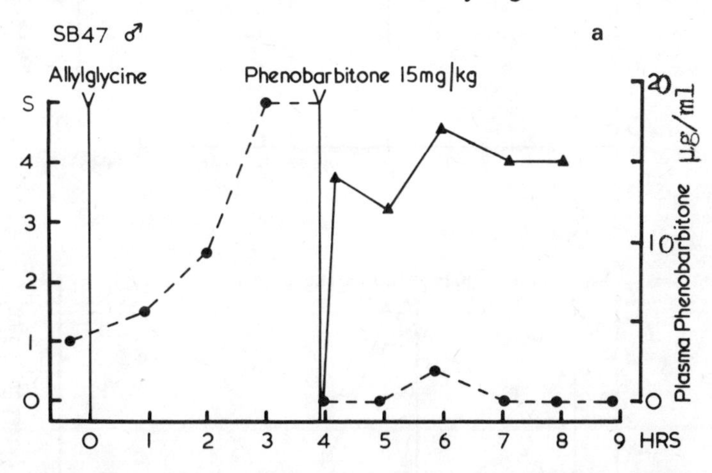

Fig. 1. *Graphs to show the relationship between epileptic response to stroboscopic stimulation (ILS) and the plasma concentration of various anticonvulsant drugs in photosensitive baboons, Papio papio* Abscissa: time following injection of 2-amino-4,pentenoic acid, 200 mg/kg i.v. Left ordinate (circles, broken lines): 0 = no response to ILS; 1 = myoclonus of eyelids during ILS; 2 = myoclonus of face and neck during ILS; 3 = myoclonus of limbs during ILS; 4 = myoclonus continuing beyond the end of ILS; S = tonic–clonic seizure after ILS. Right ordinate (triangles, solid line): plasma concentration of drug.

(*a*) Suppression of myoclonic responses in a male *Papio papio*, weight 8.5 kg, following phenobarbitone 15 mg/kg i.v.

Fig. 1. *Continued overleaf.*

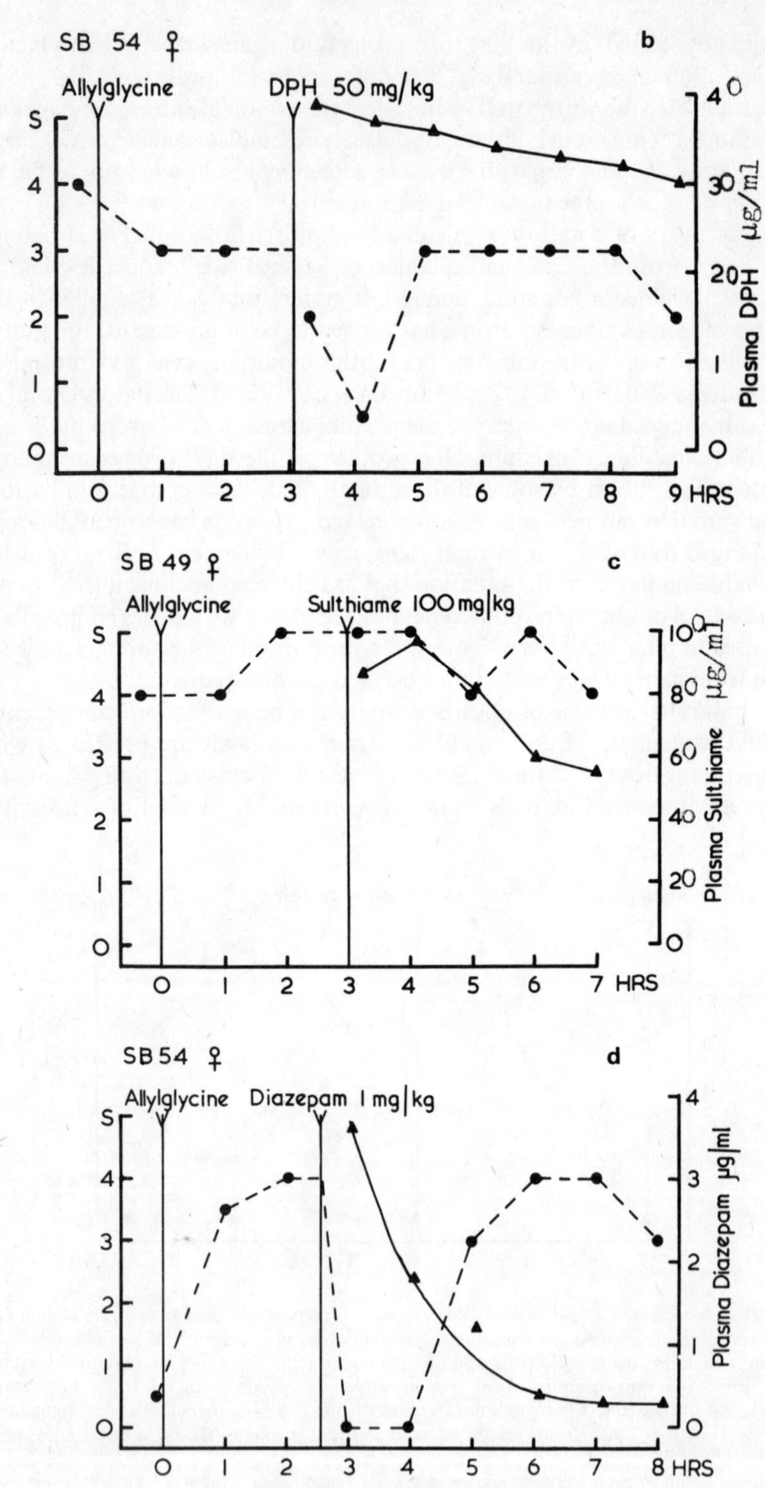

Fig. 1. *Continued*

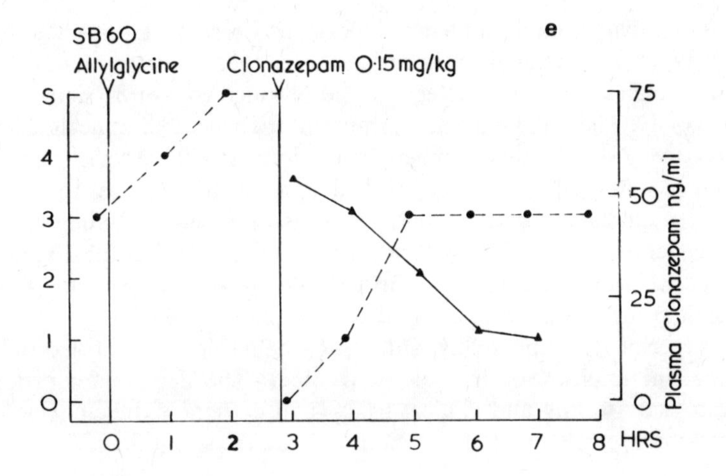

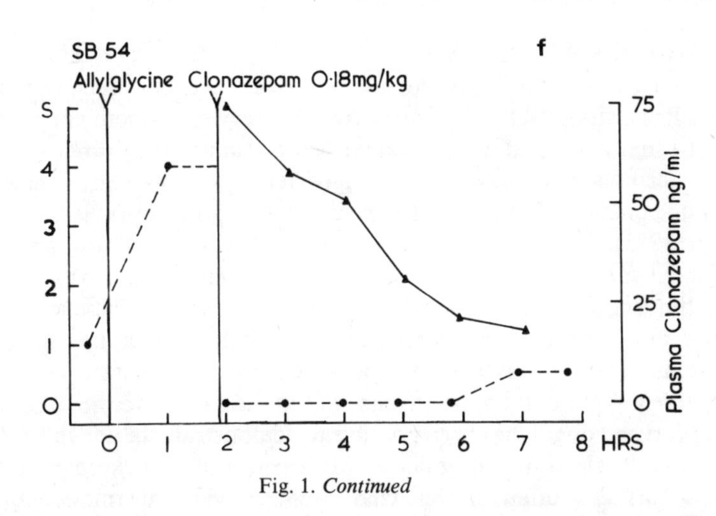

Fig. 1. *Continued*

(b) Absence of self-sustaining responses but persistence of myoclonus in a female *Papio papio*, weight 4.5 kg, following diphenylhydantoin, 50 mg/kg i.v. Signs of neurological toxicity (nystagmus, ataxia) were present throughout the test period. *(c)* Absence of therapeutic effect of sulthiame in a female *Papio papio*, weight 5 kg, of sulthiame 100 mg/kg i.v. *(d)* Brief abolition of myoclonic responses in a female *Papio papio*, weight 4.5 kg, following the administration of diazepam 1 mg/kg i.v. *(e, f)* Transient and more sustained abolition of myoclonic responses to ILS in baboons receiving clonazepam 0.15 and 0.18 mg/kg.

This figure is reproduced with permission from Meldrum *et al.* (1975*d*) [*(a)–(d)*] and from Meldrum *et al.* (1975*a*) [*(e)–(f)*].

photic stimulation, responding either with self-sustaining myoclonus or a tonic–clonic seizure when stimulation is repeated at hourly intervals (Meldrum *et al.*, 1975*d*). Anticonvulsant drugs can be administered intraveneously and the degree and duration of protection assessed in relation to the plasma concentration of drug (determined in blood samples taken at the time of the tests) (see Fig. 1). Acute neurological toxicity is apparent as nystagmus, strabismus, and ataxia. Phenobarbitone completely blocks myoclonic responses at plasma levels (7–17 mg/litre) similar to those considered therapeutic in man. Diazepam and clonazepam are also very effective, and produce little or no signs of neurological toxicity. The rapid plasma clearance of the benzodiazepines is associated with a relatively short action. Diphenylhydantoin (25–50 mg/kg i.v.) and carbamazepine (20–40 mg/kg) produce marked signs of neurological toxicity; they diminish the tendency for myoclonic responses to be self-sustaining but do not prevent generalized myoclonus during photic stimulation. Ethosuximide and sulthiame are relatively inactive in this test system. Among potential new anticonvuisant compounds supplied by drug companies a significant number are clearly superior to diphenylhydantoin, carbamazepine, ethosuximide and sulthiame but inferior to the benzodiazepines. The decision to proceed to chronic toxicity testing and clinical trials is an extremely difficult one, and is of course based partly on considerations outside the scientific field.

(2) Drugs or Procedures that Terminate Prolonged Seizures

(a) In non-epileptics

A wide variety of poisonings or metabolic disturbances can precipitate seizures in non-epileptic subjects. Such seizures are often very resistant to anticonvulsant drugs but usually respond dramatically to correction of the primary metabolic disorder. Seizures associated with hypoglycaemia, or low plasma calcium or magnesium concentration or pyridoxine deficiency come into this category (see Meldrum, 1975). However, there are a number of complex clinical states, often involving dehydration, electrolyte imbalance, and plasma hyperosmolarity that are sometimes complicated by focal or generalized seizures. These include uraemic coma, non-ketotic diabetic coma and acute dehydrating illnesses in childhood. In clinical studies it is usually impossible to assess the relative contribution of the different factors to the occurrence of seizures. However, in experimental studies in cats and dogs it has been possible to demonstrate the convulsant effects of high blood levels of urea or glucose (Maccario, 1968; Zuckerman & Glaser, 1972) and further studies of this kind could provide information of direct clinical value.

(b) In epileptics

Although the majority of seizures in epileptic patients terminate spontaneously within a few minutes of their onset, occasionally seizures are very prolonged and may lead to death or to brain damage manifest as neurological disability or intellectual deterioration. Such prolonged seizures are seen when anticonvulsant therapy is abruptly reduced or discontinued and sometimes when a brain lesion is the underlying cause of the epilepsy. The management of such cases has been enormously improved by the introduction of the benzodiazepines, which if given intramuscularly or intravenously will arrest the majority of prolonged seizures. A relatively small dose is often promptly effective if given in the first few minutes of a seizure, but massive doses may be required when a seizure has been continuing for more than 30 minutes. This appears to be a similar phenomenon to that observed in experimental animals, which show self-sustaining seizure activity once a moderate number of brief seizures have been induced in quick succession with electroshock or pentylenetetrazol. It appears that some inhibitory mechanism within the brain loses its functional capacity once seizure activity has been sustained for 20–30 min. A great many biochemical changes have been observed in the brains of experimental animals undergoing seizures, but most such studies have concerned very brief seizures and none have analysed the correlation between biochemical changes and the development of therapy resistance. It is likely that animal experiments with a model of status epilepticus could lead to a metabolically conceived therapy that would be of particular value in prolonged seizures.

(3) Prevention of Brain Damage Secondary to Prolonged Seizures

In the management of patients in status epilepticus it would be helpful to know which physiological and biochemical changes associated with seizures are responsible for epileptic brain damage. Microscopically, the brain damage takes the form of nerve cell loss and reactive gliosis in chronic epileptics. In adults or, more especially, children dying shortly after an episode of status epilepticus the predominant cytological abnormality is ischaemic cell change in the neurons of the cerebellum, neocortex, thalamus and hippocampus (see Corsellis & Meldrum, 1975; Meldrum 1975*a*). Ischaemic cell change is seen after any stress that critically limits cerebral energy metabolism (e.g. cardiac arrest, profound arterial hypotension, focal ischaemia, or hypoglycaemia (Brierley *et al.*, 1973). Neuropathologists have generally attributed epileptic brain damage to systemic hypoxia or cerebral arterial spasm or cerebral oedema leading to compression of arteries (Spielmeyer, 1927; Scholz, 1951, 1959).

The use of bicuculline to induce prolonged seizures in experimental primates enables the pattern of brain damage seen in man after status epilepticus to be reproduced in the laboratory (Meldrum & Brierley, 1973). In such experiments physiological and biochemical changes can be monitored during the seizure and correlated with the occurrence or severity of any pathology (see Table 6).

Table 6. *Physiological and biochemical changes occurring during prolonged seizures and their possible relation to subsequent brain damage*

Respiratory status	Mild *systemic hypoxia*
	Marked lactacidaemia + compensatory hypocapnia
Circulation	*Arterial hypotension*
	Raised venous pressure
	Increased or normal cerebral blood flow (but autoregulation and focal metabolic control absent or impaired)
Blood	Hyperkalaemia
	Hypoglycaemia
Metabolism	*Hyperpyrexia*
	Cerebral metabolic rate increased ($CMRo_2$ increased ×2–3)

Data from experiments with bicuculline or allylglycine in adolescent baboons (Meldrum & Horton, 1973; Meldrum & Brierley, 1973; Meldrum *et al.*, 1973; Meldrum *et al.*, 1974) except for increase in $CMRo_2$ (bicuculline, in rats—B. K. Chapman, B. S. Meldrum & B. K. Siesjö, unpublished work).

Factors which independently or in combination can lead to ischaemic cell change in cortical or cerebellar neurons are italicized.

Such prolonged generalized seizures are associated with an initial rise in systolic and diastolic blood pressure, but late in the seizure moderate arterial hypotension is usual (Meldrum & Horton, 1973). Measurement of arterial and cerebral venous oxygen tension does not indicate a severe degree of cerebral hypoxia. Initially, blood glucose concentration rises, but in some animals there is a sustained secondary hypoglycaemia. Rectal temperature rises, sometimes to a peak of 42–43°C. Although such experiments reveal a correlation between brain damage and duration of hyperpyrexia, secondary arterial hypotension, and hypoglycaemia it is not possible to evaluate the relative contributions to the pathology made by secondary peripheral effects, and any direct cerebral consequences of seizure activity. However, in baboons paralysed with a peripheral neuromuscular blocking agent and artificially ventilated, the seizure, as recorded electroencephalographically, is unmodified but the late secondary peripheral effects are greatly diminished (Meldrum *et al.,* 1973). Under these circumstances cerebellar damage is almost totally prevented, neocortical damage is reduced (for a given seizure duration), but hippocampal damage remains severe. Evidence from other experimental situations also indicates that hippocampal damage is largely a result of local consequences of seizure activity, rather than secondary systemic changes. Thus a long sequence of brief seizures, induced by 2-amino,4-pentenoic acid, associated with only mild systemic changes, leads to hippocampal lesions in the absence of evidence of neuronal loss elsewhere (Meldrum *et al.,* 1974). Prolonged seizures confined to the limbic system, induced by injection of ouabain into the septal nuclei, can also lead to neuronal loss and gliosis in the hippocampus (Baldy-Moulinier *et al.*, 1973).

Broadly, animal studies of the pathological consequences of prolonged seizures indicate that it is important to avoid or correct secondary physiological disturbances including systemic hypoxia, arterial hypotension, hyperpyrexia and hypoglycaemia. However, even when these are corrected or prevented (as by paralysis) arrest of the paroxysmal electrical activity itself is still necessary to avoid brain damage.

Preventing the development of fits or epilepsy

Our knowledge of the processes that precede the onset of epilepsy is very limited. Perinatal cerebral asphyxia, cranial trauma in later life, especially when dural penetration occurs, and focal infective or neoplastic lesions can all lead to focal or generalized seizures. However, not all patients suffering these insults develop epilepsy, and the latent interval to seizure onset is very variable. Events occurring during that interval include a loss of neurons and proliferation of reactive astrocytes and a loss of dendritic spines (reviewed in Meldrum, 1975*a*). Each of these processes could contribute to epileptogenesis, and the possibility exists of separating out such contributions in experimental models of epilepsy. The selective loss of inhibitory interneurons can be studied by conventional histology, autoradiography, histochemistry or microchemical methods in several test systems. In the cerebellum a selective loss of Purkinje cells and basket cells occurs acutely after anoxia or seizures (Meldrum & Brierley, 1973). In the neocortex smaller neurons in the third and fifth layers are selectively lost in several circumstances (Brierley, *et al.,* 1973) but proof that they are inhibitory interneurons is so far lacking.

In Golgi preparations a loss of dendritic spines can be detected in the brains of epileptic patients (De Moor, 1898; Scheibel *et al.,* 1974). Similar changes are seen in epileptogenic foci created in the neocortex of rhesus monkeys by the local application of alumina cream (Westrum *et al.*, 1965). This appearance can be the result of deafferentation (Globus & Schiebel, 1966) and may therefore follow a loss of neurons that is either local or distant. Slowly progressive changes of this kind could be occurring during the latent interval between brain damage and the development of epilepsy. They can be investigated in man only with the greatest difficulty but are susceptible to detailed experimental analysis in animal models.

Prophylactic trials are not too difficult to organize in man. The principal categories of at-risk patient available are penetrating and non-penetrating head injuries in adults (in which the risks of subsequent epilepsy are respectively 50% and 10% (Jennett 1969; Caveness, 1974)) and perinatal brain damage. However, prophylactic measures cannot be appropriately chosen until more fundamental knowledge is available.

Febrile convulsions occur in about 50 per 1000 of children between the age of 6 months and 4 years. If prolonged they may be followed by chronic epilepsy. The relation of subsequent epilepsy to brain damage produced by the seizure cannot be decided from purely clinical studies (for review see Meldrum, 1975*c*). Neuropathological studies on anterior temporal lobectomy specimens removed from children and adolescents with temporal lobe epilepsy suggest that mesial temporal sclerosis secondary to febrile convulsions is a common cause of temporal lobe epilepsy (Falconer, 1974). As similar lesions can be induced in young baboons (Meldrum *et al.,* 1974) their evolution and epileptogenicity can be investigated experimentally and possible prophylactic measured defined. Obviously prevention of the febrile convulsions themselves should be the primary goal (Lennox-Buchthal, 1973).

Conclusion: the Role of Animal Models in Basic and Applied Epilepsy Research

Animal models of epilepsy are probably best used as tools in the investigation of direct clinical problems. It is nearly always necessary to investigate a number of different animal model systems in order to establish the general validity of any observation. For example, apomorphine protects against electroshock seizures in rats, but not in mice (McKenzie & Soroko, 1972), protects against audiogenic seizures in mice (Anlezark & Meldrum, 1975) and photically induced seizures in baboons (Meldrum *et al.,* 1975*b*) and cobalt-induced spikes in rats (Dow *et al.,* 1974). However, it potentiates chemically induced seizures in rats (pentylenetetrazol; Soroko & McKenzie, 1970) and baboons (2-amino,4-pentenoic acid; Meldrum *et al.,* 1975*b*).

Animal models can also be themselves a topic for investigation. However, the justification for this is little greater than the justification for any basic research that is potentially relevant to epilepsy. Progress in fundamental understanding of cytological, biochemical and electrophysiological processes in the normal and damaged brain is essential before an adequate account of events in epilepsy can be given. Much of this research does not require an animal model of epilepsy. It can, and hopefully will, be conducted by people with no special knowledge of, or interest in, epilepsy.

References

Anlezark, G. M. & Meldrum, B. S. (1975) *Brit. J. Pharmacol.* **53**, 419–421
Baldy-Moulinier, M., Arias, L. P. & Passouant, P. (1973) *Eur. Neurol.* **9**, 333–348
Balzer, H., Holtz, P. & Palm, D. (1960) *Naunyn-Schmiedebergs. Arch. Exp. Pathol. Pharmakol.* **239**, 520–552
Bell, J. A. & Anderson, E. G. (1972) *Brain Res.* **43**, 161–169
Brierley, J. B., Meldrum, B. S. & Brown, A. W. (1973) *Arch. Neurol. Psychiat. (Chicago)* **29**, 367–374
Caveness, W. F. (1974) *Hand. Clin. Neurol.* **15**, 274–194
Chadwick, D., Jenner, P. & Reynolds, E. H. (1975) *Lancet* **i**, 473–476
Chen, G., Portman, R., Ensor, C. R. & Bratton, A. C. (1951) *J. Pharmacol. Exp. Ther.* **103**, 54–61
Chusid, J. G. & Kopeloff, L. M. (1969) *Epilepsia* **10**, 239–262
Corsellis, J. A. N. & Meldrum, B. S. (1975) in *Greenfield's Neuropathology* (Blackwood, W., ed.), Arnold, London
Curtis, D. R. & Davis, R. (1962) *Brit. J. Pharmacol.* **18**, 217–246
Curtis, D. R. & Johnston, G. A. R. (1974) *Ergeb. Physiol.* **69**, 98–188
De Moor, J. (1898) *Ann. Soc. Roy. Sci. Med. Nat. Brux.* **7**, 205–250
Dow, R. C., Hill, A. G. & McQueen, J. K. (1974) *Brit. J. Pharmacol.* **52**, 135P
Eadie, M. J. & Tyrer, J. H. (1974) *Anticonvulsant Therapy*, p. 204, Churchill Livingstone, Edinburgh
Ellaway, P. H. & Trott, J. R. (1975) *Exp. Brain Res.* **22**, 145–162
Everett, G. M. & Richards, R. K. (1944) *J. Pharmacol. Exp. Ther.* **81**, 402–407
Falconer, M. A. (1974) *Lancet* **ii**, 767–700
Fuxe, K., Corrodi, H., Hökfelt, T., Lidbrink, P. & Ungerstedt, U. (1974) *Med. Biol.* **52**, 121–132
Glaser, G. H. (1972) in *Experimental Models of Epilepsy* (Purpura, D. P., Penry, J. K., Woodbury, D. M. & Walter, R., eds.), p. 317, Raven Press, New York
Globus, A. & Scheibel, A. B. (1966) *Nature (London)* **212**, 463–465
Glötzner, F. L. (1973) *Brain Res.* **55**, 159–171
Grob, D., (1963) *Handb. Exp. Pharmakol. Suppl.* **15**, 989–1027
Grossman, R. G. & Rosman, J. L. (1971) *Brain Res.* **28**, 181–201
Haigler, H. J. & Aghajanian, G. K. (1974) *J. Pharmacol. Exp. Ther.* **188**, 688–699
Hawk, P. B. & Bergheim, O. (1937) *Practical Physiological Chemistry*, Blakiston, Philadelphia
Henn, F. A., Haljamae, H. & Hamberger, A. (1972) *Brain Res.* **43**, 437–443
Horton, R. W. & Meldrum, B. S. (1973) *Brit. J. Pharmacol.* **49**, 52–63
Jennett, W. B. (1969) in *Late Effects of Head Injury* (Walker, A. E., Caveness, W. G. & Critchley, M. D., eds.), pp. 201–214, Thomas, Springfield, Ill.
Jobe, P. C., Stull, R. E. & Geiger, P. F. (1974) *Neuropharmacology* **13**, 961–968
Killam, E. K., Matsuzaki, M. & Killam, K. F. (1973) in *The Benzodiazepines* (Garratini, S., Mussini, E. & Randall, L. O., eds.), pp. 443–460, Raven Press, New York
Knyihár, E., Kiss, J. & Csillik, B. (1971) *Res. Commun. Chem. Pathol. Pharmacol.* **2**, 395–405
Kopeloff, L. M., Chusid, J. G. & Kopeloff, N. (1954) *Neurology (Minneapolis)* **4**, 218–227
Krnjević, K. (1974) *Physiol. Rev.* **54**, 418–540
Kuffler, S. W. & Nicholls, J. C. (1966) *Ergeb. Physiol.* **57**, 1–90
Lehmann, A. (1964) *Agressologie* **5**, 311–351
Lennox-Buchtal, M. A. (1973) *Electroencephalogr. Clin. Neurophysiol. Suppl.* **32**, 1–138
Lhermitte, F., Marteau, R. & Degos, C. F. (1972) *Rev. Neurol. (Paris)* **126**, 107–114
Lidbrink, P., Corrodi, H. & Fuxe, K. (1974) *Eur. J. Pharmacol.* **26**, 35–40
Maccario, M. (1968) *Arch. Neurol. Psychiat. (Chicago)* **19**, 525–534
McKenzie, G. M. & Soroko, F. E. (1972) *J. Pharm. Pharmacol.* **24**, 696–701
Maynert, E. W. (1969) *Epilepsia* **10**, 145–162.
Medina, M. A. (1963) *J. Pharmacol. Exp. Ther.* **140**, 133–137
Meldrum, B. S. (1975*a*) in *A Textbook of Epilepsy* (Laidlaw, J. & Richens, A., eds.), Churchill Livingstone, Edinburgh
Meldrum, B. S. (1975*b*) *Int. Rev. Neurobiol.* **17**, in press.
Meldrum, B. S. (1975*c*) in *Modern Trends in Neurology* (D. Williams ed.), vol. 6, Butterworths, London

Meldrum, B. S. & Balzamo, E. (1971) *C.r. Séances Soc. Biol.* **165**, 2379–2381
Meldrum, B. S. & Balzamo, E. (1972) *Proc. 3rd, Conf. Exp. Med. Surg. Primates,* Lyon, part 2, pp. 282–288, Karger, Basel
Meldrum, B. S. & Brierley, J. B. (1973) *Arch. Neurol. Psychiat. (Chicago)* **28**, 10–17
Meldrum, B. S. & Horton, R. W. (1971) *Brain Res.* **35**, 419–436
Meldrum, B. S. & Horton, R. W. (1973) *Arch. Neurol. Psychiat. (Chicago)* **28**, 1–9
Meldrum, B. S. & Horton, R. W. (1974) in *Epilepsy. Proceedings of the Hans Berger Centenary Symposium* (Harris, P. & Mawdsley, C., eds.), pp. 55–65, Churchill Livingstone, Edinburgh
Meldrum, B. S. & Naquet, R. (1971) *Electroencephalogr. Clin. Neurophysiol.* **31**, 563–572
Meldrum, B. S., Balzamo, E., Gadea, M. & Naquet, R. (1970) *Electroencephalogr. Clin. Neurophysiol.* **29**, 333–347
Meldrum, B. S., Balzamo, E., Wada, J. A. & Vuillon-Cacciuttolo, G. (1972) *Physiol. Behav.* **9**, 615–621
Meldrum, B. S., Vigouroux, R. A. & Brierley, J. B. (1973) *Arch. Neurol. Psychiat. (Chicago)* **29**, 82–87
Meldrum, B. S., Horton, R. W. & Brierley, J. B. (1974) *Brain* **97**, 405–418
Meldrum, B. S., Anlezark, G., Balzamo, E., Horton, R. W, & Trimble, M. (1975a) *Advan. Neurol.* **10**, 119
Meldrum, B. S., Anlezark, G. & Trimble, M. (1975*b*) *Eur. J. Pharmacol.* **52**, 203–213
Meldrum, B. S., Horton, R. W. & Sawaya, M. C. B. (1975*c*) *J. Neurochem.* **24**, 1003–1009
Meldrum, B. S., Horton, R. W. & Toseland, P. (1975*d*) *Arch. Neurol. Psychiat. (Chicago)* **32**, 289–294
Merritt, H. H. & Putnam, T. J. (1938*a*) *Arch. Neurol. Psychiat. (Chicago)* **39**, 1003–1015
Merritt, H. H. & Putnam, T. J. (1938*b*) *J. Am. Med. Ass.* **111**, 1068–1073
Millichap, J. G. (1969) *Epilepsia* **10**, 315–328
Naquet, R. & Lanoir, J. (1973) in *Anticonvulsant Drugs* (Mercier, J., ed.), vol. 1, pp. 67–122, Pergamon Press, Oxford
Naquet, R, & Meldrum, B. S. (1972) in *Experimental Models of Epilepsy* (Purpura, D. P., Penry, J. K., Tower, D., Woodbury, D. M. & Walter, R., eds.), p. 373, Raven Press, New York
Papeschi, R., Molina-Negro, P., Sourkes, T. L. & Erba, G. (1972) *Neurology (Minneapolis)* **22**, 1151–1159
Pfeiffer, C. C., Jenney, E. H. & Marshall, W. H. (1956) *Electroencephalogr. Clin. Neurophysiol.* **8**, 307–315
Pollen, D. A. & Trachtenberg, M. C. (1970) *Science* **167**, 1252–1253
Putnam, T. J. & Merritt, H. H. (1937) *Science* **85**, 525–526
Reilly, R. H., Killam, K. F., Jenney, E. H., Marshall, W. H., Tausig, M. D., Apter, N. S. & Pfeiffer, C. C. (1953) *J. Am. Med. Ass.* **152**, 3117–1321
Rudzik, A. D. & Johnson, G. A. (1970) in *Amphetamines and Related Compounds* (Costa, E. & Garattini, S., eds.), p. 715, Raven Press, New York
Scheibel, M. E., Crandall, P. H. & Scheibel, A. B. (1974) *Epilepsia* **15**, 55–80
Schmidt, R. F. (1971) *Ergeb. Physiol.* **63**, 20–101
Scholz, W. (1951) *Monogr. ges. Neurol. Psych.* 75, Springer-Verlag, Berlin
Scholz, W. (1959) *Epilepsia* **1**, 36–55
Shaywitz, B. A., Cohen, D. J. & Bowers, M. B. (1975) *Neurology (Minneapolis)* **25**, 72–79
Soroko, F. E. & McKenzie, G. M. (1970) *Pharmacologist* **12**, 253
Spielmeyer, W. (1927) *Z. ges. Neurol. Psych.* **109**, 501–520
Swinyard, E. A. (1973) in *Anticonvulsant Drugs*, (Mercier, J., ed.), vol. 1, p. 47, Pergamon Press, Oxford
Swinyard, E. A. & Castellion, A. W. (1965) *J. Pharmacol. Exp. Ther.* **151**, 369–375
Swinyard, E. A., Castellion, A. W., Fink, G. B. & Goodman, L. S. (1963) *J. Pharmacol. Exp. Ther.* **140**, 375–384
Tebecis, A. K. (1974) *Transmitters and Identified Neurons in the Mammalian Central Nervous System,* p. 340, Scientechnia Publishers, Bristol
Tower, D. B. (1960) *Neurochemistry of Epilepsy,* Thomas, Springfield, Ill.
Tower, D. B. (1969) in *Basic Mechanisms of the Epilepsies* (Jasper, H. H., Ward, A. A. & Pope, A., eds.), p. 611, Little Brown, Boston, Mass.
Vuillon-Cacciuttolo, G. & Balzamo, E. (1972) *J. Pharmacol. (Paris)* **3**, 31–45
Wada, J. A. & Asakura, T. (1969) *Exp. Neurol.* **24**, 19–37
Wada, J. A., Balzamo, E., Meldrum, B. S. & Naquet, R. (1972) *Electroencephalogr Clin. Neurophysiol.* **33**, 520–526
Westrum, L. E., White, L. E. & Ward, A. A. (1965) *J. Neurosurg.* **21**, 1033–1046
Wood, J. D. & Abrahams, D. E. (1971) *J. Neurochem.* **18**, 1017–1025
Zuckerman, E. G. & Glaser, G. H. (1972) *Arch. Neurol. (Chicago)* **27**, 14–28

Chapter 14

Metal Implants as Models of Epilepsy

By P. C. EMSON

Medical Research Council Brain Metabolism Unit,
University Department of Pharmacology,
1 George Square, Edinburgh EH8 9JZ, U.K.

Since the original observation by Kopeloff *et al.* in 1942 that alumina cream applied to the frontal cortex of monkeys produced recurrent spontaneous clinical seizures, alumina cream injections and later cobalt and tungstic acid gel implants have been used to produce electroencephalographic (EEG) foci and clinical seizures in experimental animal models of epilepsy. To provide an adequate human model of epilepsy of cortical onset or of focal epilepsy, the model should mimic as closely as possible the pharmacological, physiological, biochemical and morphological properties of the human epileptogenic cortex. Table 1 compares some of the physiological and pharmacological properties of seizures of cortical onset in man with the animal models of epilepsy involving cortical heavy metal implants.

The most commonly used convulsant metals have been aluminium (as aluminium hydroxide), cobalt (as a powder in gelatine) and tungsten (as tungstic acid gel). Practical details of the preparation of these metals for implants and descriptions of the surgery involved are provided by Ward (1972). All three types of implant produce discrete brain lesions characterized by neuronal loss and gliosis. In the rat cortex cobalt diffuses rapidly from the original implant site to produce a necrotic lesion (Emson & Joseph, 1975). Beyond the zone of complete cell death gliosis occurs and ultimately forms a calcified capsule which provides a barrier to further diffusion of the cobalt, the extent of the necrotic lesion being determined by the size of the original implant and the amount of cobalt diffusing from the implants. Alumina injection into cats or monkey produces a granuloma at the site of the original injection. Histochemical staining for aluminium reveals that the majority of the alumina remains in the original implant, although aluminium can be detected in neurons around the lesion some years after the original implant (Harris, 1973). This contrasts with cobalt implantation where histochemical staining fails to reveal cobalt outside the implant once the glial capsule has been established. Tungstic acid gel has been used much less extensively than the cobalt or alumina models, and this may partly reflect the difficulty in preparing this compound (Ward, 1972). The volume of tungstic acid gel injected into the animal must be carefully controlled as the gel produces a very severe necrotic lesion and too large an injection volume will produce status epilepticus and death. EEG changes appear much sooner with tungstic acid gel than with either alumina or cobalt and can develop within four hours of injection. The histological changes are also

Table 1. *Features common to seizures of cortical onset in man and models of epilepsy involving heavy metal implants*

References: Ward (1972), Dow *et al.* (1973), Kusske *et al.* (1974), Emson & Joseph (1975), Calvin *et al.* (1973), Glotznes *et al.* (1974), Sypert & Ward (1974)

Cortical epilepsy in man	Alumina-induced focus in monkey, cat and rabbit	Cobalt-induced focus in cat, monkey and rat	Tungstic acid gel focus in cat, monkey and rat
1. Presence of abnormal high amplitude epileptic spikes in EEG		√	√
2. Spontaneous recurrent seizures	√	Myoclonic jerks but rarely full clinical seizures	Myoclonic jerks rarely full clinical seizures
3. Extracellular recording reveals presence of neurons firing in stereotyped bursts	√	Not investigated	√
4. Seizures more frequent at different periods of the day and during periods of stress	√	√	√
5. Definite epileptogenic focus not always distinguishable	Implant areas forms primary focus	Implant area forms primary focus	Implant area forms primary focus
6. Secondary foci formed in areas receiving projections from the primary focus	Secondary foci formed	Secondary foci formed	Secondary foci formed
7. Phenytoin and phenobarbitone suppress seizures	Phenytoin and phenobarbitone suppress seizures	Phenobarbitone produces epileptiform spikes in EEG of controls and epileptic animals	Not investigated

more acute. Pyknotic neurons and acute necrosis are seen within 3–6 h and by 36–48 h the gel is encapsulated by a glial capsule. If defined volumes of gel are used the model can be used as an acute convulsant agent and has the advantage over other acute topical epileptogenic agents such as penicillin or strychnine in that no systemic side effects occur.

The reasons for the epileptogenicity of aluminium hydroxide gel, cobalt and tungstic acid gels are not clear. A number of metals have been shown to produce subclinical EEG foci (Chusid & Kopeloff, 1962) but no clear periodic table relationship has been established, except that nickel, which is next to cobalt in the periodic table, is also epileptogenic but less so than cobalt (Hartman *et al.*, 1974). The epileptogenicity of the different implants varies with the animal species and with the brain region concerned. Thus alumina will not produce seizures when applied to the cortex of the rat and cobalt is most effective in producing seizures when applied to the sensorimotor or frontal cortex in the cat, primates or rats but only produces an EEG focus if applied to areas like the visual cortex (Ward, 1972). The advantages of the different models vary and their similarities to human focal epilepsy have been reviewed in Table 1. The most commonly used model at the moment is cobalt-induced epilepsy in the rat or cat, and this model has been the subject of detailed biochemical investigations (Van Gelder, 1972; Craig & Hartman 1973; Emson & Joseph, 1975). However, the disadvantages of a model involving implantation of a toxic heavy metal must always be borne in mind in reviewing biochemical changes in the resulting epileptic focus. For this reason, study of the secondary or mirror focus often resulting from the primary implant of a heavy metal into the frontal cortex may be a more valuable model (Westmoreland *et al.*, 1972), although the possiblity of axonal transport of the heavy metal into the secondary focus cannot be ruled out. In order to understand the neurochemistry of human focal epilepsy a great deal of further work must be done on human biopsy material and on the more chronic models of epilepsy in primates (Ward, 1972; Emson, 1975; Meldrum, this volume, Chapter 13). Table 2 compares some of the biochemical and morphological features found in human epileptic foci and experimental epileptic foci in animals. Common features of all these epileptic foci include evidence of deafferentiation, gliosis, changes in levels of transmitter and reduced Na^{+}/K^{+} ATPase levels, probably indicating changes in cation levels in the focus. The importance of these different features in the genesis in both human epileptic foci and in models of epilepsy involving metal implants indicate that the models of epilepsy involving metal implants provide a system sufficiently analogous to that found in man for it to be of use in testing new anticonvulsants and surgical procedures whilst allowing electrophysiological and neurochemical monitoring.

The model that has received most attention in our laboratory has been the cobalt model of epilepsy in the rat. The main reasons for this are its reproducibility, reasonably well defined time-course and comparative cheapness of rats as compared to cats or monkeys. The cobalt model in the rat has not, however, proved particularly valuable in acute pharmacological studies (Dow *et al.*, 1973). Nonetheless we have found the cobalt model particularly valuable for monitoring the biochemical changes occurring during development and re-

Table 2. *Biochemical and histological features common to cortical epileptic foci in man, and experimental heavy metal-induced cortical foci in animals*

References: (1) Scheibel, M. E., Crandall, P. H. & Scheibel, A. B. (1974); (2) Harris, A. B. (1973); (3) Emson, P. C. & Joseph, M. H. (1975); (4) Black, R. G., Abraham, J. & Ward, A. A. Jr. (1967); (5) Crapper, D. R. (1973); (6) Westrum, L. E., White, L. E. & Ward, A. A. Jr. (1964); (7) Van Gelder, N. M., Sherwin, A. L. & Rasmussen, T. (1972); (8) Tower, D. B. (1960); (9) Bradford, H. F. (this volume, Chapter 16); (10) Craig, C. R. & Hartman, E. R. (1973); (11) Van Gelder, N. M. (1972); (12) Clayton, P. R. & Emson, P. C. (1975); (13) Hunt, W. A. & Craig, C. R. (1973).

Human cortical foci	Alumina-induced cortical foci (monkey, cat and rabbit)	Cobalt-induced cortical foci (cat, rat and mouse)	Tungstic acid gel-induced cortical foci (cat, rat and monkey)
1. Glial scar. Human cortical foci show varying numbers of abnormal glial cells (1)	Alumina induces a necrotic lesion with extensive gliosis (2)	Cobalt lesion becomes surrounded by a glial capsule which ultimately calcifies (3)	Tungstic acid gel produces a severe necrotic lesion which becomes surrounded by a fibrous glial capsule (4)
2. Neuronal loss. Pyramidal cell loss is typical of cortical foci. If no definite focus then scattered neuronal loss occurs throughout the brain (1)	Alumina produces selective pyramidal cell loss. Neurons particularly affected are in cortical layers II and III and V (5)	Cobalt produces selective pyramidal cell loss. Neurons particularly affected are those in layer II and III and layer V. Histochemical staining shows these cells normally contain heavy metals (3)	Pyknotic neurons appear within hours of injection. Outside necrotic area selective nerve cell death (4)
3. Golgi staining of focal tissue reveals abnormal neurons with reduced dendritic trees and few dendritic spines (1)	Golgi staining in monkey reveals abnormal pyramidal cells in primary focus with reduced dendritic spines (6)	Golgi staining in rat reveals abnormal pyramidal cells in primary focus with reduced dendritic spines (3)	Not investigated
4. Reduced levels of transmitter amino acids and impaired ability to synthesize acetylcholine (7, 8)	Reduced levels of transmitter amino acids in primary focal area (9)	Reduced levels of transmitter amino acids and choline acetylase in primary focal area in rat (3, 10, 11)	Not investigated
5. Levels of biogenic amines and associated enzymes in focal areas not investigated	Not investigated	Reduced levels of tyrosine hydroxylase, monoamine oxidase and catechol-*O*-methyl transferase in primary and secondary foci. Increased levels of noradranaline and dopamine metabolites in cortex and caudate. (4 hydroxy-3-methoxyphenylethylene glycol and homovanillic acid) Serotonin (5-HT) levels only reduced after peak in spiking has passed. Cortical 5-HIAA and midbrain tryptophan hydroxylase are also reduced at this time (12)	Not investigated
6. Reduced levels of Na^+/K^+ ATPase activity found in focal area	Reduced levels of Na^+/K^+ ATPase activity found in focal area (8)	Reduced levels of Na^+/K^+ ATPase activity found in focal	Not investigated

gression of an EEG focus. For this purpose the absence of clinical seizures producing hypoxia and cell loss is an advantage. In these studies of the development and regression of the cobalt focus we have been greatly helped by the development of a computer technique for spike recognition (Hill & Townsend, 1973). The computer provided an estimate of spikes per minute, and a typical example of focus development is seen in Fig. 1. Spikes usually reach a peak

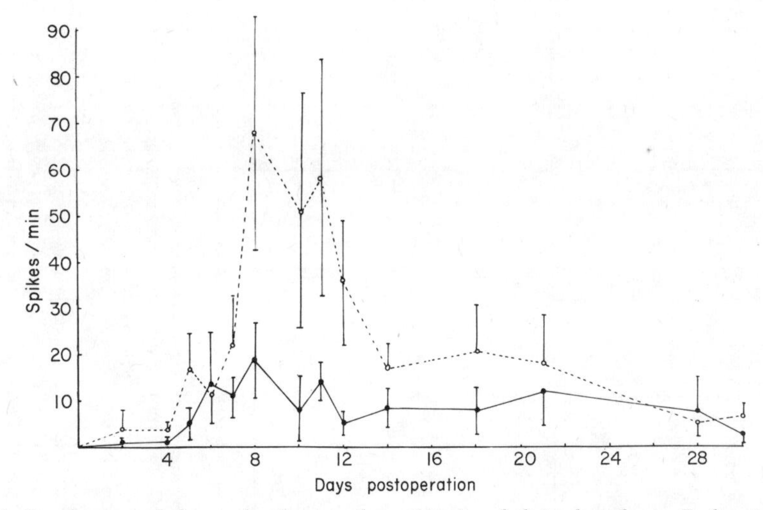

Fig. 1. *Development of abnormal epileptic spike activity in cobalt-implanted rats. Each point is the mean* ± S.D. *for primary focus (●——●) and secondary focus (○------○). The number of animals used were for days 2, 4, 5, 6, 7, 11, 12 and 28—12 animals, and for days 8, 14, 21—24 animals*

between 6–12 days postoperatively, at which time biochemical parameters of nerve cell death and terminal degeneration are maximal (Emson & Joseph, 1975), and motor manifestations of epilepsy are also maximal (contralateral forelimb and whisker twitch). Following this there is a prolonged chronic (12–100 days) phase during which biochemical parameters return to normal. However, despite the recovery of the biochemical parameters it remains to be seen whether this recovery is full in the functional sense and it is likely that the focus retains a permanently lowered threshold to convulsants. A similar study of the time-course of spike development has been made in the alumina cream-induced epilepsy in the cat (Velasco *et al.*, 1973). In the cat the numbers of spikes reach a maximum between 40–50 days postoperatively and the EEG shows no definite spikes by 80 days. No detailed biochemical observations have been made during spike development or regression in this model. The ability of monkeys to recover from the epileptogenic effects of cortical cobalt implants has been established (Marcus, 1972) and it seems likely that they will recover from alumina implants, although this has not been properly investigated (Wyler *et al.*, 1974). If the remission from seizures represents a regeneration response as seems to be the case in the rat, then it may be logical to use drugs which may encourage nervous regeneration as a prophylactic measure in cases of head injury.

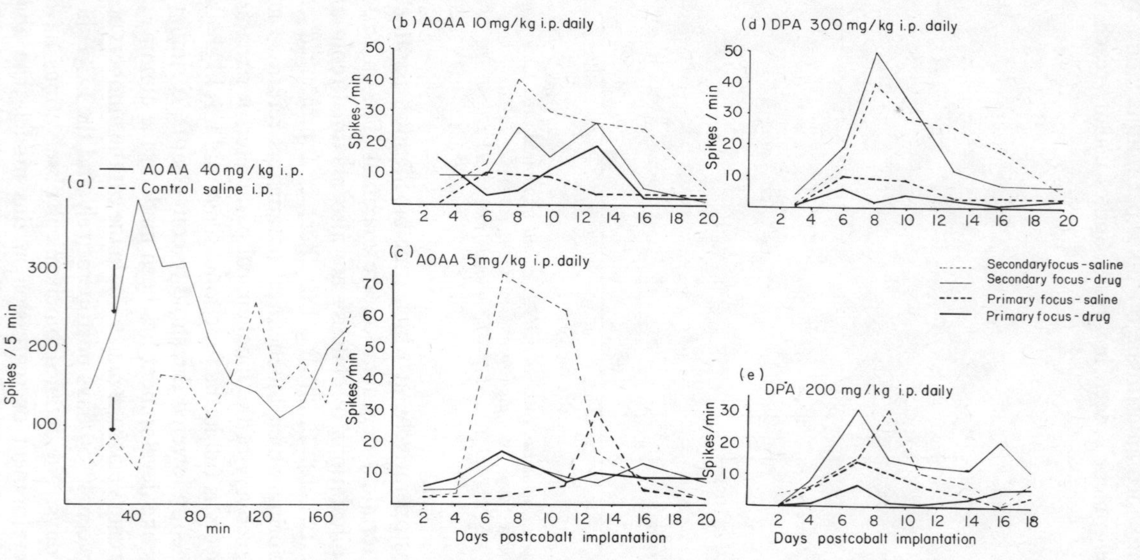

Fig. 2*a*. *The effect of an acute intraperitoneal injection of 40 mg/kg aminooxyacetic acid on the numbers of spikes in the electrocorticogram (ECoG) of a rat with a frontal cobalt implant (——). A control cobalt-treated rat received a saline injection (– – – –). There is a 30 min control period at the beginning of the experiment. ↓ indicates the time of drug or saline injection.*

Fig. 2*b–e*. *The effects of chronic treatment with aminooxyacetic acid and di-n-propylacetate on development of spiking in the primary and secondary foci of rats receiving cobalt implants in the frontal cortex. The electrocorticogram was recorded for 10 min periods at the same time of day and at two- or three-day intervals. Four rats were used in each experimental or control group.*

Having established the normal time-course of focus development in the cobalt rat we have attempted to modify its development and severity using chronic pharmacological treatments and stereotactic surgery. Our pharmacological investigations have concentrated on the role of two putative inhibitory transmitter amino acids, taurine and γ-aminobutyric acid (GABA). Both taurine and GABA levels are reduced in the primary cobalt focus when spiking is maximal and we have attempted, by chronic administration of taurine and γ-aminobutyric transaminase (GABA-T) inhibitors (raising cortical GABA levels by blocking GABA metabolism), to reverse the changes in GABA and taurine levels in the focus and hopefully reduce or suppress the activity of the focus. Figures 2*a–e* show a comparison of the effects of aminooxyacetic acid (AOAA), a well-established GABA-T inhibitor acting via interference with the pyridoxal binding site on the enzyme, and sodium di-*n*-propylacetate (DPA) a recently introduced antiepileptic agent which is claimed to be a competitive inhibitor of GABA-T. The daily administration of 1–5 mg/kg i.p. AOAA reduced spiking in both the primary and secondary foci of cobalt epileptic rats (Fig. 2*c*). At higher daily doses (10 mg/kg i.p.) of AOAA daily spiking was not reduced (Fig. 2*b*). This biphasic response to aminooxyacetic acid is also seen in experiments with acute administration of high doses of AOAA 20–60 mg/kg i.p. (Fig. 2*a*). The reason for this response probably lies in the ability of AOAA to inhibit GABA-T preferentially at low concentrations but at higher concentrations (above 10 mg/kg) to inhibit glutamic acid decarboxylase so that although brain GABA levels are still elevated the actual amount of synaptic GABA may be reduced.

In contrast to AOAA, which has a clear-cut antispike effect, neither acute nor chronic di-*n*-propylacetate reduced spiking. Study of the *in vitro* action of DPA on the GABA-T activity of human and rat brain homogenates has shown that DPA is only a weak GABA-T inhibitor, producing significant inhibition only at concentrations above 10 mM, a concentration unlikely to be approached therapeutically. This makes it unlikely that DPA is exerting any anticonvulsant action through GABA-T inhibition, although a more potent effect on succinic semialdehydes dehydrogenase ($K_i = 1.5$ mM) has recently been reported (Harvey *et al.*, 1975). However the anticonvulsant action of DPA in man seems well established (Volzke & Doose, 1973) so it may be that it is exerting its action through some as yet uncharacterized metabolite (Schobben *et al.*, 1975).

The suggestion has been made that the sulphur-containing amino acid taurine may be a naturally occurring anticonvulsant (Van Gelder, 1972). A number of observations have shown that taurine has definite anticonvulsant properties in a number of models of epilepsy including the cobalt model of epilepsy in the cat and mouse (Van Gelder, 1972), the alumina-induced focus in the cat, and penicillin and strychnine induced acute cortical foci in the rat (Adembri *et al.*, 1974). Taurine has also received attention as an antiepileptic in man with promising results, in some cases of petit mal (A. Barbeau, personal communication). Figure 3 shows the effects of acute and chronic treatment with taurine on spike development in the cobalt model of epilepsy in the rat. Neither acute taurine injections (200 mg/kg i.p. at 30 min intervals, Fig. 3*a*) nor chronic oral administration of taurine (approximately 2.4 g/kg daily, Fig. 3*b*) had any significant effects on numbers of epileptic spikes. Biochemical observations showed that the chronic treatment with taurine elevated brain taurine levels almost to normal but had no effect on the levels of other amino acids in the primary focus (M. H. Joseph & P. C. Emson, unpublished work). Because of the comparatively small increase in taurine observed we carried out a series of intracortical injections which were designed to substantially elevate brain taurine levels. Even under these conditions, which doubled brain taurine levels, we found no significant antispike or anticonvulsant activity of taurine. It seems likely that taurine is not an anticonvulsant in the cobalt epileptic rat and in this respect the cobalt model in the rat may be a poor model for human epilepsy.

The second part of our study of the chronic treatment of cobalt epileptic rats has involved investigation of the role of thalamic projection systems in inhibiting or exciting a cortical focus. Our approach has been to make coagulative lesions in individual thalamic nuclei by stereotactic surgery and to monitor the effects on the development and spike numbers of the cortical foci (K. Green & P. C. Emson, unpublished work). Our histological observations (Emson & Joseph, 1975) have shown extensive terminal degeneration in the mediodorsal and ventral thalamic nuclei following cortical implantation of cobalt. For this reason we lesioned the mediodorsal and some of the ventral thalamic nuclei. The results of a series of ipsilateral mediodorsal lesions of thalamic nuclei are

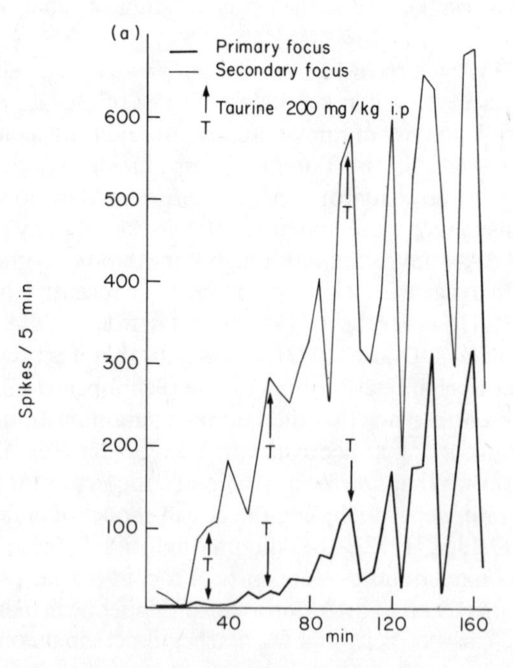

Fig. 3*a*. *The effect of a series of acute taurine injections (200 mg/kg i.p.) on epileptic spiking in a rat with a frontal cobalt implant.*

Taurine injections are indicated by (↓), the spikes in the primary focus by (—) and the spikes in the secondary focus by (——).

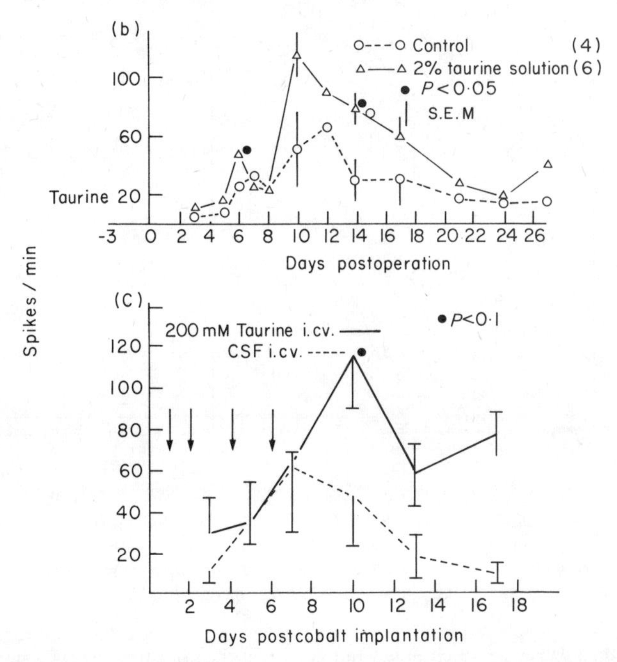

Fig. 3*b*. *The effect of chronic oral taurine administration (2% taurine in the drinking water) on the development of spikes in the ECoG of rats with frontal cortical implants (△—△)*

Control rats received normal drinking water (O- - - - -O). There were three rats in each group.

Fig. 3(*c*) *The effect of intracortical taurine injection on the development of spikes in control (- - - -) and taurine-treated (——) rats with frontal cobalt implants*

Control animals received intracortical injections of 10μl of cerebrospinal fluid (CSF) and experimental animals 10μl of 200 mM taurine dissolved in CSF. ⊥ indicates S.D.

shown in Fig. 4. Notice that the thalamic lesion significantly reduces spiking in the primary focus but has little effect on the mirror or secondary focus where the thalamic circuitry is still intact. This suggests that to some extent the secondary focus is independent of the primary focus, a suggestion supported by results from cutting the corpus callosum (the band of fibres connecting the cerebral hemispheres) which demonstrate that an established secondary focus can persist independent of the primary (Ashcroft *et al.*, 1973). Electrophysiological observations have shown that thalamic neurons can apparently trigger cortical events (Guttnick & Prince, 1975) and so it may be that in lesioning thalamic nuclei we are removing a type of positive feedback to the cortical focus. Stereotactic lesioning of thalamic nuclei is used in the treatment of intractable human cortical epilepsy (Sano, 1974) so that the ability to investigate this systematically in a model of epilepsy should have great value. If lesioning of specific brain nuclei can be combined with detailed neurochemical and pharmacological scrutiny of the focus then we should be in a much better position to understand the mechanisms of cobalt-induced epilepsy and hopefully to suggest new rational therapies for human epilepsy.

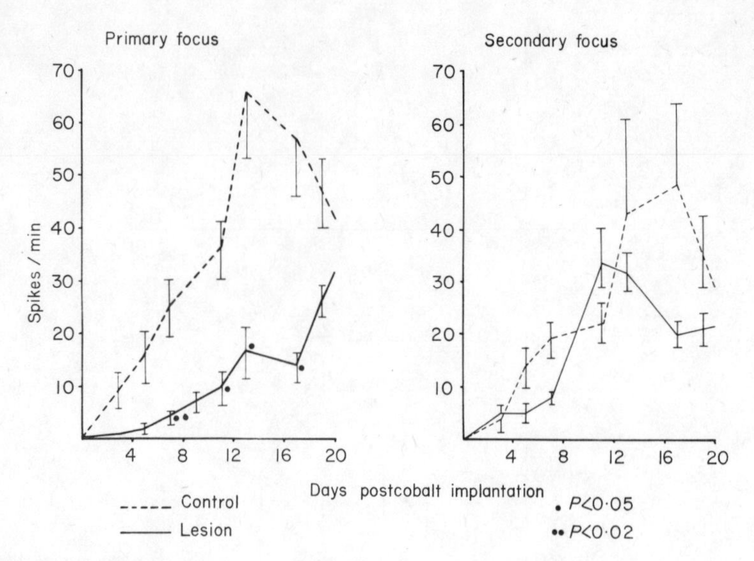

Fig. 4. *The effects of lesioning the mediodorsal thalamic nucleus on the development of epileptic spiking in the primary and secondary foci of cobalt epileptic rats*

The thalamic nucleus was lesioned one week prior to insertion of cobalt into the cortex. Control rats without thalamic lesion received cobalt implants at the same time. There were three rats in the experimental and three in the control group.

Acknowledgements

I am indebted to all the members of the Epilepsy group within the MRC Brain Metabolism Unit for their support and encouragement. In particular I should like to thank Mrs. Kathryn Green (supported by the Epilepsy Research Fund) and Dr. M. H. Joseph for permission to include our unpublished data in this review.

References

Adembri, G., Bartolini, A., Bartolini, R., Giotti, A. & Zilleti, L. (1974) *Brit. J. Pharmacol.* **52**, 439–440

Ashcroft, G. W., Dow, R. C., Harris, P., Hill, A. G., Ingleby, J., McQueen, J. K. & Townsend, H. R. A. (1973) in *Proceedings of the Fifth European Symposium on Epilepsy,* London

Black, R. G., Abraham, J. & Ward, A. A. Jr. (1967) *Epilepsia* **8**, 58–63

Calvin, W. H., Ojemann, G. A. & Ward, A. A. Jr. (1973) *Electroencephalogr. Clin. Neurophysiol.* **34**, 337–351

Chusid, J. G. & Kopeloff, L. M. (1962) *J. Appl. Physiol.* **17**, 696–700

Clayton, P. R. & Emson, P. C. (1975) *Biochem. Soc. Trans.* **3**, 261–263

Craig, C. R. & Hartman, E. R. (1973) *Epilepsia* **14**, 409–414

Crapper, D. R. (1973) *Electroencephalogr. Clin. Neurophysiol.* **35**, 575–588

Dow, R. C., Forfar, J. C. & McQueen, J. K. (1973) *Epilepsia* **14**, 203–212

Emson, P. C. (1975) *Int. J. Biochem.* (In press)

Emson, P. C. & Joseph, M. H. (1975) *Brain Res.* **93**, 91–110

Glötzner, F. L., Feltz, E. E. & Ward, A. A. Jr. (1974) *Exp. Neurol.* **42**, 502–528

Guttnick, M. J. and Prince, D. A. (1975) *Exp. Neurol.* **46**, 418–431

Harris, A. B. (1973) *Exp. Neurol.* **38**, 33–63

Hartman, E. R., Colstanti, B. K. & Craig, C. R. (1974) *Epilepsia* **15**, 121–129

Harvey, P. K. P., Bradford, H. F. & Davison, A. N. (1975) *FEBS Lett.* **52**, 251–254
Hill, A. G. & Townsend, H. R. A. (1973) *Int. J. Bio-Med. Computing* **4**, 149–156
Hunt, W. A. & Craig, C. R. (1973) *J. Neurochem.* **20**, 559–567
Kopeloff, L. M., Barrera, S. E. & Kopeloff, N. (1942) *Am. J. Psychiat.* **98**, 881–902
Kusske, J. A., Wyler, A. R. & Ward, A. A. Jr (1974) *Exp. Neurology* **42**, 587–592
Marcus, E. M. (1972) in *Experimental Models of Epilepsy* (Purpura, D. P., Perry, J. K., Woodbury, D. M., Tower, D. B. & Walter, R. D., eds.), pp 113–146, Raven Press, New York
Sano, K. (1974) in *Epilepsy. Proceedings of the Hans Berger Centenary Symposium*, pp. 215–221 (Harris, P. & Mawdsley, C., eds.), Churchill Livingstone, Edinburgh
Scheibel, M. E., Crandall, P. H. & Scheibel, A. B. (1974) *Epilepsia* **15**, 55–80
Schobben, F., van der Kleijn, E. & Cabreels, F. J. M. (1975) *Eur. J. Clin. Pharmacol.* **8**, 97–106
Shaywitz, B. A., Cohen, D. J. & Bowers, M. C. *Neurology (Minneapolis)* **25**, 72–79
Sypert, G. W. & Ward, A. A. Jr (1974) *Exp. Neurol.* **45**, 19–41
Tower, D. B. (1960) *Neurochemistry of Epilepsy*, Thomas, Springfield, Ill.
Van Gelder, N. M. (1972) *Brain Res.* **47**, 157–165
Van Gelder, N. M., Sherwin, A. L. & Rasmussen, T. (1972) *Brain Res.* **40**, 385–393
Velasco, M., Velasco, F., Estrada Villaneuva, F. & Olvera, A. (1973) *Epilepsia* **14**, 3–27
Volzke, E. & Doose, H. (1973) *Epilepsia* **14**, 185–193
Ward, A. A. Jr. (1972) in *Experimental Models of Epilepsy* (Purpura, D. P., Perry, J. K., Tower, D. B., Woodbury, D. M. & Walter, R. D., eds.), pp. 13–36, Raven Press, New York
Westmoreland, B. F., Hanna, G. R. & Bass, N. H. (1972) *Brain Res.* **42**, 485–499
Westrum, L. E., White, L. E. & Ward, A. A. Jr. (1964) *J. Neurosurg.* **21**, 1033–1046
Wyler, A. R., Fetz, E. E. & Ward, A. A. Jr (1974) *Exp. Neurol.* **44**, 113–125

Discussion Papeı

Epilepsy—A Clinician's View

By H. R. A. TOWNSEND

Department of Surgical Neurology,
Western General Hospital, Crewe Road, Edinburgh EH4 2XO, U.K.

The trouble with epilepsy is that most of the time it isn't there! As far as the patient is concerned most of the time he is fit and well, alert and active; yet he cannot drive a motor car, fly an aeroplane, or even go swimming by himself. The patient's parents are torn between the desire to protect their child and the desire that he or she should live a normal life. From the point of view of the doctor the efficacy of treatment is almost impossible to assess. If the patient has attacks, then treatment has failed. If the patient has no attacks, then would he have had any even without treatment?

All this is because epilepsy is a disorder, not of structure, but of function, and the malfunction is not consistent but only intermittent. *Fits* are intermittent, dramatic interruptions of consciousness and disturbances of behaviour. *Epilepsy*, however, is the likelihood of having a fit. Any medical diagnosis is in some sense a prognosis, an attempt to peer into the future, but nowhere is this more obvious than in the diagnosis of epilepsy where the questions always uppermost are: Will he have another fit? When will he have another fit? Will the attacks go away, or will they keep on coming back in the future?

The attempt to find out about epilepsy, its cause and possible cure, starts with an attempt to classify the manifestations of the disease.

The classical epileptic major fit is the symptom of grand mal epilepsy. The term 'aura' derives from an indescribable feeling that many epileptic patients have for some time—hours, even days—prior to a major fit, a feeling of heightened awareness, unnatural clarity of perception, difficult, if not impossible, to describe but clearly recognizable. The fit comes suddenly. All the muscles contract fiercely and simultaneously. Perhaps involuntarily crying out or biting the tongue, the patient falls to the ground. The muscles remain contracted, respiration stops. Gradually the tonic contraction breaks into shock-like clonic jerks becoming slower and slower until the fit passes off and the exhausted patient passes into a coma from which recovery is gradual.

It is not surprising that this condition was interpreted in olden days as 'possession by the devil'. Sufferers from epilepsy have always been regarded with a mixture of awe and fear. This is the disease of beggars and of kings, of slaves and prophets.

The drama of a major fit is quite unlike the manifestations of the other form of epilepsy classically recognized as 'petit mal'. This is a disease of children and teenagers. The attacks may go almost unnoticed except by an acute observer. They consist simply of interruptions of the stream of consciousness, a simple activity—like walking—may be continued automatically without discernible interruption. Speech may be inhibited or become repetitive or jumbled, but simple tasks like counting out loud or reciting a well known poem may be unaffected. The outward signs of consciousness are difficult to quantify but something, particularly about the eyes, seems to tell us when consciousness is even momentarily absent. Children who have these little spells many times a day inevitably suffer as a result of the breaks in the thread of consciousness, but there is no dramatic fit, no period of confusion, it is just as if consciousness were switched off and then on again. The condition is usually self-limiting and disappears at about the end of puberty leaving no trace behind, but often, too, petit mal and grand mal co-exist and as the child grows older the major fits supplant the minor ones.

In practice, of course, particular individuals who show pure major or minor fits, like those I have described, are rare. Mixtures of patterns seem to be the rule and even in one patient the attacks are rarely identical on every occasion. A practical classification considers the frequency of the attacks. If epileptic fits occur rarely, perhaps only one or two in a lifetime, can this ready be called epilepsy? A young person who has a single attack may never have another one. It seems pointless to try and treat a disease which may not exist.

The clinical problem is the individual with relatively frequent attacks. These may be monthly, weekly or even daily. Once a patient has been observed for long enough to be able to make some sort of estimate of the frequency of attacks without treatment, it is possible to institute some form of rational therapy. Nevertheless it takes very understanding relatives or a strong-willed physician to resist the temptation to institute therapy immediately, or at all events after the second attack. Having made the decision to give anticonvulsants it is very difficult to be sure when treatment is being successful. On the one hand it is essential to maintain a sufficient dosage to minimize the possibility of another attack, on the other hand not only is it tedious (and expensive) to take anticonvulsants but it may be positively detrimental, not to say dangerous.

Electroencephalography (EEG) is of some help in this situation, parodoxically because the phenomena which can be seen in the EEG, and which appear to be of epileptic origin, are for practical purposes, not affected at all by the doses of anticonvulsants given in clinical practice.

Major attacks tend to occur in bouts, and as far as I know the reason for this tendency is quite unknown. Sometimes probable causes may be fairly obvious: overindulgence in alcohol is a fairly sure way to bring on a bout of fits, so too is any minor infectious illness, even a common cold, or a tummy attack. Frequently, excitement or overexertion can be indicted. Children in hospital may have an attack after a visit from their parents. All these are reactions to an external environment, but there may be internal determinants; attacks often occur during sleep, women frequently have fits associated with their periods.

Status epilepticus, in which fits follow each other without respite, is a major clinical emergency as the patient's life is at risk and even in the best staffed and equipped units, it may not always be possible to stop the discharges. The condition may be provoked by a brain injury or poisoning, or by the sudden withdrawal of a drug. There is something strange at work here, because the ordinary epileptic fit is followed by a sort of refractory period, during which the epileptic process cannot restart until the brain has had a certain amount of time to recover. In status epilepticus this mechanism of exhaustion fails to operate.

A less dangerous, but nevertheless extremely interesting form of status epilepticus is minor status. This condition presents clinically as a disorder of consciousness of abrupt onset, which may last for hours or even days, and then ceases equally abruptly and unpredictably. Without recourse to the electroencephalogram it is almost impossible to distinguish from an hysterical fugue state. In this case, whatever the mechanism that normally stops the minor fit and restores consciousness to normal functioning, simply fails to operate.

The great conceptual advance which followed the development of electroencephalography was to recognize the importance of a localized malfunction or *focus* which could occasionally interfere with the functioning of the whole brain. One hundred years ago Hughlings Jackson suggested that epilepsy was caused by a discharge in the grey matter of the cortex, and that those manifestations of epilepsy, now called Jacksonian epilepsy, in which the twitching or clonic movements slowly spread from one region of the body to another were tracing the progress of a localized discharge gradually moving across the surface of the brain. The advent of electrical recordings enables us to demonstrate the discharge that Jackson imagined, and led to the recognition, as forms of epilepsy, of other phenomena which had hitherto either not been recognized or whose epileptic nature had been debatable. The most dramatic of these are the remarkable disorders involving organized behaviour or complex perception which are found when epileptic discharges involve the temporal lobe.

So we have three aspects of this curious condition—epilepsy.

The classical contrast between the major fit with muscular jerking and subsequent exhaustion, and the minor fit which appears to be a pure disturbance of consciousness. Could these contrasting phenomena be due to disturbances of a biochemical mechanism or are they more likely to be due to differences in anatomical site?

Consideration of the frequency of occurrence of attacks and the temporal patterns in which they occur leads us to many questions. Is the epileptic attack an accident, a particular random combination of inputs to a part of the cortex, or is it determined by some chemical substance which gradually builds up until released in the discharge of the clinical fit? There is, for example, evidence that some patients actually feel worse if their attacks are stopped, and some psychiatric patients suffer a psychotic relapse when their epilepsy is quiescent and vice versa. On the other hand epilepsy is a manifestation of the function of the brain, which also has volition as one of its functions. Patients will tell you, and I see no reason to disbelieve them, that they can, by 'taking thought', control and suppress attacks, so long as they are not distracted at the crucial moment. In the case of severe epilepsy or status epilepticus, there is relatively little room for psychology, Here, surely, biochemical and gross structural factors must play the major part.

The surgery of epilepsy has mainly concentrated on focal epilepsy, for obvious reasons. If a focus can be shown to exist then its removal should at least ameliorate the condition. There has been a tendency to try and group all forms of epilepsy into varieties of focal epilepsy so that those which are not focal in the periphery must be focal in the centre (i.e. centrencephalon epilepsy). There have even been neurosurgical assaults on the centrencephalon itself!

To me the surprising fact is not that we may have fits, but that we usually don't. Perhaps the fact that we are close relatives of monkeys has something to do with it. Monkeys, wild ones at any rate, do not suffer from epilepsy—the first fit is the last! In the course of evolution the central nervous system must have built up a series of checks and balances—inhibitory mechanisms—at every level to prevent the system from getting out of control.

At the cell membrane there is some evidence that epileptic spikes are caused by abnormal, non-decrementing, conduction along apical dendrites. At the synapse inhibitory and excitatory transmitters balance each other. At the control level there is an hierarchic system of reflexes each providing local stability or homoeostatis at its own level. Subsystems, for example the hemispheres, are so constructed that damage to one has only minimal effect on the functioning of the other and many functions seem to be partially duplicated. Finally the whole brain although limited in its powers of physical regeneration is almost unbelievably plastic in its ability to redeploy its higher level functional units.

No wonder the treatment of epilepsy is so difficult and we are so often reduced to—as my mentor Dr. Nevin often used to say—'leaving the patient less brain to be epileptic with'.

Discussion Paper

Lesions in Human Epilepsy

By C. E. POLKEY

The Neurosurgical Unit of Guy's, Maudsley and King's College Hospitals, The Maudsley Hospital, De Crispigny Park, Denmark Hill, London SE5 8AF, U.K.

In the initiation and propagation of abnormal electrical activity in the brain which leads to the clinically observed events described as an epileptic fit, it is possible to recognize three separate links in a chain of events and any of these links may, if modified, either alter the nature of an attack, abort it, or even cause it not to appear at all. It is therefore appropriate to describe the lesions which may be found in human epilepsy in relation to this chain of events. Such lesions may be seen at many stages and by different observers, paediatricians, clinical neurophysiologicsts, neurologists, neurosurgeons and pathologists, each of whom may seen only a part of the process, and this review attempts to amalgamate these various viewpoints. The experimental aspects of epilepsy have already been discussed and therefore these will not be dealt with except in so far as they bear directly upon the lesions found in human epilepsy.

An epileptic fit must originate in normal neuronal tissue which has been induced to behave in an abnormal way. This is a basic concept which was well known in the 19th century to Hughlings Jackson and his predecessors (see Taylor, 1931). Therefore the first of the three stages or links in the chain must be the pathological change itself, which may vary from a gross organic lesion such as a tumour, abscess or similar change affecting the cerebral tissue, through more subtle change such as that induced by cerebrovascular disease or other degenerative disease to the effects of purely biochemical processes such as anoxia or hypoglycaemia. These in their turn set up the second link: this is a local effect which alters the environment of one or several groups of neurons, and this alteration in the environment which gives rise to focal abnormal electrical behaviour constitutes the second link in the chain. The third link in the chain relates to the properties of the neuronal population itself. It is doubtful whether a group of neurons, till it attains a certain numerical size, is capable of concerted action. De Lisle Burns (1958), in a series of elegant isolating and undercutting experiments, showed that a small number of cortical neurons was electrically silent and in order to maintain the regular alterations of the expiratory and inspiratory neurons in the medulla, for example, a minimal number of interneuronal connections was necessary (Burns & Salmoiraghi, 1960). In a similar fashion a certain number of neuronal pathways, possibly in a state of altered excitability, are necessary for the spread and propagation of the original minor disturbance, and this forms the third link in the chain.

Although in terms of neurochemistry it is probably useful to concentrate on the second and third stages in the chain, it is necessary for an understanding of these later events, and to give some idea of the sort of material in which the lesions in human epilepsy may be studied, to spend a brief time on certain selected items in the first link. A clinician, in assessing a patient with epilepsy, is naturally anxious, wherever possible, to find a cause. Therefore it is from the results of clinical appraisal and special investigations that one would expect to find the basic causative lesions in human epilepsy. By this means an answer which may be reasonably accurate can be obtained in a proportion of cases. In a smaller number of cases surgery is recommended and a specimen is obtained which may be examined by the pathologist. The value of this specimen depends upon the extent and techniques of the surgical resection. Thus in cases of predominantly temporal lobe epilepsy a realistic view is obtained only if the mesial temporal structures are resected *en bloc* (Falconer, 1969). Finally, patients may die and their brains become available for detailed pathological examination and reference is made later to results obtained in this way. It must be said that the results from these last two sources, although definitive, are obtained from a small and selected part of the epileptic population.

Early surveys are not explicit regarding the precise cause of chronic epilepsy, and Lennox & Lennox (1960*a*), quoting several early series from mental institutions in the United States published about 1915, gives three figures, varying between 34% and 89%, for cases in which an organic diagnosis was possible. He himself states that in a series of 2500 chronic epileptics there was thought to be a genetic factor in 35%, an acquired factor in 16%, both were thought to be together in 11%, and in 38% there was no causative factor discernible (Lennox & Lennox 1960*b*). This last figure is worthy of attention since it is a constant feature of every list of causes of chronic epilepsy whatever the source of the material. There is a recent survey of adult temporal lobe epilepsy from the London Hospital in which cases referred to both neurologists and neurosurgeons were combined and assessed (Currie *et al*, 1971). In this series there were 666 patients and among these a definite cause, even on clinical grounds, could only be found in 25%. This included patients with gross organic lesions who had presented with chronic epilepsy but not those with known lesions such as tumour in which seizures formed part of their more florid clinical picture. Only 62 of these patients were subjected to surgery and a definite pathological diagnosis was obtained in 29 patients (47%).

When we look at series of surgical material from centres specializing in the treatment of chronic epilepsy, the number of different lesions decreases considerably although the type of lesion described is of more interest in the study of the pathophysiology of chronic epilepsy. Thus Rasmussen (1972), from Montreal, describing 1456 patients operated upon between 1928 and 1968, lists the causes as follows:

Tumours 19%, arteriovenous malformations 1%, birth trauma, anoxia or compression 25% [this group includes cases which would be classified by us as mesial temporal sclerosis, which were placed by Earle *et al.* (1953) in the group of incisural sclerosis], postnatal trauma 19%, postinflammatory scarring 12%, miscellaneous 5%, and unknown 19%.

The nature of the pathological lesion associated with some forms of chronic epilepsy has been gradually described over many years. Over 100 years ago, Bouchet & Cazauviehl (1825) in France, described chronic induration in the hippocampus of nine out of 18 epileptics. Its microscopical features were subsequently described by Sommer (1880) and Bratz (1899). The lesions which these pathologists described were then ignored for 50 years. Meanwhile surgeons were becoming interested in the prospect of curing epilepsy by operation, and Horsley in 1886 began operating upon cases of chronic epilepsy following trauma (Northfields, 1968). This problem had also interested Foerster, and together with Penfield (Foerster & Penfield, 1930) investigations were begun on the pathological nature of cerebral scarring. They showed that in an injured area a scar was formed by fibrous astocytes and that this scar became richly vascularized and attached thereby to the overlying meninges; subsequently this scar tissue underwent contraction and as it did so a tethering cicatrix was formed. Adjacent to this cicatrix, and between it and the grey matter, was an intermediate zone in which there was gradual destruction of neuronal tissue over a long period. This was in some way responsible for the epileptogenic focus, and thus they reasoned that if this scar were excised out to the gyral margins, the artificial cavity and minimal scarring which would follow would be less likely to provoke epileptic seizures. Penfield subsequently began to remove scars on this basis with considerable success and this method is now used in neurosurgical practice.

The precise properties of these scars which make them epileptogenic still remains unelucidated but we shall see later that gliosis, which is a process whereby glial cells (the specialized connective tissue cells of the central nervous system) infiltrate diseased and damaged parts of the brain, is common to several of these epileptogenic lesions.

The relationship between epilepsy and the gross organic lesions such as abscesses, intracranial haemorrhage and intracranial tumours both intra- and extracerebral, although clinically very evident, is in terms of the precise effect whereby they cause the remaining neuronal population to act as an epileptogenic focus, necessarily obscure. Neither is it always constant; thus some patients with a tumour will present with fits whereas others may be free until it has been removed. Therefore in studying the relationship between the organic lesion and the epileptogenic focus it is more profitable to concentrate upon the type of discrete lesion seen in the surgical and postmortem material available from chronic epileptics.

It is at this point that we must return to the gliotic lesions of the hippocampus described by Sommer in 1880. Stauder (1935) described how he found similar lesions in the brains of 36 out of 53 epileptics at postmortem examination, and that 33 of the 36 with lesions had been diagnosed as having temporal lobe epilepsy in life. Further papers appeared, but pertinent to the present review is that by Margerison & Corsellis (1966). They examined the brains of 55 chronic epileptics dying in hospital, and were able to identify in 36 of these brains a process of nerve cell loss and glial proliferation occurring in the hippocampus. In its more severe form it resembled the hippocampal lesion described by Sommer in which there is neuronal loss and gliosis in the Sommer sector (H1 field) together with a similar process in the end folium and dentate gyrus described by Bratz in 1899. This is the lesion described as Ammon's Horn, or mesial temporal sclerosis. In addition, in some of these brains there was a less severe form in which these processes affected the end folium and dentate gyrus only and this they called end-folium sclerosis. They were able to catalogue these effects separate from major structural abnormality, the results of injury, and the effects of acute anoxia in patients dying in a fit or status epilepticus. They also found on studying the clinical details and electroencephalographic findings a correlation, positive in 21 cases and negative in seven cases, between all three features sufficient to associate these sclerotic lesions with chronic temporal lobe epilepsy.

In surgical material obtained by Falconer (1974) from approximately 300 cases of temporal lobectomy for drug-resistant temporal lobe epilepsy the sort of pathological lesion which is found falls into four main groups. In about a half, the lesion of Ammon's Horn, or mesial temporal sclerosis, is seen. In a fifth, hamartomas, which are congenital malformations of glial and nerve cells or small tumours of glial or mixed glial and neural origin, are seen, in a tenth there are scars of infarcts and in the remainder, about another fifth, there is no definite lesion. In a small number of cases an interesting focal dysplasia, thought to be a 'forme fruste' of tuberose sclerosis, is seen (Taylor *et al.*, 1971).

That there is a relationship between epilepsy and inheritance is known. The exact nature of it is described by Gastaut (1969), who, adding his own observations to those of Lennox and Metrakos, feels that there is evidence for a weak dominant gene of irregular penetrance which transmits a generalized interictal electroencephalographic discharge. Its effect declines with age. Gastaut also feels that there is evidence for a recessive gene transmitting a more severe form of disorder. Inherited factors also proved of interest in studying the characteristics of young chronic epileptics. Ounstead *et al.* in 1966 described the findings in a group of 100 children aged between 3 and 15 years in whom a definite diagnosis of temporal lobe epilepsy had been made. He was able to divide them into three roughly equal groups. In one group there had been a record of a definite acute or chronic cerebral insult, in a second group there was a definite history of a severe febrile convulsion, or even status epilepticus, occurring before the age of 3. In this group there was a higher incidence of familial epilepsy than in the other two groups. The second group corresponds in many particulars to those patients in Falconer's material in whom the lesion of mesial temporal sclerosis was found (Falconer, 1970). You will already have heard today about a possible experimental parallel with this course of events, but the clinical story does show that there are two factors which must be considered, both the lesion-producing mechanism and the type of neuronal material which is insulted.

This leads naturally into the third link in the chain, the neuronal population and parts of the brain upon which the epileptogenic process is working. It is known that the structures of the limbic system are peculiarly sensitive to this process. There is clinical evidence that lesions remote from the anterior and medial parts of the temporal lobe can produce fits and findings upon special investigation identical with those produced by a lesion in that area (Falconer *et al.*, 1962). It is well known in the experimental situation that the production of a discrete discharging focus, for example, with chemical agents such as penicillin powder or alumina cream, will result in the appearance of secondary foci of abnormal electrical activity, and such area will persist for some while after the primary focus is removed.

Evidence of a similar mechanism is hard to find in human cases, partly because the presence of only one discrete organic lesion is difficult to establish with absolute certainty and also because EEG evidence of mirror foci in humans is hard to come by. However, Falconer & Kennedy (1961) describe seven cases in which removal of a unilateral lesion resulted in the disappearance of bilateral foci. Gastaut *et al.* (1960), in describing the HHE syndrome, tells how after an acute insult to one cerebral hemisphere in a child, which may be ushered in by hemiconvulsions, there may develop a hemiplegia and chronic epilepsy. During the acute episode and the first week thereafter the EEG abnormalities are largely unilateral but the patient may subsequently develop bilateral foci. Gastaut *et al.* (1960), in describing the HHE syndrome, tells how structures on the affected side, but it was noted by Krynauw (1950) that the EEG in the good hemisphere will often return to normal after the affected one has been removed. Finally, in describing his recent work in chronic bilateral stimulation of the anterior cerebellum for the relief of chronic epilepsy, Irving Cooper notes that there may be a time lag between the onset of stimulation and the lessening of fits frequency and this same time lag is seen when there is an accidental breakdown of the system. This suggests that there is some medium length alteration in neuronal properties accompanying chronic epileptic events in humans.

In summary, there are three points on which the clinician may look to the neurochemist for help. First, could there be some biochemical explanation for the 20% of patients in whom no definite cause is implicated? Second, what is the precise chemical abnormality in the gliotic areas, so common to many organic lesions found in chronic epilepsy, which causes the abnormal behaviour in the adjacent neurons giving rise to an epileptogenic focus? Third, and regrettably barely touched upon in this meeting, what is the nature of the secondary changes in neurons involved in the spread of the discharge widely sometimes through both hemispheres?

Acknowledgements

I am pleased to acknowledge the debt I owe to Mr. Murray Falconer, whose original work on temporal lobe epilepsy forms a large part of this review and under whose personal supervision I have learnt to appreciate the problems associated with the surgical treatment of focal epilepsy.

References

Bouchet & Cazauviehl (1825) *Arch. Gen Med. (Paris)* **9**, 510–542
Bratz, I. (1899) *Arch. Psychiat. Nervenkrankh.* **31**, 820–836
Burns, B. D. (1958) *The Mammalian Cerebral Cortex,* pp. 1–21, Arnold, London
Burns, B. D. & Salmoiraghi, G. C. (1960) *J. Neurophysiol.* **23**, 27–46
Cooper, I. (1974) *The Cerebellum, Epilepsy and Behaviour* (Cooper, I. S., Riklan, M. & Snider, R. S. eds.) Plenum Press, New York
Currie, S., Heathfield, K. W. G., Henson, R. A. & Scott, D. F. (1971) *Brain* **94**, 173–190
Earle, K. M., Baldwin, M. & Penfield, W. (1953) *Arch. Neurol. Psychiat.* (*Chicago*) **69**, 27–42

Falconer, M. A. (1969) in *Current Problems in Neuropsychiatry*, pp. 95–101, Headley Bros., Ashford, Kent
Falconer, M. A. (1970) *J. Neurosurg.* **33**, 233–252
Falconer, M. A. (1974) *Lancet* **ii**, 767–770
Falconer, M. A. & Kennedy, W. A. (1961) *J. Neurol. Neurosurg. Psychiat.* **24**, 205–212
Falconer, M. A., Driver, M. V. & Serafetinides, E. A. (1962) *Brain* **85**, 521–534
Foerster, O. & Penfield W. (1930) *Brain,* **53**, 99–119
Gastaut, H. (1969) *Epilepsia* **10**, 3–6
Gastaut, H., Poirier, F., Payan, H., Salamon, G., Toga, M. & Vigouroux, M. (1960) *Epilepsia* **1**, 418–447
Horsley, 1886
Krynauw, R. A. (1950) *J. Neurol. Neurosurg. Psychiat.* **13**, 243–267
Lennox, W. G. & Lennox, M. A. (1960*a*) *Epilepsy and Related Disorders,* vol. 2, pp. 576–577, Little Brown, Boston
Lennox, W. G. & Lennox, M. A. (1960*b*) *Epilepsy and Related Disorders,* vol. 1, p. 57, Little Brown, Boston
Margerison, J. H. & Corsellis, J. A. N. (1966) *Brain* **89**, 499–530
Northfields, D. W. (1968) *Brit. Med. J.* **2**, 471
Ounstead, C., Lindsay, J. & Norman, R. M. (1966) *Biological Factors in Temporal Lobe Epilepsy*, pp. 50–81, Heinemann, London
Rasmussen, T. (1972) in *Scientific Foundations of Neurology*, pp. 101–108, Heinemann, London
Sommer, W. (1880) *Arch. Psychiat. Nervenkrankh.* **10**, 631–675
Stander, H. K. (1935) *Arch. Psychiat. Nervenkr.* **104**, 181–212
Taylor, E. (1931) in *Selected Writings of John Hughlings Jackson*, vol. 1 *Epilepsy and Epileptiform Convulsions*, pp 217–238, Hodder and Stoughton, London
Taylor, D. C., Falconer, M. A., Bruton, C. J. & Corsellis, J. A. N. (1971) *J. Neurol. Neurosurg. Psychiat.* **34**, 369–377

Chapter 15

Observations on the Mode of Action of Antiepileptic Drugs

By C. D. RICHARDS

The National Institute for Medical Research,
Mill Hill, London NW7 1AA, U.K.

The epilepsies are a group of neurological disorders which are characterized by recurrent, abnormal paroxysmal discharges of many neurons. The overt symptoms vary according to the type and extent of the disorder but include brief, recurrent, lapses of consciousness (petit mal), bizarre stereotyped behavioural patterns (psychomotor epilepsy) and intense tonic–clonic seizures (grand mal). There are many known causes of epilepsy including congenital defects in the brain, local damage due to trauma or infections of central nervous tissue (e.g. meningitis) and degenerative disease of the brain. Equally, many factors are able to precipitate an epileptic attack in a susceptible subject: flickering light, hyperventilation, stress and sudden sounds, for example. For further description of the clinical aspects of the epilepsies see the articles by Harris (this volume, Chapter 12) and Townsend (this volume, pp. 175–178). The simplest possible case for analysis is presented here. It is assumed that there is a single lesion giving rise to a primary focus which generates paroxysmal discharges to be propagated throughout the rest of the brain via neural pathways or the release of nerve cell excitants. Of course, any actual case is likely to be more complex than this, but the essential elements remain, abnormal neural tissue providing additional inputs to networks of normal nerve cells.

An antiepileptic drug is a substance that reduces or prevents the occurrence of epileptic seizures. However, such a definition is too broad, as any central depressant would presumably prevent the propagation of seizure discharges. In practice, clinically useful antiepileptic drugs reduce or prevent the occurrence of epileptic seizures without causing overall depression of the nervous system.

To be able to understand the mode of action of antiepileptics, we need to be able to answer the following questions:

(1) How are antiepileptics taken up by and distributed in the central nervous system (CNS)?
(2) With what sites do they interact?
(3) What biochemical changes are produced by their interaction with the target site?
(4) How do the biochemical changes modify the seizure activity of the CNS?

The main discussion in this chapter is concerned with some electrophysiological aspects of the final question listed above. Antiepileptic drugs can act in two ways to prevent a seizure, either they can modify the activity of the focus to make it less liable to produce the pattern of discharge capable of eliciting a seizure or they can act on essentially normal nerve cells to prevent them propagating the seizure discharge. A combination of both these mechanisms is also possible.

There is little evidence about the action of antiepileptic drugs on the paroxysmal activity of neurons in an epileptic focus, but Morrell *et al.* (1959), Musgrave & Purpura (1963) and Escueta *et al.* (1974) all report that antiepileptic drugs can reduce the discharge frequency of primary foci. Morrell *et al.* (1959) concluded that phenobarbitone and trimethadione depress focal activity while diphenylhydantoin does not; Escueta *et al.* (1974) and Musgrave & Purpura (1963), found that diphenylhydantoin does reduce the activity of discharging foci, and Musgrave & Purpura found no effect of phenobarbitone on the discharge frequency of a focus. The reasons for these discrepancies are not clear, as all groups used a freezing technique to produce their lesions. None of these groups found that antiepileptic drugs completely suppress the activity of discharging foci. Moreover, Morrell *et al.* (1959) found that the clinical efficacy of various antiepileptics correlated best, not with effects on discharging foci, but with their ability to prevent the spread of epileptiform activity. Therefore, at least part of their antiepileptic activity must result from effects on normal nerve cells and the nature of such effects will be discussed.

The susceptibility of a network of nerve cells to epileptiform discharge could be diminished either by decreasing the excitation or by increasing the inhibition of the neurons within the network. The decrease in excitation could be achieved in at least five different ways:

(i) a decrease in the excitatory synaptic transmission within the nerve network;
(ii) an increase in the threshold depolarization required for spike discharge;
(iii) a decrease in the post-tetanic potentiation (PTP) of excitatory synaptic transmission;
(iv) an increase in postactivation depression (PAD) of excitatory synaptic transmission;
(v) a decrease in the sensitivity of neurons to excitatory substances released during paroxysmal activity (e.g. glutamate).

To illustrate my thesis I propose to examine which of these various mechanisms contribute to the antiepileptic activity of the barbiturates. For this analysis, it is clearly desirable to have an *in vitro* preparation of mammalian brain tissue which can be maintained under well defined circumstances. There are now several such preparations available and the one used for the work to be described was the isolated olfactory cortex of the guinea-pig originally described by Yamamoto & McIlwain (1966). These preparations are made by taking a thin tangential slice from the surface of the olfactory cortex. They include the superficial 300–500 μm of the olfactory tubercle, prepiriform cortex and anterior piriform cortex together with one clearly defined afferent pathway—the lateral olfactory tract. Evoked potentials arise in response to a synchronous volley to the lateral olfactory tract and consist of the compound action potential of the olfactory tract fibres, a negative field potential generated by the synchronous depolarization of the superficial cortical synapses (population excitatory postsynaptic potential, EPSP) and one or more positive peaks superimposed on the population EPSP. These positive peaks reflect the synchronous discharge of the cortical cells in response to the afferent volley and so are called population spikes.

Evidence for this interpretation of the evoked potentials is given in the papers by Richards & Sercombe (1968, 1970) and Richards & ter Keurs (1971). The relationship between the anatomical connexions, the field potentials and the events occurring inside a neuron in the prepiriform cortex is illustrated in diagrammatic form in Fig. 1.

It is well established that the barbiturates can depress excitatory synaptic transmission in ganglia (Larrabee & Posternak, 1952), spinal cord (Somjen, 1963; Somjen & Gill, 1963; Weakly, 1969), the olfactory cortex (Richards, 1971, 1972) and the hippocampus (Bliss & Richards, 1971; Richards, 1974). This depression appears to result primarily from interference with chemical transmission. However, the doses of barbiturate required to cause such depression are high, and lie within the anaesthetic range (see Richards 1972*b* for a detailed discussion of this point). Furthermore, a generalized decrease in excitatory

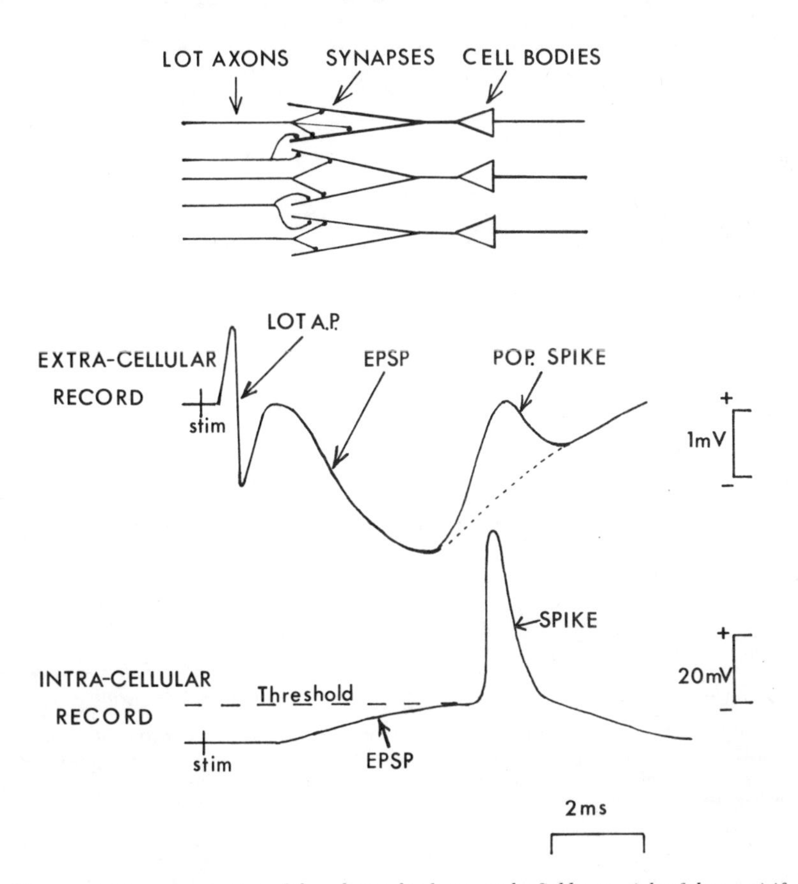

Fig. 1. *Diagrammatic representation of the relationship between the field potentials of the prepiriform cortex evoked by stimulation of the lateral olfactory tract (LOT), the anatomical organization of the cortex and the changes of potential occurring inside a 'typical' neuron in the prepiriform cortex after LOT stimulation*

The calibrations of amplitude and time are given as a guide to the relative magnitude of intracellular and extracellular potential changes. AP, action potential; EPSP, excitatory postsynaptic potential; POP. spike, population spike.

synaptic transmission by the barbiturates could only be of clinical value in extreme situations such as *status epilepticus,* and it is noteworthy that intravenous injections of thiopentone are used clinically for status epilepticus. Such generalized effects can have little to do with the antiepileptic activity of phenobarbitone, which is generally unaccompanied by sedation.

It is frequently stated that antiepileptics increase the cortical threshold for convulsions induced by electroshock. Such a statement implies that antiepileptics may increase the threshold depolarization required for spike generation in cortical neurons. This mechanism would indeed make the epileptic discharge

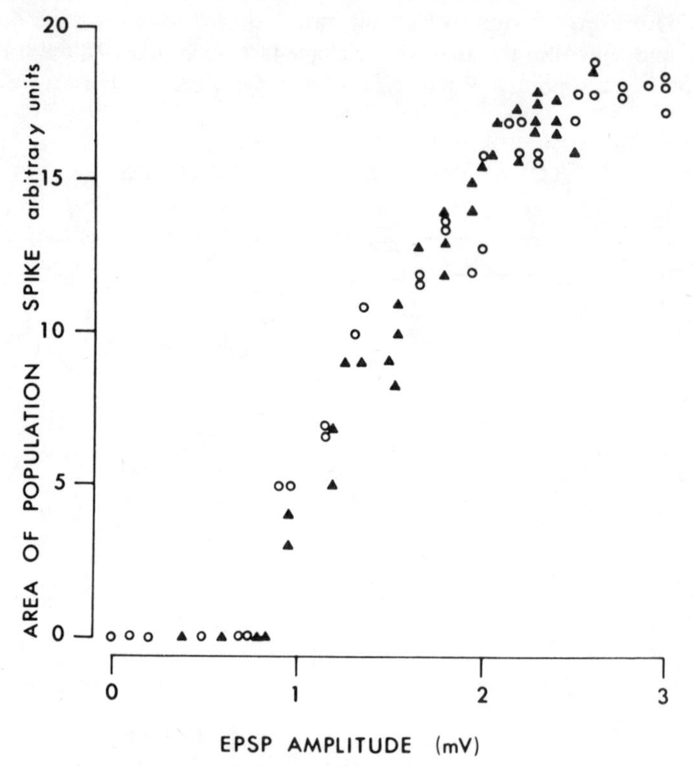

Fig. 2. *The absence of any effect of pentobarbitone on the relationship between the area of the population spike and EPSP amplitude. Reproduced from J. Physiol, with permission*

This relationship is determined by varying the intensity of the lateral olfactory tract volley. If the threshold of the cortical cells had been increased by pentobarbitone then the relationship should have been shifted along the abscissa to the right. ○, control; △, 0.2 mM pentobarbitone. (See also Richards, 1972*b*.)

of normal cortical cells less likely, but would also tend to cause generalized depression of the nervous system. Examination of the relationship between the population EPSP and population spike shows no shift towards a higher threshold even when anaesthetic levels of pentobarbitone are administered (Fig. 2). Similarly, pentobarbitone did not alter the electrical threshold for the fibres of the olfactory tract (see Richards 1972*b*). These results suggest that the electrical threshold of individual cortical neurons is not increased by the barbiturates and that this mechanism cannot account for their antiepileptic properties.

One of the characteristics of neurons in epileptic foci is continual high frequency discharges (Ward, 1969). Furthermore, Pinsky & Burns (1961) have shown that trains of stimuli at 50 Hz are the most efficacious stimuli for eliciting epileptiform after discharge in neurologically isolated slabs of the cat's cerebral cortex; Calvin (1975) and Porter & Muir (1971) have also found that burst firing of cortical cells provides potent stimuli for the discharge of neurons. It is known that repetitive stimulation of the lateral olfactory tract at high frequencies for very short periods of time is associated with the depression of the population EPSP evoked a few seconds (0.5–10 s) after the initial conditioning train. Conversely, prolonged repetitive stimulation (10–30 s) at high frequencies (20–100 Hz) is associated with the potentiation of the population EPSP evoked 20–200 s after the conditioning train (Richards, 1972*a*). The first of these phenomena is called postactivation depression (PAD) and the second post-tetanic potentiation (PTP). Similar depression and potentiation of postsynaptic potentials has been found at other central synapses, motoneurons receiving group 1a afferents from muscle for example (Curtis & Eccles, 1960).

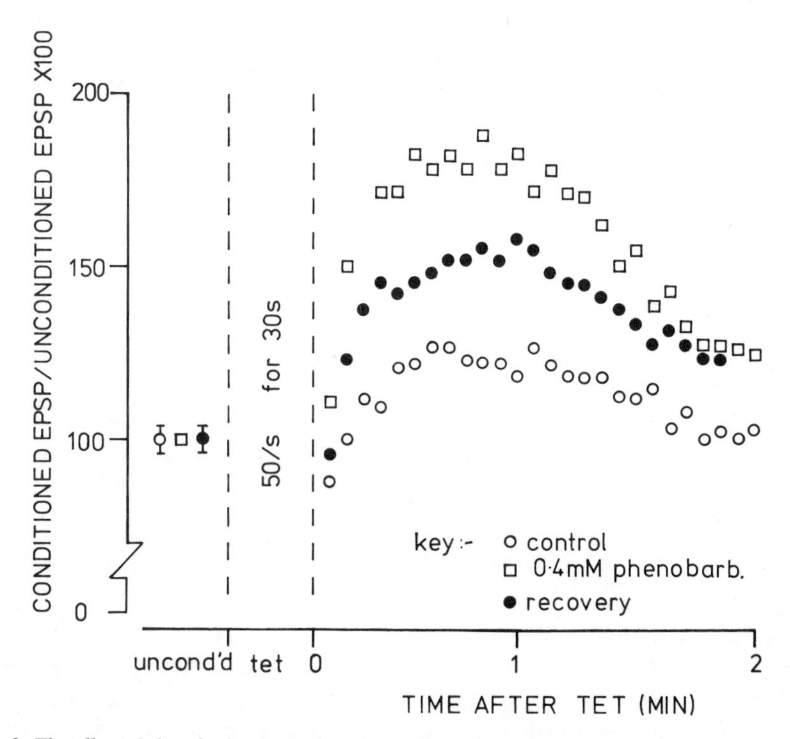

Fig. 3. *The effect of phenobarbitone on the post-tetanic potentiation (PTP) of the population EPSPs evoked at* 0.2 *Hz after a tetanic train at* 50 *Hz for* 30 *s*

All values for the amplitude of the EPSPs are expressed as a percentage of the mean of the values found for the 30 s preceding the tetanic train. The maximum variation in the amplitude of the control values is given by the vertical bars above and below the symbols on the left. There was little variation of the EPSP in 0.4 mM phenobarbitone.

If antiepileptics increased PAD or decreased PTP without effect on normal synaptic transmission, then they could limit the spread of a seizure discharge without depression of normal cerebral function. The effects of diphenylhydantoin and trimethadione on the post-tetanic potentiation of monosynaptic reflex discharge in the spinal cord have been examined by Esplin (1957) and Esplin & Curto (1957). Diphenylhydantoin decreased PTP of the monosynaptic reflex without affecting normal reflex activity but trimethadione was without effect. In the olfactory cortex both pentobarbitone and phenobarbitone (Fig. 3) fail to depress the PTP of the population EPSP even at concentrations sufficient to depress the population EPSP by half. Thus depression of PTP cannot, even in part, account for the antiepileptic activity of the barbiturates.

In contrast to this, pentobarbitone consistently enhanced the PAD seen after a train of four impulses (see Fig. 4). This result was obtained in five preparations

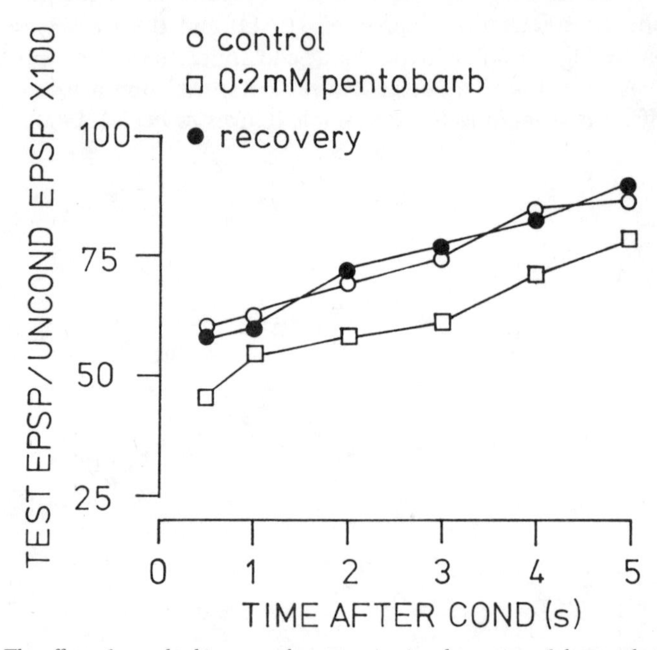

Fig. 4. *The effect of pentobarbitone on the postactivation depression of the population EPSP*

The magnitude of the test EPSP is expressed as a percentage of the first EPSP of the conditioning train. This percentage is then plotted against the interval between the conditioning train and the test EPSP. Each point is the mean of six separate determinations. The tests were carried out at 15 s intervals.

of the olfactory cortex, It is tempting to conclude that this could help to account for the antiepileptic action of the barbiturates but it should be remembered that these results were obtained with anaesthetic levels of pentobarbitone (0.2–0.3 mM). Nonetheless, further experiments of this kind with antiepileptic drugs are clearly needed, partly to see how far such a mechanism actually does contribute towards antiepileptic activity and partly to elucidate the underlying mechanism.

So far I have only considered the possibility that seizures are propagated through normal neural pathways. However, it is possible that some progressive forms of seizure—Jacksonian epilepsy for example—may be the result of chemical stimulation of neurons by released excitants such as glutamic acid. The march which is characteristic of such seizures would thus be envisaged as the result of a leakage of the excitant to adjacent areas of the cortex and so to activate groups of neurons not directly connected to the focus. This thesis is discussed in more detail by Bradford (this volume, Chapter 16). It follows from this analysis that antiepileptics effective against such disorders could act either to prevent leakage of neurons to the excitant or to decrease the sensitivity of neurons to the excitant. One substance likely to be involved in such a spread of excitation is glutamate, which is a very potent nerve cell excitant (Curtis *et al.* 1960). Bradford (this

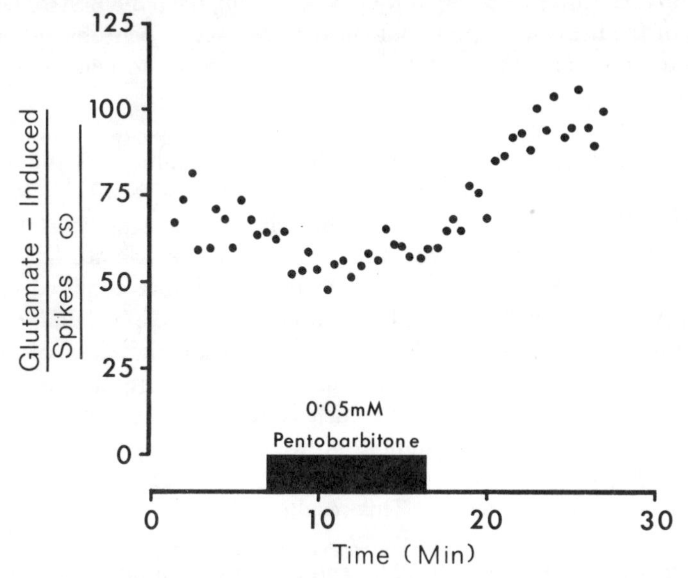

Fig. 5. *The action of* 0.05 mM *pentobarbitone on the sensitivity of a cell in the prepiriform cortex to iontophoretically applied* L-*glutamate*

Each point represents the number of spikes produced by a 8 nA pulse of glutamate for 10 s duration. The tests of glutamate sensitivity were made at 20 s intervals. (By courtesy of J. C. Smaje.)

volume, Chapter 16) discusses the evidence concerning the leakage of glutamate from active foci, here I present evidence to suggest that barbiturates decrease the sensitivity of cortical neurons to glutamate.

Crawford & Curtis (1966) and Crawford (1970) have shown that the sensitivity of cortical cells to iontophoretically applied glutamate is depressed by barbiturates but the interpretation of these results is complicated by the action of anaesthetic levels of barbiturates on the activity of the brain as a whole. The nerve cells in the *in vitro* preparations of the olfactory cortex have no spontaneous electrical activity; therefore the interpretation of similar experiments on the olfactory cortex is less likely to be complicated by the action of the barbiturates on sites remote from the cell under investigation.

The procedure used was as follows: separate barrels of multibarrelled electrodes were filled with sodium glutamate and sodium chloride. One sodium chloride barrel served as a recording electrode and another as a current control barrel. Glutamate was ejected continuously from the electrode during its advance through the tissue until the spike activity of an individual nerve cell could be recorded free from noise and generalized background spike activity. Thereafter, the glutamate was applied at regular intervals (30 s–1 min) for a fixed period (10–30 s) throughout the experiment.

In agreement with earlier results, pentobarbitone depressed the glutamate sensitivity of cortical neurons (Fig. 5 and Richards & Smaje, 1974) although the glutamate sensitivity of other cells was not always as readily depressed as that shown in Fig. 5. As evoked field potential studies have shown that pentobarbitone does not affect the threshold depolarization for impulse generation of neurons in the olfactory cortex (Fig. 2 and Richards, 1972*b*), the reduction in the responsiveness of the neurons to glutamate in the presence of pentobarbitone can be ascribed to a decrease in their sensitivity to glutamate, i.e. to a change in receptor sensitivity.

Now, an anticonvulsant dose of phenobarbitone is about 15 mg/kg i.v. for the rat (Barnes & Eltherington, 1966), which gives a value of 20–40 mg/litre in blood, or approximately 0.08–0.16 mM (Glasson & Benakis, 1973). Of this total, about 40–50% will be bound to serum protein (Svensmark, 1973), thus about 0.04–0.1 mM phenobarbitone would be in the extracellular fluid. This correlates well with the minimum concentration of pentobarbitone required to depress the sensitivity of cortical neurons to iontophoretically applied glutamate.

To round off this discussion it is necessary to consider the action of antiepileptic drugs on inhibitory pathways within the CNS. There are two inhibitory processes: post synaptic inhibition which is analogous to excitatory synaptic transmission; and presynaptic inhibition, which has no excitatory counterpart but is thought to result from the depolarization of afferent fibres caused by the activity of axo-axonal synapses. This results in failure of afferent impulses to invade the nerve terminals. Barbiturates prolong both postsynaptic inhibition (Larson & Major, 1970; Nicoll, 1972) and presynaptic inhibition (Eccles *et al.*, 1963; Miyahara *et al.*, 1966) but, as these effects have only been produced by anaesthetic doses (10–20 mg/kg pentobarbitone), the relevance of these observations to the antiepileptic activity of the barbiturates is a matter of conjecture.

To conclude, the part of the antiepileptic activity of the barbiturates that cannot be ascribed to action directly on the focus seems to be attributable to an increase in postactivation depression and a decrease in the sensitivity to endogenously released nerve cell excitants such as glutamic acid. In addition, increases in both postsynaptic and presynaptic inhibition may occur. It is likely that the total effect relies on the summation of a number of subtle effects which are insufficient to cause overall depression of the nervous system but which stabilize normal nerve networks against seizure discharge.

References

Barnes, C. D. & Eltherington, L. G. (1966) *Drug Dosage in Laboratory Animals*, University of California Press, Los Angeles

Bliss, T. V. P. & Richards, C. D. (1971) *J. Physiol. (London)* **214**, 7P

Calvin, W. H. (1975) *Brain Res.* **84**, 1
Crawford, J. M. (1970) *Neuropharmacology* **9**, 31
Crawford, J. M. & Curtis, D. R. (1966) *J. Physiol. (London)* **186**, 121
Curtis, D. R. & Eccles, J. C. (1960) *J. Physiol. (London)* **150**, 374
Curtis, D. R., Phillis, J. W. & Watkins, J. C. (1960) *J. Physiol. (London)* **150**, 656.
Eccles, J. C., Schmidt, R. & Willis, W. D. (1963) *J. Physiol. (London)* **168** 500.
Escueta, A. V., Davidson, D., Hartwig, G. & Reilly, E. (1974) *Brain Res.* **86**, 85.
Esplin, D. W. (1957) *J. Pharmacol. Exp. Ther.* **120**, 301
Esplin, D. W. & Curto, E. M. (1957) *J. Pharmacol. Exp. Ther.* **121**, 457
Glasson, B. & Benakis, A. (1973) in *Anticonvulsant Drugs* (Mercer, J., ed.), vol. 1, p. 244, Pergamon Press, Oxford
Larrabee, M. G. & Posternak, J. M. (1952) *J. Neurophysiol.* **15**, 91
Larson, M. D. & Major, M. A. (1970) *Brain Res.* **21**, 309
Miyahara, J. T., Esplin, D. W. & Zaslicka, B. (1966) *J. Pharm. Exp. Ther.* **154**, 119–127
Morrell, F., Bradley, W. & Ptashne, M. (1959) *Neurology (Minneapolis)* **9**, 492
Musgrave, F. S. & Purpura, D. P. (1963) *Electroencephalogr. Clin. Neurophysiol.* **15**, 921
Nicoll, R. A. (1972) *J. Physiol. (London)* **223**, 803
Pinsky, C. & Burns, B. D. (1961) *J. Neurophysiol.* **25**, 359
Porter, R. & Muir, R. B. (1971) *Brain Res.* **34**, 127
Richards, C. D. (1971) *J. Physiol. (London)* **217**, 41P
Richards, C. D. (1972*a*) *J. Physiol. (London)* **222**, 209
Richards, C. D. (1972*b*) *J. Physiol. (London)*. **227**, 749
Richards, C. D. (1974) in *Molecular Mechanisms in General Anaesthesia* (M. J. Halsey, Millar, R. A. & Sutton, J. A., eds.), pp. 90–11, Churchill Livingstone, Edinburgh
Richards, C. D. & ter Keurs, W. J. (1971) *Brain Res.* **26**, 446
Richards, C. D. & Sercombe, R. (1968) *J. Physiol. (London)* **197**, 667
Richards, C. D. & Sercombe, R. (1970) *J. Physiol. (London)* **211**, 571
Richards, C. D. & Smaje, J. C. (1974) *J. Physiol. (London)*. **239**, 103P
Somjen, G. G. (1963) *J. Pharmacol. Exp. Ther.* **140**, 396
Somjen, G. G. & Gill, M. (1963) *J. Pharmacol. Exp. Ther.* **140**, 19
Svensmark, O. (1973) in *Anticonvulsant Drugs* (Mercier, J., ed.), vol. 1, p. 297, Pergamon Press, Oxford
Ward, A. A. (1969) in *Basic Mechanisms of the Epilepsies* (Jasper, H. H., Ward, A. A. & Pope, A.. eds.), pp. 263–278, Little Brown, Boston
Weakly, J. N. (1969) *J. Physiol. (London)* **204**, 63
Yamamoto, C. & McIlwain, H. (1966) *J. Neurochem.* **13**, 1333

Chapter 16

On Amino Acid Involvement in Basic Mechanisms of the Epilepsies

By H. F. BRADFORD

Department of Biochemistry,
Imperial College of Science and Technology, London SW7 2AZ, U.K.

Amino Acids and Neurotransmission

Currently a great deal of interest is centred on the possible involvement of certain amino acids in the mechanisms of epilepsy. This is partly the result of an awakening in neurophysiology to the paradox that a range of simple and ubiquitous amino acids which are involved in a wide range of metabolic pathways in the central nervous system (CNS) are likely also to function as major synaptic transmitters in the brain and spinal cord. At present a list of these compounds would include glutamate and aspartate as excitatory agents, and γ-aminobutyric acid (GABA), glycine and possibly taurine as inhibitory agents (for reviews see Krnjević, 1974; Curtis & Johnston, 1974). It is this wider involvement of amino acids in mainstream biochemistry rather than their involvement in synaptic transmission which is the factor determining their tissue concentration, and consequently it is not surprising to find that they are not specially concentrated on the presynaptic side of the synapse since relatively small amounts are required for transmission purposes. A potent, high-affinity, uptake system operates to keep amino acid levels in the extracellular space at a minimum, and this seems to be present in most cell types. In the cerebral cortex as in many other brain regions most neurons will respond with firing to glutamate iontophoresed from micropipettes, and such iontophoresis is commonly employed by electrophysiologists to detect the presence of active cells. Concentrations as low as 10–100 μM glutamate are probably sufficient to initiate action potential discharge, and the glutamate ion exerts its effect through interaction with glutamate receptors in glutamate-operated synapses. Similarly GABA is a very potent inhibitory agent. Tissue levels of glutamate and GABA may be 3–10 mM, and therefore a very large concentration gradient exists across the cell membrane. Since these amino acids are being continually released during activity at the synapse, and possibly across other membranes, both during activity and at rest (Weinreich & Hammerschlag, 1975), great importance must be attached to the maintenance of low levels of these compounds in the extracellular space, particularly in the region of the synapse. Glutamine on the other hand, which is readily formed from glutamate and ammonia, is physiologically inactive, and, in addition, readily moves between cells and their extracellular space being the most highly concentrated amino acid in cerebrospinal fluid (300–500 μM). It may in fact be providing a means of recycling glutamic acid which is released from the presynapse, or other sites, and subsequently absorbed by perikarya and glial cells. The high level of glutamine commonly found in CSF is usually attributed to glutamine synthetase activity directed towards removing ammonia, but the removal of the potent excitatory glutamate ion must be regarded as an equally important activity for this enzyme.

The Amino Acid Content of Epileptic Foci

Since the early observation by Tower and his co-workers that glutamate and GABA are diminished in concentration in slices of excised human temporal lobe foci incubated in nutrient salines, the picture has enlarged to include specific changes in the levels of several other amino acids in epileptogenic cortex both in humans and in a wide range of experimental epilepsy (Fig. 1; Tables 1 and 4).

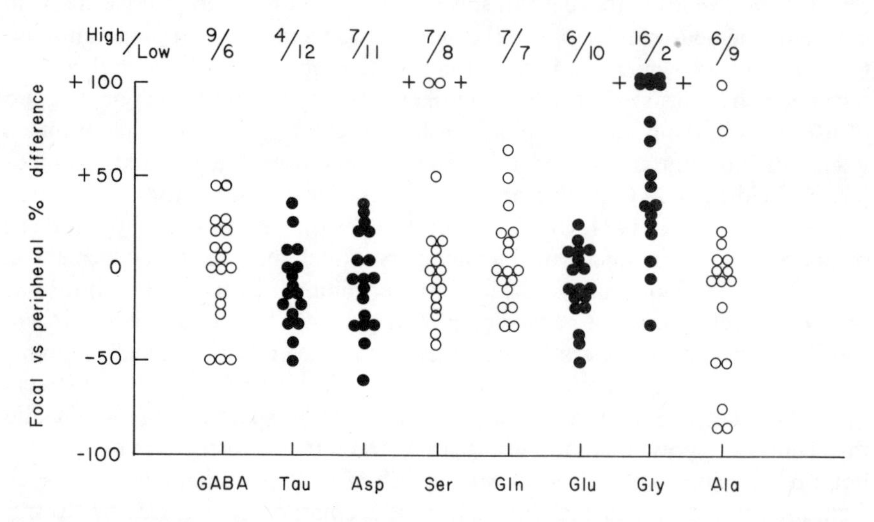

Fig. 1. *A comparison between focal epileptogenic nervous tissue and surrounding cortex with respect to amino acid content*

Data refer to the percentage by which the concentration of an amino acid in the focus differs from its matching peripheral sample (focal-peripheral)/(peripheral) $\times$ 1000). The results are therefore independent from possible erroneous estimation of concentrations in normal human cortex. Solid circles (●) call attention to amino acids having a consistently differing concentration in the focal area as compared to surrounding regions (over 50%). Ratios indicate the number of focal samples in which the level of a particular amino acid was either above or below the concentration found in the matching peripheral sample (from Van Gelder *et al.*, 1972).

These include losses of glutamate, glutamine, aspartate and taurine and increases in glycine and sometimes in serine in tissue samples isolated from the spiking focus and compared with normal tissues from the same preparation. Not all these changes have been consistently observed by all investigators but where they have been detected, most have been in the direction indicated (Perry *et al.*, 1972).

Table 1. *Amino acid changes in various experimental primary cortical epileptogenic foci*

Focus	Animal	GABA	Glutamate	Glutamine	Aspartate	Taurine	Serine	Glycine	Alanine	Reference
			(% change from control value)							
Cobalt	Mouse	18↓	12↓	a	a	15↓	60↑	32↑	69↑	Van Gelder (1972)
Cobalt	Cat	50↓	28↓	54↑	20↓	a	57↑	49↑	—	Van Gelder (1972)
Cobalt	Cat	75↓	50↓	a	32↓	60↑	43↑	53↑	29↑	Koyama (1972)
Cobalt	Rat	50↓	50↓	a	60↓	60↓	—	15↑	—	Emson & Joseph (1974)
Freeze lesion	Cat	a	50↓	50↓	—	—	—	—	—	Berl *et al.* (1959)
Penicillin	Cat	33↓	50↓	—	—	—	—	—	—	Gottsfeld & Elazar (1972)

Values are those from maximum convulsive stage: a = value unchanged or less than 10% changed from control.

Table 2. *Some common systemic chemical convulsants and their likely mode of action*

Convulsant	Chemical structure	Proposed mode of action	References
3-Mercaptopropionic acid	SH—CH_2—CH_2—COOH	Inhibition of glutamate decarboxylase and enhanced GABA amino transferase	Lamar (1970); de Lorez Arnaiz *et al.* (1971, 1973)
Allylglycine	CH_2 = CH . CH_2 . CH . NH . COOH	Inhibits glutamate decarboxylase	Alberici *et al.* (1969); de Lorez Arnaiz *et al.* (1971)
Pentylenetetrazole (metrazole)	CH_2 CH_2 / N—N | | ‖ CH_2 N—N CH_2 CH_2		Stone (1957); Hawkins & Sarett (1957)
Methionine sulphoximine	CH_3 . S . CH_2 . CH_2 . CH . NH_2—COOH ‖ NH	Inhibits glutamine synthetase and protein synthesis	Hrebicek *et al.* (1971); Sellinger & Weiler (1963)
Thiosemicarbazide	NH_2 . CS . NH . NH_2	Pyridoxal phosphate antagonists glutamate decarboxylase inhibition	Wood & Peesker (1974); Killam & Bain (1957); Roa *et al.* (1964); Collins (1973)
Semicarbazide, and other pyridoxal antagonists	NH_2 . CO . NH . NH_2		
Strychnine	Alkaloid	Blocks postsynaptic inhibition at glycine receptors?	Curtis *et al.* (1968*a*, *b*); Haber & Saidel (1948)
Picrotoxin	Alkaloid	Blocks presynaptic inhibition or GABA receptors?	Curtis *et al.* (1968*a*, *b*)
Bicuculline	Alkaloid		

Table 3. *Changes in whole-brain amino acid levels during experimental seizures induced by systemic convulsants*

Convulsant	Animal	Dose (mg/kg)	GABA	Glutamate	Glutamine	Aspartate	Taurine	References
Metrazole	Rat	100	—	13↓	13↑	—	—	(1)
Metrazole	Rat	100	17↑	—	—	—	—	(2)
Metrazole	Rat	100	a	a	—	a	a	(9)
Metrazole	Rat	70	15↑	25↓	—	15↓	—	(4)
Thiosemicarbazide	Dog	20	34↓	a	—	—	—	(5)
Thiosemicarbazide	Rat	7·5	20↓	a	—	—	—	(10)
Semicarbazide	Rat	200	32↓	a	—	a	a	(9)
Semicarbazide	Rat	500	38↓	—	—	—	—	(1)
β-Mercaptopropionic acid	Rat	150	31↓	a	a	a	—	(3)
Allylglycine								
Strychnine	Rat	4	15↑	—	—	—	—	(1)
Strychnine	Rabbit	0.1	21↑	18↑	13↓	—	—	(6)
Strychnine			—	30↓	—	—	—	(8)
Picrotoxin	Rat	10	a	—	—	—	—	(2)
Picrotoxin	Rabbit	3	a	a	13↓	—	—	(6)
Picrotoxin	Mouse*	2–3	16↑	a	27↑	—	—	(6)
Hyperbaric oxygen	Rat	6 atm	35↓	a	—	a	—	(7)
Hyperbaric oxygen	Rat	5 atm	38↓	—	—	—	—	(2)

* Intracerebral injection.

Values are those from maximum convulsive stage; ↑, level raised; ↓, level lowered; — not measured. a = value unchanged or less than 10% changed from control. References: (1) De Ropp & Snedeker(1961); (2) DeFeudis & Elliott (1968); (3) de Lorez Arnaiz *et al.* (1972, 1973); (4) Nahorsky *et al.* (1970); (5) Roa *et al.* (1964); (6) Saito & Tokunaga (1967); (7) Wood & Watson (1963); (8) Haber & Saidel (1948); (9) Killam & Bain (1957); (10) Collins (1973).

Amino Acid Changes in Whole Brain

Many systemic convulsant agents are able to change whole-brain levels of certain of these amino acids in the same direction as occurs in foci (Tables 2 and 3). A lowering of GABA and sometimes of glutamate is most commonly observed. It is not surprising that agents such as allylglycine, 3-mercaptopropionic acid, semicarbazides, hydrazides and other antipyridocal agents which inhibit glutamic acid decarboxylase (GAD) cause lowered GABA levels (Table 3) but this is sometimes accompanied by lowered glutamate levels (Alberici *et al.*, 1969; de Lorez Arnaiz *et al.*, 1971, 1972, 1973). A similar linkage

Table 4. *Glutamate metabolism in incubated slices of human cerebral cortex*

Conditions	Incubation time (min)	(μM/g) Glutamic acid level	Glutamine level	γ-Aminobutyric acid level
Normal	0	8.1	2.7	4.75
	60	10.0	4.55	6.1
Epileptogenic	0	7.05	3.0	4.35
	60	5.15	5.1	3.25
+ L-Asparagine or γ-aminobutyrate	60	9.5	4.75	—
Patient C.G., ♂, 29				
Normal	0	7.35	2.2	
	60	10.35	3.75	
Epileptogenic	0	7.35	2.85	
	60	6.0	4.2	

From Tower (1960).

between GABA and glutamate depletion is seen in whole-brain following seizures induced by methionine sulphoximine, a compound which prevents glutamine formation and interferes with methionine metabolism, including protein synthesis (Hrebicek *et al.*, 1971). Tower (1960) has argued that since transamination of GABA leads to glutamate production via α-oxoglutarate, diminished GABA levels and lowered flux through this 'GABA shunt' pathway would necessarily lead to a fall in glutamate concentrations (Fig. 2). However, a loss of glutamate under conditions where GABA was substantially reduced was not reported by Roa *et al.* (1964) and may not be a consistent correlation. Other convulsant agents administered systemically which have been reported to change brain GABA, glutamate, glutamine and aspartate levels are metrazole, picrotoxin, strychnine and oxygen at high pressure (Table 3), although in many cases the changes are detectable only after onset of seizure and cannot therefore be initiating factors. Many of these amino acids are important intermediates in mainstream carbohydrate and energy metabolism in addition to their role in protein synthesis and other biosynthetic pathways. For this reason it is important when attempting to connect changed amino acid levels with seizure initiation or propagation that aberrations in these processes are given equal consideration alongside any direct influence amino acids may have on excitation and inhibition.

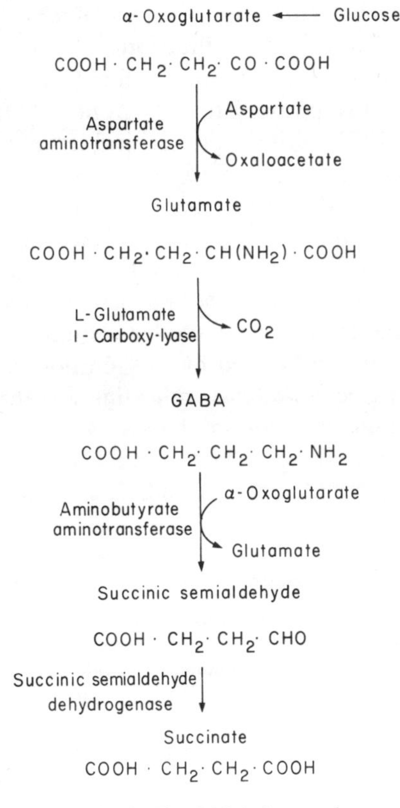

Fig. 2. *The GABA shunt pathway*

Amino Acid Changes: Metabolic Implications

Abnormal pool sizes of glutamate, aspartate, GABA and glutamine usually indicate changes patterns of carbohydrate metabolism. In the epileptic brain this appears to be occurring both locally in the spiking focus and in a more general way throughout the brain during and after seizures. In the spiking focus these changes might bear a causal relationship with the hyperactivity, for instance, by diminishing the availability of GABA for inhibition, or of ATP for ion transport. Deficiencies could arise through a decreased incorporation of glucose carbon into amino acids, and available evidence indicates that although glucose uptake and incorporation into alanine and lactate is stimulated during seizures, incorporation into other amino acids is reduced (Yoshino & Elliott, 1970; Dunn & Guiddita, 1971). Pool sizes, however, are either unchanged, rise, or show a slight fall (Sactor *et al.*, 1966; Nahorsky *et al.*, 1970; Leonard & Palfryman, 1972; King *et al.*, 1974). Of course these are seizures induced by electroshock or drugs in otherwise normal animals, and measurement of the response in chronically epileptic animals would give a more relevant answer. Restricted availability of glucose and other substrates, or of oxygen in epileptogenic tissue, could lead to the observed fall in pool size of glutamate, aspartate and GABA since they would continue to supply Krebs cycle intermediates without being replaced (Tower, 1960; Bradford, 1968). However, slices of human temporal lobe, and slices of the spiking foci from cobalt epileptic rats show no obvious change in respiratory rates from normal (Pappius & Elliott, 1954; Colastani & Craig, 1973), although the critical test of performance would be under conditions accelerating activity to rates closer to those found *in vivo* and this was not examined. Moreoever, since any hyperactivity in the focus would not persist *in vitro*, an abnormality may not be detected. However, Tower and his co-workers have shown that tissue levels of glutamate and GABA in excised temporal lobe foci fall during an hour's incubation while normal tissue shows these substances increasing under similar conditions. Significantly, the fall in glutamate is apparently prevented by the addition of 10 mM GABA or asparagine to the nutrient glucose–salines, indicating that the pathways linking GABA, asparagine and glutamate via the Krebs cycle and GABA shunt are sufficiently active to replenish the glutamate losses (Table 4 and Fig. 2). Intravenous injections of GABA (125 mg/kg) and, less effectively, of glutamate

Table 5. *Changes in amino acid content of freeze-focus following systemic GABA and glutamate administration*

	(% of control level)			
	Glutamate	Glutamine	GABA	No. of experiments
Untreated	54	51	95	6
After glutamate injection (100 mg/kg)	80	81	121	5
After GABA injection (125 mg/kg)	73	70	244	6

Values are means for the number of experiments quoted. Cats were the animals employed. Injections were via an indwelling femoral vein cannula. From Berl *et al.* (1961).

(100 mg/kg), into cats with active freeze-foci which showed lowered pool sizes of glutamate and GABA, suppressed the epileptoform activity but also raised the amino acid levels at the focus to within the normal range. There was no correlation in time between the two events. Suppression of abnormal electrical activity occurred 15 s after GABA administration but recovery of amino acid levels in the focus was not detectable until a further 15 min had elapsed. Following serial infusion, levels of GABA in the focus were three–fourfold higher than control. The authors interpret the changes in glutamate levels following GABA administration in terms of increased GABA shunt activity, and the effects on electrical activity of the focus in terms of a direct inhibitory action of GABA rather than via any readjustment in metabolism it might cause. There is a local breakdown of the blood–brain barrier in the freeze-focus and GABA would therefore gain entry to the brain (Table 5, Fig. 3; Berl *et al.*, 1959; Berl *et al.*, 1961). Since GABA, in the presence of glutamate, is an extremely good substrate for ATP synthesis by brain mitochondria but is ineffective for liver and probably other tissue except kidney, the effects of GABA in refurbishing amino acid levels and suppressing seizures could well be through a special metabolic role this substance might perform (Scriver & Whelan, 1969; Lee *et al.*, 1974).

In exploring the possible causal link between diminished amino acid levels and seizure initiation, some authors have emphasized the inhibitory effect which such depletion might have on the synthesis of proteins such as the S100 protein which could be important in controlling neuronal activity. It is suggested

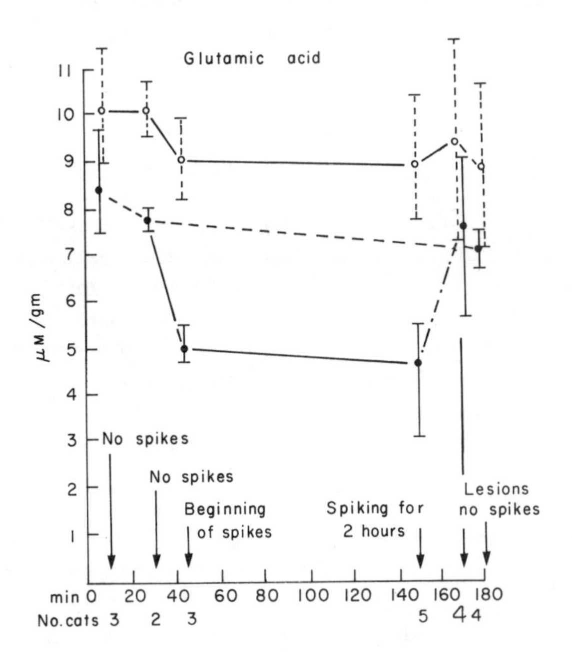

Fig. 3. *Changes in glutamate levels in freeze-focus of cats*

Values are mean ±S.D. for number of cats indicated. Note that onset of spiking correlates with the fall in tissue content of the amino acids (from Berl *et al.*, 1960).

that the changes in glycine and taurine imply alterations in the patterns of protein synthesis at the focus (Van Gelder *et al.*, 1972; Van Gelder & Courtois, 1972). Electrical stimulation both *in vitro* (Orrego & Lipman, 1967; Jones & McIlwain, 1971) and *in vivo* (Dunn *et al.*, 1972; Essman, 1973) causes lowered protein synthesis rates in whole brain or brain slices though this is not associated in these experiments with lowered pool size of key amino acids and is probably irrelevant to present arguments. Indeed, at the synapse itself, protein synthesis rates are substantially increased by electrical stimulation (Wedege & Bradford, 1975).

Lowered glutamate level could on its own lead to lowered GABA and aspartate levels because of the interconversion of these amino acids which occurs. Lowered GABA levels would probably reduce the flux through the GABA shunt pathway, but since this is anyway likely to be making a small (10%) contribution to the overall economy of carbohydrate and energy metabolism this is not likely to be a major influence on rates of oxidative metabolism (Machiyama *et al.*, 1965). Agents such as allyglycine, 3-mercaptopropionic acid and antipyridoxal agents with known actions on glutamate decarboxylase activity (GAD) are a special case. They show that the lowering brain GABA levels correlates closely with the precipitation of convulsions and convey special significance for epilepsy occurring during vitamin B_6 deficiency. However, unless lowered GABA levels or impairment of the glutamate decarboxylase enzyme system, including pyridoxal deficiency, can be established as a feature of tissues, including foci, from patients with epilepsy, no compelling connexion exists to link GABA changes with the disease. It must remain one of several candidates in this respect.

Glutamate, GABA: Excitation and Inhibition

It now seems fairly certain that synapses carrying postsynaptic receptors for GABA or for glutamate, and which mediate inhibition and excitation respectively, occur widely in the CNS together with a more restricted distribution of equivalent sites for glycine and aspartate. The synapses which are operated by these amino acids are likely to constitute a large proportion of the total in the CNS, which means that the high concentrations and ubiquitous presence of these compounds, particularly glutamate, aspartate and GABA leads to a high risk of anomalous contact between the amino acids and their synapses.

GABA and its enzymes

The three enzymes glutamate decarboxylase (GAD), GABA-transminase (GABA-T) and succinic semialdehyde dehydrogenase (SSDH) are responsible for the production and removal of GABA (Fig. 2). Agents which reduce GAD activity often markedly lower brain GABA concentrations (e.g. 40%), and in parallel cause seizures, e.g. hydrazides, allyglycine, 3-mercaptopropionic acid (Killam & Bain, 1957; de Lorenz Arnaiz *et al.*, 1971, 1972, 1973) (Table 3). Although there is no clear correlation between brain GABA content and seizure onset, one research group have developed an equation relating excitable state to both the level of GABA and the degree of GAD inhibition, where hydrazine and its derivatives are the convulsant agents (Wood & Peesker, 1974). Because of the likely involvement of GABA in both metabolism and inhibitory neurotransmission, changes in compartmentation of GABA must be judged as more critical than overall changes in tissue content; changed location could occur without detectable change in tissue content. Inhibition of GABA-T or SSDH activity leads to a rise in brain GABA level, and simultaneously protects against seizures, examples being α-amino oxyacetic (AAOA; Collins, 1973; Snodgrass & Iversen, 1973; Perry *et al.*, 1974) and di-*n*-propylacetate (Epilim; Simler *et al.*, 1968, 1973; Harvey, *et al.*, 1975). Whilst AOAA is unfortunately toxic when administered to humans, and has convulsant action above a certain dosage (Tapia *et al.*, 1969; Perry *et al.* 1974), Epilim, whose passage across the blood–brain barrier is facilitated by its fatty nature, is proving a most useful anticonvulsant in the clinic (Jeavons & Clark, 1974). There is evidence that some commonly employed barbiturate anticonvulsants raise brain GABA levels (Saad *et al.*, 1972), which, if substantiated, should be taken into account in assessing mechanism of actions of these substances. Conflicting with this generalization is the report that AOAA at doses which more than doubled brain GABA gave no protection against audiogenic seizures during barbiturate withdrawal (Crossland & Turnbull, 1972).

Restricted availability of GABA could reduce inhibitory transmission, reduce the rate of flow through the 'GABA shunt', or both. The anticonvulsant Epilim blocks the shunt pathway and elevates brain GABA levels, and the implication appears to be that raising brain intracellular GABA above normal reduced the incidence of fits perhaps by increasing GABA output at its synapse. This anticonvulsant could equally well produce its effects by blocking GABA uptake and increasing extracellular GABA levels in regions containing GABA operated inhibitory synapses. Since there is evidence that AOAA and Epilim are both inhibitors of GABA uptake, this feature of their action might be additional to any effect due to raised cellular levels of GABA (Snodgrass & Iversen, 1973; Johnstone & Balcar, 1974; Harvey *et al.*, 1975). The potent inhibition of neuronal discharge produced by extracellular GABA can be seen following its iontophoresis locally in cerebral cortex and elsewhere *in vivo* (Krnjević & Phillis, 1963) or when it is present in salines bathing slices of cerebrellar cortex *in vitro* (Okamoto & Quastel, 1973). There has been a report of the successful clinical use of GABA as an anticonvulsant against petit mal seizure in particular (Tower, 1960), but this has not been pursued further probably because GABA does not appear to penetrate normal brain very well. However, where the blood–brain barrier is ineffective, local entry of GABA could be expected, e.g. at foci (see also Hayashi, 1966).

Glutamate and its enzymes

The considerations brought to bear on GABA apply equally to glutamate, although its more direct relationship with the Krebs cycle makes distinction between its metabolic and transmission involvement much more difficult. In contradistinction to GABA, glutamate is a potent excitatory agent and readily causes firing of most neurons when iontophoresed into cerebral cortex and other CNS regions (for review see Johnston, 1974). It increases the firing rate of neurons present in cerebrellar slices *in vitro* (Okamoto & Quastel, 1973) and can be shown to depolarize and increase the sodium permeability of cells in cerebral cortex slices (Gibson & McIlwain, 1965; Bradford & McIlwain, 1966). Glutamate (together with smaller amounts of aspartate, lysine and taurine) is released from the cortical surface, following stimulation of the reticular formation, suggesting that it could be the transmitter which mediates reticulocortical activation (Jasper & Koyama, 1969).

The glutamate ion is, therefore, well qualified to be considered as an excitatory agent in the seizure-initiating mechanisms. Small amounts of the substance introduced intracerebrally readily produce convulsions (Hayashi, 1954; Weichert & Gollnitz, 1968; Hennecke & Weichert, 1970). Systemically administered glutamate produces seizures in 10-day-old rats (Curtis & Johnston, 1972) and also, at higher doses, in adult rats (Bhagaram *et al.*, 1971; Bradford & Dodd, 1975). Raised whole-brain levels of glutamate and of glutamate dehydrogenase have been reported in the preconvulsive state before onset of seizures initiated by a range of agents (Weichert & Gollnitz, 1968, 1969), and a release of glutamate has been observed to occur from cobalt-foci during the onset of seizures as observed either by EEG or as motor response (Figs. 4 and 5; Koyama, 1972; Dodd & Bradford, 1975). In our experiments this appears to an effect specific to glutamate and not due therefore to a generalized tissue degeneration. Thus amino acids such as glutamine which are more concentrated in the tissue than glutamate are not released, and the glutamate released to the

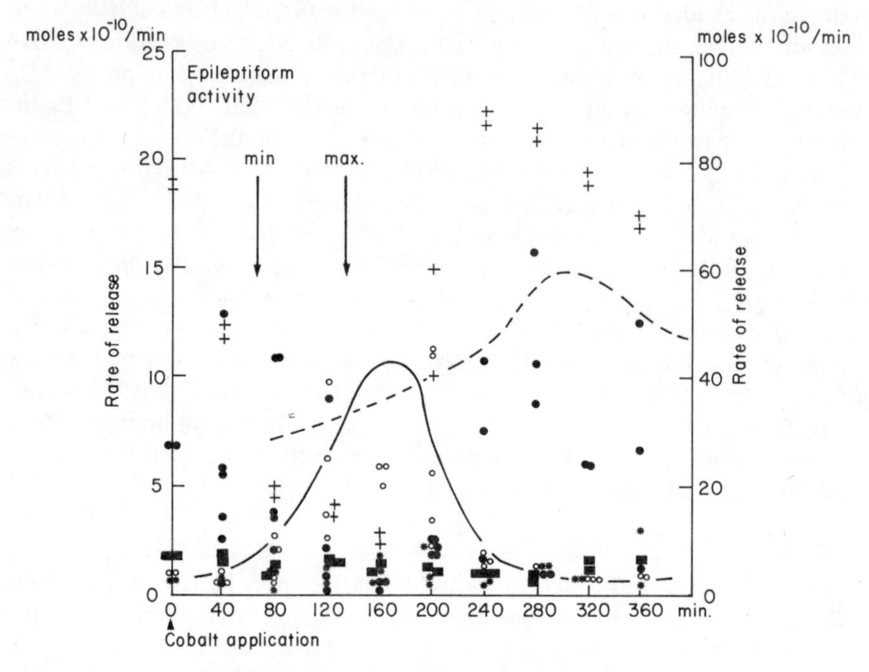

Fig. 4. *Glutamate release from cobalt-focus in vivo*

——, Glutamate pattern; . . . . . , glutamine pattern; ■ taurine; * aspartic acid; ● glutamine, ○ glutamic acid; and + urea in moles × 10^{-10} min. Ordinates indicate rate of release in moles × 10^{-10} min (left-hand side is used for values of taurine, aspartic acid, glutamine and glutamic acid; right-hand side is used for urea) and abscissa indicates the time from cobalt application in minutes (adapted from Koyama, 1972).

superfusion fluid is increased two–threefold both in absolute quantity and as a proportion of total amino acid release (Fig. 5). The complementary change was observed to occur in the freeze-focus, namely a reduction in glutamate content correlating with the time of onset of epileptiform EEG activity (Fig. 3; Berl *et al.*, 1959). In view of their high concentration and their soluble state, intermittent leakage of glutamate or other amino acids from the focus could account for part or all of their lowered tissue levels, and leakage from incubated human focal tissue to the glucose–saline might partly explain the inability of the human focal tissue to maintain or increase these levels *in vitro* (Table 4). The authors did, however, consider this possibility and reported amino acids to be absent from the incubation medium, so the 'leakage' theory would require that the relatively small amount of amino acid involved, having been released to the extracellular space, and in contact with receptors, was reabsorbed by the tissue into compartments where it is rapidly metabolized, GABA and asparagine (added to 10 mM), which readjusted tissue levels, could exert their effect through acceleration of the activities of the Krebs cycle and the shunt pathway, to rates of glutamate and GABA formation in excess of the rate of loss from the tissue. In the experiments of Berl *et al.* (1961), glutamate as well as GABA prevented much of the epileptiform activity of the freeze-focus in cats when administered systemically (100 mg/kg) (Table 5). However, the effect often required several injections, and involved a delay period not seen with GABA. Moreover, it was often preceded

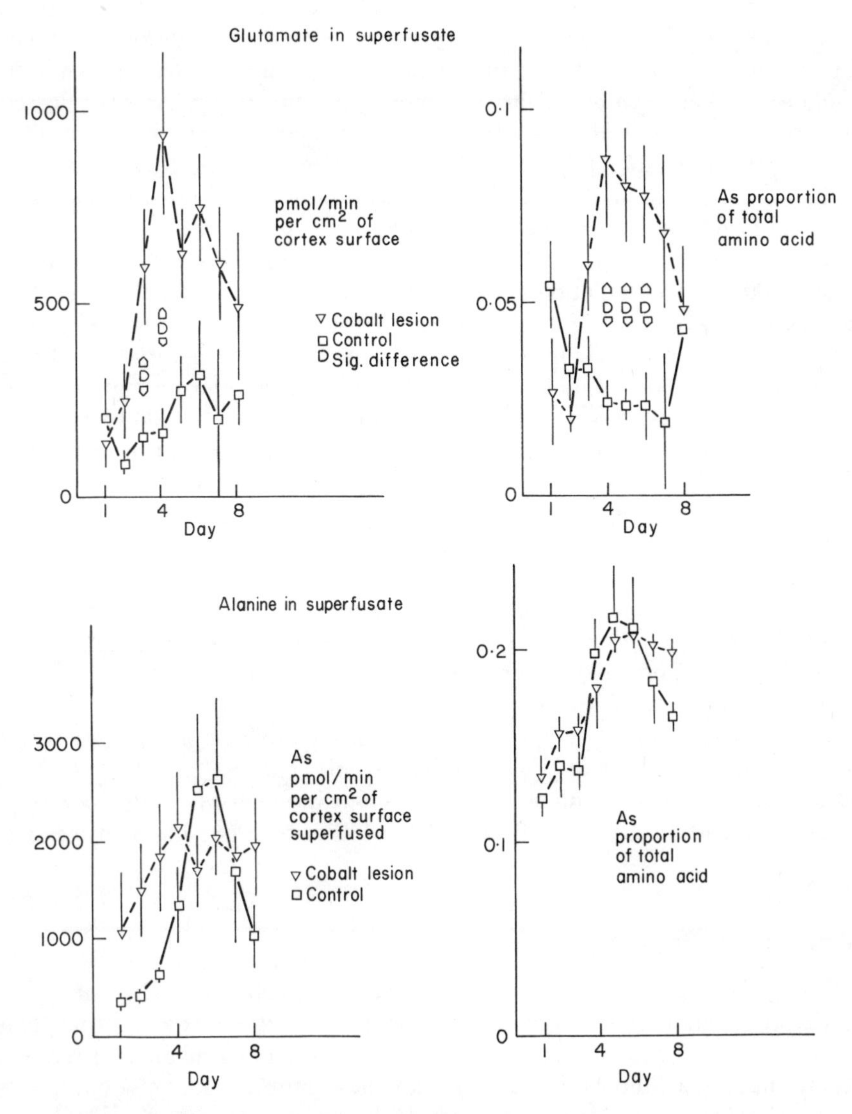

Fig. 5. *Patterns of release of glutamate and alanine* in vivo *from cobalt focus and cerebral cortex of rat*
The animals were awake and unrestrained and collection was via a superfusion system employing artificial CSF (flow rate 5 ml/h; see Dodd *et al.*, 1974; Dodd & Bradford, 1975).

by activation of the focus. Similarly, cobalt-foci in rats may be activated by sub-seizures-threshold doses of glutamate administered systemically (Bradford & Dodd, 1975). Arguing against a leakage of glutamate as a basic excitatory mechanism is the work of Brown & Stone (1973), who could not detect an increase of glutamate or aspartate in fluid samples withdrawn from the subarachnoid space of dogs experiencing convulsions induced by thiosemicarbazide, metrazole, picrotoxin or methionine sulphoximine. However, a rather wide range of control glutamate levels (measured before injections of convulsant) were reported for the group of animals studied, making firm conclusions difficult. In addition, superfusion systems may have to be employed to overcome the great efficiency of the glutamate reuptake system. The theory also requires that excitation by released glutamate would predominate over inhibition by any released GABA. This could be encompassed by the fact that more glutamate than GABA is released, and by the possibility that GABA functions primarily at picrotoxin-sensitive presynaptic (axo-axonic) sites modulating the release of excitatory transmitters including glutamate. In the latter arrangement the action of *extracellular* glutamate could not be damped by the presence of *extracellular* GABA in regions where glutamate synapses predominated (e.g. cerebral cortex). Abnormally high glutamate accumulation in the extracellular space might well be reflected as raised levels of this amino acid in the CSF of epileptic patients, and several reports which substantiate this expectation now exist in the literature (see Plum, 1974, for refs.). Caution must, however, be exercised in interpreting these data because of the ease with which glutamine present in CSF (in 100-fold greater abundance than other amino acids) is converted to glutamate during extraction and analysis (Blanshard *et al.,* 1975). Potassium elevation in CSF has also been reported to occur during convulsions in monkeys (Meyer *et al.,* 1970) and may be linked with the rise in CSF glutamate.

If glutamate release were to occur as readily from central nerve fibre tracts as is reported for peripheral nerves and spinal nerve roots (Wheeler *et al.,* 1966; DeFeudis, 1971; Weinreich & Hammerschlag, 1975) and its accumulation were to ensue to the appropriate extent, then some participation of glutamate in the slower mechanisms of seizure propagation could be envisaged. In addition, extracellular glutamate may mediate the considerable brain oedema which accompanies convulsion (e.g. see de Lorez Arnaiz *et al.,* 1972, 1973) since incubating neural tissue with this amino acid has long been known to promote tissue swelling (Terner *et al.,* 1950; Bradford & McIlwain, 1966; Van Harreveld & Fiftkova, 1971.

Glutamate and Spreading Depression

Strong evidence is accumulating which suggests that glutamate is also involved in the propagation of spreading depression in cerebral cortex and retina. This condition results from strong stimulation at one point and manifests itself as a general neuronal inhibition together with the disappearance of spontaneous electrical activity, and it spreads slowly (3 mm/min) in all directions from the initial point (Leao, 1944). It appears paradoxical that a convulsant such as glutamate should mediate a state of general inhibition, and seems inconsistent with a likely excitant role in epilepsy. However, the depression appears to result initially from gross depolarization of the neurons, and its propagation is associated with the release of both potassium and glutamate at the advancing front; this causes further depolarization and thereby triggers further release. The neurons recover their membrane potential in about 2 min but the depression of excitability continues for 10 or 15 min. There is evidence that GABA is released during the period of depolarization and may accumulate in the extracellular space causing the more prolonged period of inhibition (Van Harreveld & Fiftkova 1970, 1973; Phillis & Ochs, 1971; Van Harreveld, 1972; Kaczmarek & Adey, 1974). In this way both glutamate and GABA would appear to be central to the mechanisms of propagation and maintenance of spreading depression. The brief burst of action potentials which is often recorded in the advancing front would be expected if rapid and gross depolarization were lowering the membrane potential to the critical firing level before causing a brief period of cathodal block. Thus a role of glutamate in the spread of this essentially inhibitory condition is compatible with its established excitatory action, but suggests that spreading depression rather than convulsion should follow intracerebral injection of glutamate or its leakage in an epileptic focus. This anomaly remains to be fully answered but is probably related to the extracellular concentrations of glutamate present in the different situations, and to the secondary involvement of GABA or other inhibitory agents in spreading depression. The paroxysmal depolarization shifts (PDS) seen in many experimental foci do, after all, involve a massive depolarization, with a brief discharge of action potentials followed by cathodal block much as is seen during the invasion of spreading depression before the second and much longer period of inhibition commences. Also, it is notable that the condition is much easier to evoke in small mammals such as rat and rabbit with smooth cortices and a high neuronal population density, than in cats and primates which have the opposite kind of cortical structure, and spreading depression has yet to be demonstrated in man (Phillis & Ochs, 1971). Thus in primates and man it may not be possible for glutamate to set in train the events which lead to spreading depression.

Conclusions

It is the amino acids rather than acetylcholine or the monoamines which are currently emerging as the compounds of special interest in relation to epilepsy. Highlighted among these are glutamate, GABA and taurine, and speculation continues on their mode of involvement. Our own view has for some years focused on the excitatory properties of glutamic acid.

Its high tissue content appears to have stimulated the evolution of high-affinity transport processes which ensure a very low concentration of glutamate in the region of synapses operated by this amino acid. Changes which reduce the efficiency of this transport activity through inhibition or oversaturation will lead first to raised levels of extracellular glutamate and then to neuronal hyperactivity followed by cathodal block (cf spreading depression) depending on the levels reached. Golgi staining reveals the presence of dendritic deafferentiation in both human and experimental epilepsy and this could occur on the hyperactive neurons detected in these foci. Should such deafferentiation involve the widespread appearance of active glutamate receptors on the dendritic surfaces of epileptogenic cortical neurons, as occurs in denervated insect muscle (Cull-Candy & Usherwood, 1973), supersensitivity to glutamate would ensue and might explain the abnormally high synaptic activity which appears to generate the paroxysmal depolarization shifts commonly observed. This situation would allow even low levels of extracellular glutamate to become convulsive, and should folate accumulation in epileptic foci be a widespread phenomenon glutamate transport block could promote glutamate accumulation locally. The capacity of non-synaptic regions of nerves to release glutamate suggests that this ion may be held within neurons by the membrane potential and briefly effluxed from both non-synaptic and synaptic regions during depolarizing wave of the action potential. If this were so, the apparently random and episodic occurrence of many epileptic fits might be explained by the glutamate-releasing effect of certain activity patterns in the region of the epileptic focus Damaged neurons present in the active focus and displaying lowered membrane potential would more readily reach the threshold level at which glutamate was released. However, so simple a model does not adequately encompass many other factors, including the observed changes in taurine and aspartate, and one is forced to keep in the picture the central importance of these amino acids in mainstream metabolism and protein synthesis in the brain when formulating and searching for 'seizure-initiating' mechanisms in which they might perform a causal role.

Finally, it must be emphasized that drugs which interfere with a specific transmitter system (e.g. GABA) and cause convulsions, do not indicate a causal role for that system in human epilepsy, but simply that it is susceptible to assault by these chemical agents. Equally, anticonvulsant drugs which raise the levels of endogenous inhibitory substances (e.g. GABA raised by Epilim) do not indicate deficiency in the levels of that substance in human epilepsy, they might well be suppressing the convulsive symptoms rather than extinguishing the trigger mechanism which is at the root of the disease. What they do indicate is that a search is necessary to establish whether aberrations in these systems occur in human epileptogenic brain tissue.

References

Alberici, M. de Lorez Arnaiz, R. G. & de Robertis, E. (1969) *Biochem. Pharmacol.* **18**, 137–143

Berl, S., Purpura, D. P., Girado, M. & Waelsch, H. (1959) *J. Neurochem.* **4**, 311–317

Berl, S., Purpura, D. P., Gonzales-Monteagudo, O. & Waelsch, H. (1960) in *Inhibition in the CNS and GABA* (Roberts, E., ed.), pp. 445–453, Pergamon

Berl, S., Takagaki, G. & Purpura, D. P, (1961) *J. Neurochem.* **7**, 198–209

Blanshard, K. C., Bradford, H. F., Dodd, P. D. & Thomas, A. J. (1975) *Anal. Biochem.* **67**, 233–244

Bhagaran, H. N., Coursin, D. B. & Stewart, C. N. (1971) *Nature (London)* **232**, 275–176

Bradford, H. F. (1968) in *Applied Neurochemistry* (Davison, A. N. & Dobbing, J., eds.), pp. 222–250, Blackwell, Oxford

Bradford, H. F. & Dodd, P. R. (1975) In preparation

Bradford, H. F. & McIlwain, H. (1966) *J. Neurochem.* **13**, 1163–1177

Brown, D. J. & Stone, W. E. (1973) *Brain Res.* **54**, 143–148

Colasanti, B. K. & Craig, C. R. (1973) *Neuropharmacology* **12**, 221–231

Collins, G. G. S. (1973) *J. Neurochem.* **22**, 101–111

Crossland, J. & Turnbull, M. J. (1972) *Neuropharmacology* **11**, 733–738
Cull-Candy, S. G. & Usherwood, P. N. R. (1973) *Nature (London)* **246**, 62–64
Curtis, D. R. & Johnston, G. A. R. (1974) *Ergeb. Physiol.* **69**, 97–188
Curtis, D. R., Hosli, L., Johnston, G. A. R. (1968a) *Exp. Brain Res.* **6**, 1–18
Curtis, D. R., Hosli, L., Johnston, G. A. R. & Johnston, I. H. (1968*b*) *Exp. Brain Res.* **51**, 235–258
DeFeudis, V. (1971) *Exp. Neurol.* **30**, 291–296
DeFeudis, V. & Elliott, K. A. C. (1968) *Can. J. Physiol. Pharmacol.* **46**, 803–804
de Lorez Arnaiz, R. G., Alberici, de Canal, M. & de Robertis, E. (1971) *Int. J. Neurosci.* **2**, 137–144
de Lorez Arnaiz, R. G., Alberici de Canal, M. & de Robertis, E. (1972) *J. Neurochem.* **19**, 1379–1386
de Lorez Arnaiz, R. G., Alberici de Canal, M., Robiolo, B. & Mistrorgio de Pacheo, M. (1973) *J. Neurochem.* **21**, 615–624
De Ropp, R. S. & Snedeker, E. H. (1961) *Proc. Soc. Exp. Biol. Med.* **106**, 696–700
Dodd, P. R. & Bradford, H. F. (1974) *J. Neurochem.* **23**, 289–292
Dodd, P. R. & Bradford, H. F. (1975) *Brain Res.* in press.
Dodd, P. R., Pritchard, M. J., Adams, R. C. F., Bradford, H. F., Hicks, G. & Blanshard, K. C. (1974) *J. Physics E: Scient. Instrum.* **7**, 897–901
Dunn, A. & Guiddita, A. (1971) *Brain Res.* **27**, 418–421
Dunn, A., Guiddita, A. & Pagluicca, N. (1972) *J. Neurochem.* **18**, 2093–2099
Emson, P. C. & Joseph, M. H. (1974) *Biochem. Soc. Trans.* **2**, 287
Essman, W. B. (1973) *The Neurochemistry of Electroshock*, Spectrum, Flushing, N.Y.
Gibson, M. & McIlwain, H. (1965) *J. Physiol. (London)* **176**, 261–283
Gottsfeld, Z. & Elazar, Z. (1972) *Nature (London)* **240**, 478–479
Haber, C. & Saidel, L. (1948) *Fed. Proc. Fed. Am. Soc. Exp. Biol.* **7**, 47
Hawkins, J. E. Jr. & Sarett, L. H. (1957) *Clinica Chim. Acta* **2**, 481
Harvey, P. K. P., Bradford, H. F. & Davison, A. N. (1975) *FEBS Lett.* **52**, 251–154
Hayashi, T. (1954) *Keio J. Med.* **3**, 183–192
Hayashi, T. (1966) In *Enzymes in Mental Health* (Martin, G. J. & Kisch, B., eds.) pp. 160–170, J. B. Lippincott & Co. Philadelphia
Hennecke, A. & Weichert, G. (1970) *Epilepsia* **11**, 327–331
Hrebicek, J., Kolousek, J., Wederman, M. & Charamza, O. (1971) *Brain Res.* **28**, 109–117
Jasper, H. H. & Koyama, I. (1969) *Can. J. Physiol. Pharmacol.* **47**, 889–905
Jeavons, P. M. & Clark, J. E. (1974) *Brit. Med. J.* **ii**, 584–586
Johnston, G. A. R. (1972) *Biochem. Pharmacol.* **22**, 137–139
Johnston, G. A. R. & Balcar, V. J. (1971) *J. Neurochem.* **22**, 609–610
Jones, D. A. & McIlwain, H. (1971) *J. Neurochem.* **18**, 41–58
Kaczmarek, L. K. & Adey, W. R. (1974) *J. Neurobiol.* **5**, 231–241
Killam, K. F. & Bain, J. A. (1957) *J. Pharmacol. Exp. Ther.* **119**, 255–262
King, L. J., Carl, J. L. & Lao, L. (1974) *J. Neurochem.* **22**, 307–210
Koyama, I. (1972) *Can. J. Physiol. Pharmacol.* **50**, 740–752
Krnjević, K. & Phillis, J. W. (1963) *J. Physiol. (London)* **165**, 274–304
Krnjević, K. (1974) *Physiol. Rev.* **54**, 418–540
Lamar, C. (1970) *J. Neurochem.* **17**, 165–170
Leao, A. A. P. (1944) *J. Neurophysiol.* **7**, 359–390
Lee, L. W., Liao, G. L. & Yatsu, F. M. (1974) *J. Neurochem.* **23**, 721–724
Leonard, B. E. & Palfryman, M. G. (1972) *Biochem. Pharmacol.* **21**, 1206–1209
Machiyama, Y., Balazs, R. & Julian, T. (1965) *Biochem. J.* **96**, 688
Meyer, J. S., Kanda, T., Shinohara, Y. & Fukunchi, Y. (1970) *Neurology (Minneapolis)* **20**, 1179–1184
Nahorsky, S. R., Roberts, D. J. & Stewart, G. G. (1970) *J. Neurochem.* **17**, 621–631
Okamoto, K. & Quastel, J. H. (1973) *Proc. Roy. Soc. B* **184**, 83–90
Orrego, F. & Lipman, F. (1969) *J. Biol. Chem.* **242**, 665–671
Pappius, H. & Elliott, K. A. C. (1954) *Can. J. Biochem. Physiol.* **32**, 484–490
Perry, T. L., Hansen, S., Sokoi, M. & Wade, J. A. (1972) *Clin. Res.* **20**, 949
Perry, T. L., Urquart, N., Hansen, S. & Kennedy, J. (1974) *J. Neurochem.* **23**, 443–446
Phillis, J. W. & Ochs, S. (1971) *Exp. Brain Res.* **12**, 132–149
Plum, C. N. (1974) *J. Neurochem.* **23**, 595–600
Roa, P. D., Tews, J. K. & Stone, W. E. (1964) *Biochem. Pharmacol.* **13**, 477–487
Saad, S. F., Elmasry, A. M. & Scott, P. M. (1972) *Eur. J. Pharmacol.* **17**, 386–392
Sacktor, B., Wilson, J. E. & Tiekert, C. G. (1966) *J. Biol. Chem.* **241**, 5071–5075
Saito, S. & Tokunaga, Y. (1967) *J. Pharm. Exp. Ther.* **157**, 546–554
Schriver, C. R. & Whelan, D. T. (1969) *Ann. N.Y. Acad. Sci.* **166**, 83–96
Sellinger, O. Z. & Weiler, P. (1963) *Biochem. Pharmacol.* **12**, 989–997

Simler, S., Randrianariosa, H., Lehmann, A. & Mandel, P. (1968) *J. Physiol.* (*Paris*) *Suppl.* **60**, 547–551
Simler, S., Ciesieisli, L., Maitre, M., Randrianariosa, H. & Mandel, P. (1973) *Biochem. Pharmacol.* **22**, 1701–1708
Snodgrass, S. R. & Iversen, L. L. (1973) *J. Neurochem.* **20**, 431–439
Stone, W. E. (1957) *Am. J. Phys. Med.* **36**, 222–255
Tapia, R., Perez, de la Mora, M. & Massieu, H. G. (1969) *Ann. N.Y. Acad. Sci.* **166**, 257–266
Tower, D. B. (1960) *The Neurochemistry of Epilepsy*, Thomas, Springfield, Ill.
Terner, C., Eggleston, L. V. & Krebs, H. A. (1950) *Biochem. J.* **47**, 139–149
Van Gelder, N. M. (1972) *Brain Res.* **40**,
Van Gelder, N. M. & Courtois, A. (1972) *Brain Res.* **43**, 477–484
Van Gelder, N. M., Sherwin, A. L. & Rasmussen, T. (1972) *Brain Res.* **40**, 385–393
Van Harreveld, A. & Fiftkova, E. (1970) *J. Neurobiol.* **2**, 13–29
Van Harreveld, A. & Fiftkova, E. (1971) *J. Neurochem.* **18**, 2145–2154
Van Harreveld, A. (1972) *Structure and Function of Nervous Tissue* (Bourne, G. H., eds.), Academic Press, New York and London
Van Harreveld, A. & Fiftkova, E. (1973) *J. Neurobiol.* **4**, 375–387
Wedege, E. & Bradford, H. F. (1975) *FEBS Lett.* on press
Weichert, P. & Gollnitz, P. (1968) *J. Neurochem.* **15**, 1265–1270
Weichert, P. & Gollnitz, P. (1969) *J. Neurochem.* **16**, 689–693
Weinreich, D. & Hammerschlag, R. (1975) *Brain Res.* **84**, 137–142
Wheeler, D. D., Boyarsky, L. L. & Brooks, W. H. (1966) *J. Cell. Comp. Physiol.* **67**, 141–147
Wood, J. D. & Peesker, S. J. (1974) *J. Neurochem.* **23**, **703**–712
Wood, J. D. & Watson, W. J. (1963) *Can. J. Biochem. Physiol.* **41**, 1907–1013
Yoshino, Y. & Elliott, K. A. C. (1970) *Can. J. Biochem. Physiol.* **48**, 228–235

Chapter 17

Amino Acid Transmitters: Pharmacological and Electrophysiological Aspects

By D. W. STRAUGHAN

Department of Pharmacology, The School of Pharmacy, University of London, 29/39 Brunswick Square, London WC1N 1AX, U.K.

The first part of this chapter will review the current state of knowledge on amino acid transmitters with particular respect to their release *in vivo* and to drugs which affect their postsynaptic actions. The extent to which changes in the functional activity of amino acid transmitters are involved in experimental seizure states produced by lesions or drugs, and provide a locus of action for anticonvulsant drugs, will then be considered.

Amino Acid Transmitters

The structural and electrophysiological features of the bulk of the synapses in the mammalian central nervous system (CNS) suggest central synaptic transmission is mediated through the release of chemical transmitter substances from the nerve endings and not by simple electrical coupling. Clearly, the identification of these chemical transmitters, the pathways involving them and the characteristics of their release and actions on postsynaptic membranes is fundamental to our basic understanding of how the brain works. Additionally, such knowledge provides a substrate for the understanding of drug action in the CNS and a physiological basis for neurological and psychiatric disease. For a full account of this subject, reference should be made to recent comprehensive reviews by Krnjević (1974) on central and peripheral chemical synaptic transmitters in vertebrates and by Curtis & Johnston (1974) on amino acid transmitters in the mammalian CNS.

Despite impressive evidence regarding the location and functions of the different monoamines in the CNS, it is worth emphasizing that only a small fraction of all the excitatory and inhibitory nerve endings in brain appear to utilize monoamines or acetylcholine (ACh) as transmitters. The possibility that amino acids such as glutamate and γ-aminobutyric acid (GABA) might be concerned in the control of neuronal excitability in addition to a role in cerebral metabolism was first proposed by Hayashi (1954, 1956). He showed that microinjections of glutamate into mammalian brain had a direct excitatory effect and that GABA and related neutral amino acids were effective depressants. The application of amino acids into the extracellular environment of single neurons by the technique of iontophoresis showed glutamate and GABA to have reproducible, potent, short-lived excitatory and inhibitory effects respectively, on cortical neurons and in other areas of CNS. Because of their actions and their high concentrations, Krnjević & Phillis (1963) suggested that glutamate and GABA might be natural transmitters in the brain.

Subsequently, extensive investigations have been undertaken with respect to a variety of amino acids having a neurotransmitter role in the mammalian CNS. L-Glutamic and L-aspartic acid are present in substantial amounts and widely distributed throughout the CNS but small regional differences have led to proposals that they operate as excitatory transmitters in different areas. Thus in the spinal cord glutamate has been suggested as the excitatory transmitter released by primary afferent fibres, because it is concentrated more in dorsal grey matter than ventrally. Aspartic acid has been proposed as an excitatory transmitter released from spinal interneurons because it is more concentrated in the ventral grey than dorsally and following anoxic lesions there is a correlation between the decrease in concentration of aspartic acid and loss of interneurons.

GABA, glycine and taurine are found in substantial amounts in the CNS and this location is unique for GABA, which is not found in cerebrospinal fluid, plasma or in peripheral tissues. In most parts of the CNS there is a relatively good correlation between the synthetic enzyme for GABA, glutamic acid decarboxylase (GAD) and GABA content; and high levels of GAD and GABA have been found selectively in regions of the CNS where there is a high density of inhibitory nerve terminals such as the hippocampus (Storm-Mathisen & Fonnum, 1971), Deiter's nucleus (Fonnum *et al.*, 1970; Otsuka *et al.*, 1971) and cerebellum (Fonnum & Walberg, 1973). It is of interest that the highest GABA levels are found in the substantia nigra.

While glutamate, aspartate, GABA, glycine and taurine in brain can be shown to be present in nerve-ending particles, it has not yet been possible to demonstrate a preponderant location in the synaptic vesicle fractions. This might be attributed to methodological difficulties or to the vesicle hypothesis being applicable only to some but not all central transmitters. However, it should be noted that there are very substantial amounts of taurine in glial cell cultures, and taurine has a very slow turnover (Collins, 1974). This may make a transmitter function unlikely or very difficult to demonstrate convincingly. It should also be noted that appreciable quantities of the other putative transmitter amino acids also occur in glial cultures and in glial-enriched fractions from brain.

Release Experiments

The release of amino acids from CNS *in vivo* has been investigated in a number of laboratories. If it can be shown that electrical stimulation or potassium-induced depolarization causes a specific Ca^{2+}-dependent release of particular amino acids, such experiments would provide important evidence supporting a transmitter role. However, though they may be simple in concept, the experiments involve many difficulties in interpretation, particularly with regard to the precise source of the material that is released. The amounts of endogenous amino acid released may be extremely small, and provide for difficulties of measurement, but this has been largely overcome in recent years by the application of dansylation techniques allowing the measurement of the whole spectrum of amino acids. Alternatively, labelled amino acids can be applied exogenously—these are taken up through specific carriers to label endogenous stores. The efflux of radioactivity from the CNS can then readily be measured by the sensitive techniques of liquid scintillation spectrometry. However, it is clear that these labelled amino acids can be accumulated in glial cells as well as in the nerve endings. Also depolarizing stimuli can release labelled GABA from peripheral glial cells in a calcium-dependent fashion (Minchin & Iversen, 1974), although a Ca^{2+}-independent release of GABA and ACh has been noted in other studies (Bowery & Brown, 1974*b*; Dennis & Miledi, 1974).

Thus it would seem wise to place the major emphasis on the release of endogenous amino acids in this presentation, although it is not certain that this release comes from nerve terminals rather than cell bodies or glia. These endogenous release experiments are summarized in Tables 1 and 2.

In pioneer experiments, Jasper *et al.* (1965) applied plastic cups filled with warm saline to the exposed cerebral cortex of unanaesthetized cats and analysed the amino acid effluent with respect to the EEG. They noted an inverse relationship between GABA and glutamate release and between GABA release and cortical activation as judged by the EEG. Thus preparations showing an activated EEG showed a 40% increase in glutamate release, and a decrease in GABA release of approximately two-thirds compared with those from cortex with a synchronized 'sleeping' EEG. These results were confirmed by Jasper & Koyama (1969), who showed in addition that electrical stimulation of the midbrain reticular formation caused significant increases in the release of glutamate, aspartate and glycine, while substantial increases in taurine release were noted in some experiments. None of these changes could be simply attributed to changes in the concentrations of these amino acids in cerebrospinal fluid. The possibility that these changes in amino acid release with arousal were due to changes in systemic blood pressure appeared to be ruled out, since similar results were seen when the release from one hemisphere which had been deafferented and showed a sleeping EEG was compared with the release from the opposite hemisphere which was aroused by electrical stimulation of the reticular formation On some occasions the spectrum of amino acids released by stimulation was sufficiently broad (including histidine and threonine) to suggest non-specific release. Additionally, it might be argued that the effects of reticular stimulation might not be confined purely to excitatory neurons but might cause increased activity in recurrent collateral inhibitory pathways with the possibility of an increased release of GABA. Indeed, a simple correlation between behavioural or EEG arousal and predominant activity in excitatory or inhibitory neurons might be surprising. It should also be noted from Table 1 that the rates of GABA release from sleeping cortex in Jasper and colleagues' experiments are considerably higher than the resting rates for anaesthetized animals obtained by all subsequent groups. This might be due to pial puncturing, assay problems, differences between unanaesthetized and anaesthetized animals or perhaps a combination of these.

In an elegant series of experiments, Iversen and his colleagues measured the release of GABA from the surface of the visual cortex of Dial-anaesthetized cats by means of a sensitive enzymic fluorimetric assay procedure (Iversen

Table 1. *Spontaneous (resting) release of endogenous amino acid from cerebral cortex in vivo*

Species		Anaesthetic	Glutamate	Aspartate	GABA	Glycine	Taurine	Alanine	Glutamine	Ref.
Cat		None Cerveau isolé	0.7	0.2	0.3				0.1	(1)
Cat	6	None Enceph. isolé	2.5	0.7	(2.5)	2.2	2.3	4.0	4.7	(2)
Cat	3	Dial	5.4	2.6	(0.28)	6.8	—	5.7	—	(3)
Cat	8	Pentobarb.	2.5	1.8	0.7	13.7	—	10.5	5.0	(4)
Rat	31–43	Ureth.	4.4	2.7	0.3	0.7	1.0	0.8	6.7	(5)

Values are all in terms of the same units, i.e. 10^{-10} mol/min/cm^2.

Table 2. *Evoked release of endogenous amino acids from cerebral cortex in vivo*
(Resting level of release = 1)

Procedure	No.	Glutamate	Aspartate	GABA	Glycine	Taurine	Alanine	Glutamine	Ref.
Sensory arousal	1	6.28	6.23		6.25	10.9	14.07	4.86	(2)
R.F. stim. 100–200 Hz Enceph. isolé	6	3.04*	2.26*		1.59*	1.82	1.35	1.0	
Cort. stim. 200 Hz 1.5 mA 1 ms	3	1.18	1.11	10.0†*	0.79	—	1.0	—	(3)
Cort. stim. 100 Hz 2.5 mA 1 ms		4.9*	1.4	9.0*	1.2	3.6	1.6	1.3	(6)
Cort. stim 5 mA		6.9*	4.6*	17.8*	2.2*	5.1*	2.9*	1.5	(6)
Topical K^+ 50 mM	7	0.8	1.2	2.8*	1.05	2.2*	0.9	0.42†*	(5)

Reference key: (1) Jasper, *et al.* (1965); (2) Jasper & Koyama (1969); (3) Iversen, *et al.* (1971); (4) Crowshaw, *et al.* (1967); (5) Clark & Collins (1975); (6) Collins (1974); Collins unpublished observations.
* Statistically significant; † Ca-dependent.

et al., 1971). The release of endogenous GABA was increased between three and seven times by high frequency stimulation of the brain surface or lateral geniculate nucleus sufficient to produce prolonged inhibition of the firing of cortical neurons. In a small number of experiments (see Table 2) they examined the total amino acid content of cortical samples using an amino acid analyser and detected no significant increase in the release of aspartic acid, glutamic acid, glycine or alanine, suggesting that the increased release of GABA observed during inhibitory stimulation was specific. Furthermore, when the cortex was exposed to a calcium-free collection fluid, the normal increase in GABA release evoked by epicortical stimulation was prevented though the resting release of endogenous GABA was not affected. In recent experiments Clark & Collins (1975, and unpublished observations) have applied sensitive dansylation techniques to the analysis of endogenous amino acid release from the cerebral cortex of urethane anaesthetized rats and rabbits. The resting release of glutamate, aspartate and GABA was similar to that seen in Iversen *et al.*'s experiments in the cortex though the glycine and alanine release was much lower. Epicortical stimulation with currents of 2.5 mA selectively increased the release of glutamate, GABA and taurine but not that of aspartate, glycine, alanine or glutamine, while larger stimulating currents also evoked the release of aspartate and glycine but not glutamine. Bathing the cortical surface with 50 mM K^+ also increased the release of GABA and taurine selectively, the release of aspartate, glutamate or glycine being unaffected. Further, Collins was able to show that the electrically evoked release of glutamate and GABA but not taurine was calcium dependent. These experiments would appear to support a transmitter role for glutamate and GABA in the cerebral cortex and might suggest a 'non-transmitter' role for taurine, perhaps released from glia by depolarization, in modulating cortical excitability.

Calcium-dependent release provides strong support though not absolute evidence of 'neurosecretion'. Thus on occasions the evoked release of GABA from glial cells may be calcium dependent; A reduced synaptic input by Ca^{2+} depletion might reduce release of amino acids from neurons indirectly as a consequence of reduced activity.

The release of endogenous amino acids from other sites has been demonstrated; thus Obata & Takeda (1969) showed that stimulation of the cerebellar cortex increased the release of GABA into the fluid perfusing cat fourth ventricle and Crawford & Connor (1973) showed that prolonged entorhinal stimulation increased the release of glutamate from the surface of the hippocampal cortex. The specificity and significance of these results remains to be determined since the release of other amino acids and tests of calcium dependence were not made. However, Roberts (1974*a*) has studied the release of a range of both endogenous and exogenous putative amino acid transmitters from the superfused rat dorsal column nuclei. He reported that stimulation of dorsal column tract fibres (200–500 Hz) caused a statistically significant increase over resting in the release of endogenous glutamate (1.24×) and GABA (1.26×). His results also showed a substantial increase over resting in the release of glycine and alanine with dorsal column stimulation and even with lemniscal stimulation in some experiments.

Postsynaptic Actions of Amino Acid Transmitters and Synaptic Transmission

Excitatory amino acids

When applied into the environment of neurons by iontophoresis, L-glutamate and L-aspartate consistently and powerfully excite neurons and are approximately equipotent, e.g. spinal cord, Curtis & Watkins (1963); cerebral cortex, Krnjević & Phillis (1963). Intracellular recording techniques reveal that this glutamate excitation is associated with a depolarization and a decrease in the resistance of the postsynaptic membrane. This is thought to involve particularly an increase in Na^+ permeability but an increase in K^+ permeability probably also occurs. To date, it has not proved possible to show that the equilibrium potential for the naturally released excitatory transmitter in cerebral cortex and spinal cord is identical with that for glutamate applied iontophoretically. But, the intracellular microelectrode in the cell soma will not readily 'see' the changes induced in the dendrites by release of endogenous excitatory transmitters whereas the iontophoretic application of glutamate may act locally, perhaps on extra-junctional receptors in the soma. Thus this non-identification can be explained by technical factors rather than by assuming that glutamate is not the excitatory transmitter.

Clearly the use of selective antagonists would help in the confirmation of glutamate and/or aspartate as excitatory synaptic transmitters, but none of the agents presently available has proved to be completely satisfactory. 1-Hydroxy-3-amino-pyrrolid-2-one (HA 966) reduces the sensitivity of cortical and spinal neurons to acidic amino acids before affecting excitation by ACh and also reduces the sensitivity of cuneate neurons to glutamate (Curtis *et al.*, 1973; Davies & Watkins, 1973), and this action of HA 966 is not apparently mediated through GABA or glycine receptors. Figure 1 shows HA 966 shifting the L-glutamate dose-response curve for excitation of the cortical neuron at a time when the ACh dose-response curve is unaffected; in contrast atropine selectively blocks ACh responses and GABA non-selectively shifts the curves for both excitant agonists (Clarke *et al.*, 1974). HA 966 is also a moderately effective depressant of synaptic excitation in the cuneate. However, HA 966 does not distinguish between the excitations produced by L-glutamate, L-aspartate or DL-homocysteic acid. Moreover, the safety margin for HA 966 in blocking L-glutamate is small, compared with its block of a totally unrelated agonist like ACh acting through different receptors and different conductance effects. Concentrations of HA 966 adequate to abolish Renshaw cell excitation by the amino acids reduced excitation by ACh to less than half of the control value (Curtis *et al.*, 1973). Also, doubling the application of HA 966 needed to reduce L-glutamate excitation of cortical neurons by 50% reduced the ACh responses to a similar extent—some local anaesthetics act similarly (Clarke, 1975). In contrast, applications of atropine have to be increased about eightfold over those affecting ACh in order to produce a comparable reduction in glutamate firing of cortical neurons. This difficulty in blocking glutamate 'receptors' may reflect an atypical nature. Krnjević (1974) has suggested that the depolarizing action of glutamate involves the displacement of Ca^{2+} from critical sites in the membrane, although the excitant effect cannot simply be attributed to chelation. Such an action on membrane

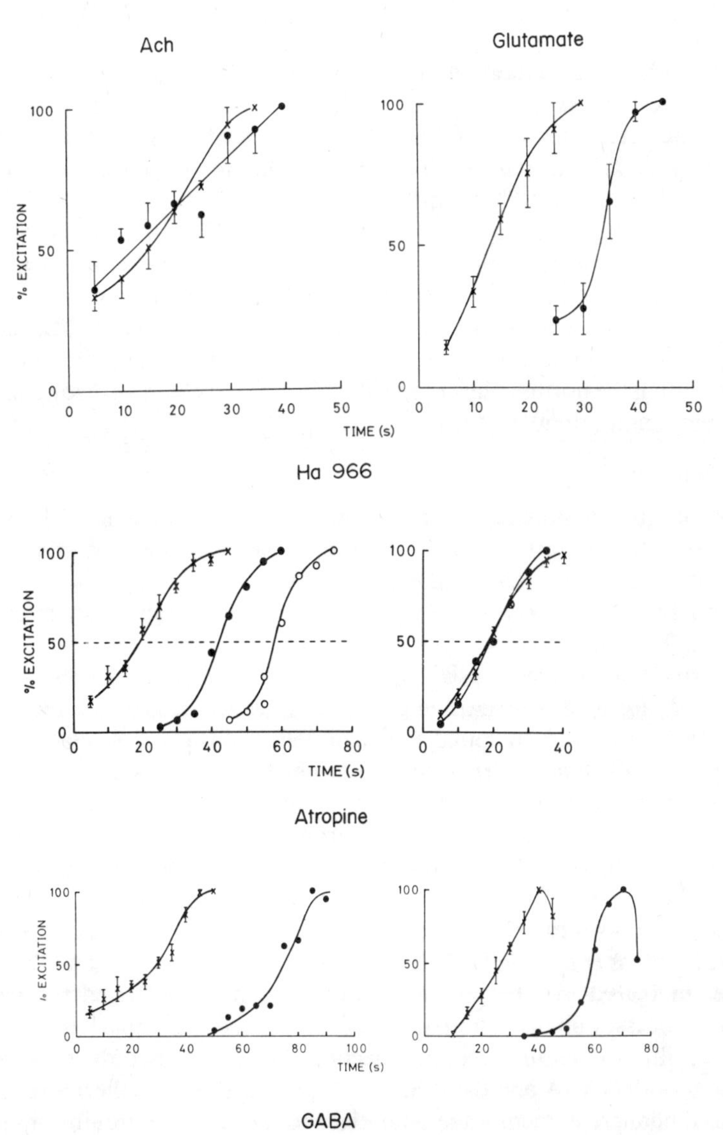

Fig. 1. *Data plotted from a cortical neuron in an N_2O-halothane anaesthetized cat illustrating the effect on excitations by iontophoretic ACh and glutamate of an additional iontophoretic application of HA 966, atropine and GABA. With short applications of agonist at a fixed current level the response changes rapidly with time allowing cumulative time to be used as a measure of dose in the construction of dose-response curves*

Top records: Control responses (×) to ACh 100 nA and L-glutamate 50 nA, and 150 s after start of an application of HA 966 50 nA (●). Each curve is the mean ±S.E.L. of 3 responses.

Middle records: Control responses (×) to ACh 35 nA and L-glutamate 40 nA (mean ±S.E.L. 3 responses) and following atropine 5 nA. For ACh record (●) is 35 s after start of atropine, (○) 70 s after start of atropine; and for the glutamate record (●) is 100 s after start of atropine.

Bottom records: Control responses (×) to ACh 50 nA and L-glutamate 50 nA (means ±S.E.L. 3 responses) and following GABA 5 nA. The ACh record was 150 s and the glutamate 200 s after the start of the GABA application.

Abscissae: Percentage maximum excitation which was for top records ACh 40–45 Hz, glutamate 55–60 Hz, for middle records ACh 260–300 Hz, glutamate 310–320 Hz, and bottom records ACh 34–40 Hz and glutamate 40–48 Hz.

Ordinate: Time in seconds.

This figure is reproduced with permission from Clarke *et al.* (1974).

Ca^{2+} might make glutamate excitations particularly vulnerable to the effect of local anaesthetics. Nevertheless, evidence for the existence of an excitatory amino acid receptor comes from observations of a differential sensitivity to acidic amino acid in spinal cord neurons: L-glutamate excites dorsal horn interneurons more powerfully than L-aspartate, while L-aspartate excites Renshaw cells more powerfully than L-glutamate (Duggan, 1974). In addition, studies with synaptic membrane fractions isolated from brain, have demonstrated a high affinity binding of L-glutamate which could be inhibited by L-aspartate and DL-homocysteric acid but not by D-glutamate (Roberts, 1974; Michaelis *et al.*, 1974). This glutamate binding was thought to be to specific sites possibly representing the normal physiological receptor, and could be distinguished from the high affinity uptake systems.

Inhibitory amino acids

The inhibitory potency of various putative transmitter neutral amino acids applied extracellularly by microiontophoresis has been examined in many laboratories. In all areas of the CNS, GABA was an effective depressant of neuronal firing, but there were regional differences in sensitivity to glycine and taurine. Thus, in the cortex, GABA is a more powerful depressant than glycine or taurine (on a current basis GABA is seven times more depressant than glycine, and taurine is weaker than glycine). However, in Deiter's nucleus and the cuneate nucleus and on spinal Renshaw cells GABA and glycine appear to be equipotent. Glycine is generally somewhat more depressant than GABA on spinal motor neurons and interneurons, and taurine may resemble glycine in potency. Further, tests on cortical neurons in which intracellular recording is combined with extracellular iontophoresis of the most potent depressant GABA, revealed close similarity between the effects of GABA and those of the natural inhibitory transmitter released by epicortical stimulation (Krnjević & Schwartz, 1967; Dreifuss *et al.*, 1969). Thus both GABA and the natural inhibitory transmitter increased membrane conductance and hyperpolarized the membrane. These processes involve similar ionic changes and particular an increased Cl^- permeability so that intracellular injections of Cl^- reversed the hyperpolarizing effect of both GABA and the IPSP to a depolarization. Furthermore, artificially induced changes in membrane potential showed that the equilibrium potentials for GABA and the endogenous inhibitory transmitter were very similar. Thus the evidence from physiological tests of identification that GABA is the major inhibitory transmitter in the cerebral cortex is very strong. However, in subcortical areas, e.g. Deiter's nucleus and spinal cord, although the effects of the inhibitory amino acids on membrane properties closely resembled those produced by the natural inhibitory transmitter, the effects of GABA could not be distinguished from glycine (Curtis *et al.*, 1968; Bruggencate & Engberg, 1971). Fortunately, pharmacological antagonists separate the rival candidates and give powerful evidence confirming the identity of the inhibitory transmitters. Thus postsynaptic or direct inhibition in the spinal cord and the effects of glycine (and 'glycine-like' α or β amino acids including taurine) are readily blocked by systemic or iontophoretic injections of strychnine which do not affect GABA (and 'GABA-like' γ and higher ω amino acids—Curtis *et al.*, 1968). There are also some strychnine-sensitive inhibitions in the medullary reticular formation (Tebecis & Di Maria, 1972). Thus physiological and pharmacological tests of identity provide strong evidence for glycine being the main postsynaptic inhibitory transmitter in the spinal cord.

The pharmacological evidence that GABA is the major postsynaptic inhibitory transmitter in the brain rests on the fact that synaptic inhibitions and the effects of GABA in most areas studied are not readily blocked by strychnine but are blocked by picrotoxin, and bicuculline applied iontophoretically which is less active against glycine (see Curtis & Felix, 1971; Curtis & Johnston, 1974). However, in the cerebral cortex, where potentiation instead of block of synaptic inhibition may sometimes be seen, iontophoretic bicuculline reduces the effects of taurine and β-alanine as well as GABA, and the safety margin or selectivity between block of GABA in preference to glycine may not be high, particularly in the rat (Biscoe *et al.*, 1972). Particular caution has to be applied in interpreting the apparent block of postsynaptic inhibition sometimes seen following the intravenous injection of GABA antagonists (Hill *et al.*, 1972). These agents are powerful convulsants and the effects observed may be due to actions at sites remote to the neuron under study as is discussed later.

Snyder's group have provided powerful evidence for the existence of distinct GABA and glycine receptors. They studied the binding of [^{3}H]GABA to synaptic membrane fragments from rat brain, in sodium-free media (Zukin *et al.*, 1974). The affinity constant of binding for cold GABA was 0.1 μM and [^{3}H]GABA binding was inhibited by bicuculline (IC_{50}5 μM) and strychnine (IC. 50 μM) but not by picrotoxin, glycine or taurine. These results suggest the antiGABA action of picrotoxin in intact cells is perhaps due to an effect on the ion channel rather than on the receptor *per se* and that bicuculline is a weak antagonist. They also imply that taurine does not act on the GABA receptor in brain. In contrast when the affinity of [^{3}H]strychnine binding to proposed glycine receptors from spinal cord was measured, the affinity constant of strychnine at 0.03 nM was about 300 times greater than that of glycine—consistent with the neuropharmacological evidence that it is a very powerful glycine antagonist (Young & Snyder, 1973).

The superior cervical ganglion preparation rather conveniently contains GABA receptors, and GABA increases Cl^- permeability just as it does in the CNS. These receptors can be blocked by bicuculline (IC_{50} 15 μM) and picrotoxin (IC_{50} 37 μM) (Bowery & Brown, 1974*a*). Bicuculline is more selective than picrotoxin with a discrimination ratio between the GABA and muscarinic receptors of 0.06 compared with 0.2 for picrotoxin. Unfortunately this preparation contains no glycine receptors so that glycine has a relative depolarizing activity of less than 10^{-5} compared to that with GABA. In addition, taurine has only 0.1% of the activity relative to GABA.

Presynaptic Inhibition

The processes discussed so far have involved nerve terminals ending on the dendrites or soma of the postsynaptic cell, and postsynaptic actions of transmitters. There is morphological evidence for another type of synapse involving axo-axonic contacts at a few sites particularly in the afferent or sensory pathway, e.g. dorsal horn of the spinal cord, the dorsal column (cuneate and gracile) nuclei, the lateral geniculate nucleus and the thalamus. These synapses are believed to be on the terminals of excitatory nerves and provide the morphological basis for presynaptic or 'remote' inhibition, whereby the presynaptic inhibitory transmitter depolarizes the excitatory terminals, reduces their resistance and the magnitude of the action potential, and thereby reduces the amount of excitatory transmitter released. This inhibition is accompanied by a reduction in EPSPs without observable direct actions in the postsynaptic membrane of motor neurons. This presynaptic inhibition correlates well with changes in terminal excitability, and the depolarization is reflected in the recording of slow surface positive potentials (P-waves) and positive spinal dorsal root potentials (see Schmidt, 1971).

There is a substantial amount of indirect evidence that the presynaptic inhibitory transmitter is GABA; thus Davidson & Southwick (1971) found an increase in terminal excitability when they superfused the dorsal column nuclei with solutions containing GABA, and in the amphibian spinal cord GABA depolarizes dorsal root fibres very strongly. Furthermore, semicarbazide reduces spinal GABA levels in acute spinal cats and suppresses positive dorsal root potentials (a correlate of presynaptic inhibition), monosynaptic reflexes and presynaptic inhibition (Bell & Anderson, 1972). In contrast with the weak effects of systemic GABA, antagonists on postsynaptic inhibitions, systemic picrotoxin and bicuculline readily block spinal and supraspinal presynaptic inhibition and the depolarization of afferent terminals by GABA in mammals and amphibians (see Curtis *et al.*, 1971; Schmidt, 1971; Davidoff, 1972*a*; Levy, 1974).

The ionic mechanisms underlying the action of GABA on presynaptic terminals are far from clear. Studies in frog spinal cord (Barker & Nicoll, 1973) suggested that this action of GABA involved changes in Na^+ but not Cl^- permeability. In contrast, however, a more recent study in the cuneate nucleus has suggested that the presynaptic action of GABA involves an increase in Cl^- conductance (Davidson & Simpson, 1975) as it does in frog cord (Nishi *et al.*, 1974). The situation is far from simple; indeed, some workers consider that the phenomenon of presynaptic inhibition with its associated depolarization of primary afferent terminals can be attributed in part or whole to the accumulation of K^+ around nerve terminals as a consequence of neuronal activity. Schmidt (1971) has given an excellent account of the physiology of presynaptic inhibition and its pharmacology—aspects of this latter process will be considered in the next section.

Synaptic Mechanisms and the Seizure Focus

Clearly, the excitatory and inhibitory amino acids and the synaptic mechanisms they are thought to mediate will exert powerful effects on neuronal excitability and might well show a primary abnormality in epilepsy. The ways in which this could occur are similar to those shown in Table 3. From a variety of biochemical studies it is known that the levels of glutamate, glutamine, aspartate, taurine and GABA may be decreased in both acute and chronic seizure foci, while glycine levels may increase (Koyama, 1972; Van Gelder *et al.,* 1972). By themselves such measurements do not reveal in which compartment changes take place and in particular if extraneuronal levels of neuroactive amino acids are affected, nor do they show how neuronal excitability overall will be affected. Additionally there is the problem of determining whether the changes cause seizures, or are a result of them.

Table 3
Potential mechanisms for convulsant drug action

BLOCK OF INHIBITION	INCREASED EXCITATION
Transmitter release	
Decreased release of inhibitory transmitters e.g. allyglycine, thiosemicarbazide, tetanus toxin	Increased release of excitatory transmitter *per se*, or synchronization of excitatory feedback
Receptors	
Block of postsynaptic inhibitory receptors, e.g. strychnine	Stimulation of postsynaptic excitatory receptors *per se*, or enhanced activity of excitatory transmitter
Block of presynaptic inhibitory transmitter, e.g. picrotoxin, bicuculline, leptazol, bemigride	
MEMBRANE	
Block of membrane changes initiated by receptor activation, e.g. D-tubocurarine, TETS	Direct increased excitability of postsynaptic membrane

One of the favoured methods of producing acute epileptic activity in laboratory animals involves the local injection of penicillin into the cortex to give a model of focal epilepsy, or the intravenous injection of penicillin to give a generalized epileptiform activity which has many similarities with human myoclonic petit mal (Prince & Farrell, 1969; Gloor & Testa, 1974). Does this involve a change in the efficacy of inhibitory mechanisms particularly GABA? In the acute penicillin focus the concentration of GABA decreases, and there is also a decrease in GAD levels of 18% in the focus and 13% in the homotropic point in the contralateral cortex. However, any decrease in the synaptic release of GABA might be counteracted by the 50% decrease in GABA uptake which is perhaps attributable to the increased extracellular K^+ levels rather than to penicillin *per se* (Gottesfeld & Elazar, 1975). Another possibility is that penicillin induces seizures in mammalian brain through a blockade of postsynaptic GABA receptors. Indeed, when applied by iontophoresis, penicillin can reduce GABA but not glycine depression of spinal and cunate neurons (Curtis *et al.,* 1972; Davidoff, 1972*b*; Hill *et al.,* 1975) and reduces the effects of GABA on cerebral cortical neurons (Curtis *et al.,* 1972; Hill *et al.,* 1973*a*). However, penicillin is a weak GABA antagonist in iontophoretic experiments and for the ganglionic GABA receptor the IC_{50} is in the region of 2 mM (Bowery, 1973; and unpublished work). In addition, when GABA antagonists are infused intravenously in subconvulsant or convulsant doses, it is difficult to show block of postsynaptic GABA receptors in the cortex or nucleate nucleus (Hill *et al.,* 1972, 1973*b*).

Intracellular studies provide further evidence for the view that postsynaptic inhibitory mechanisms are not selectively blocked either in the acute penicillin epileptic focus or in chronic foci produced by diverse agents including cobalt. The characteristic neuronal event in the epileptic focus, coincident with the cortical paroxysm or interictal spike, is a sudden large depolarization (paroxysmal depolarizing shift, or PDS). In both neocortex and hippocampus this PDS is usually followed by a hyperpolarization which can last for up to 2 seconds. Both the PDS and the subsequent hyperpolarization are believed to represent synaptic events, and it is postulated that they are in fact giant recurrent EPSP–recurrent IPSP sequences. This hyperpolarization involves an increased Cl^- conductance and can be reversed to a depolarizing potential by the intracellular injection of Cl^- (Prince, 1968). It is of course possible that inhibition is quantitatively abnormal, but these studies suggest no absolute abnormality in excitatory or inhibitory mechanisms in penicillin and chronic seizure foci. If one accepts this view, then the development of the focus and synchronous discharges requires not one, but a group of neurons and perhaps enhanced positive feedback of recurrent excitation through some defect in temporal organization and co-operation (Ayala *et al.*, 1973). It is interesting to note that the cerebellar cortex fails to develop interictal spikes and Ayala *et al.* suggest that this is because it lacks the recurrent excitatory system of cerebral or hippocampal cortex.

The role of extracellular cortical K^+ in epileptiform activity is complex—substantial increases in extracellular K^+ occur as a result of epileptic activity and, through excessive depolarization of afferent fibre terminals in the thalamus, could cause powerful antidromic excitation of cortical neurons. Also K^+ accumulation in the extracellular space (occurring as a result of neuronal activity during interictal discharges) is believed to have a causative role in the transition from interictal discharges to major seizures. However, Lux (1974) has suggested that these increases in K^+ may be transient and trigger a longer lasting decrease in extracellular K^+ perhaps by stimulating uptake. This would have the same end result as blocking presynaptic inhibition with drugs—an absolute or relative hyperpolarization of afferent terminals, an increase in terminal action potential amplitude and an increase in the amount of transmitter released. This would easily contribute to the formation of the giant EPSP in individual neurons which underlie the paroxysmal spike in the ECoG.

Convulsant drugs and inhibitory processes

A large number of drugs are able to cause convulsions in experimental animals and man and the potential mechanisms are very diverse (see Table 3). Although many of these mechanisms at the single cell level do not appear to be involved in acute and chronic epileptic foci, an analysis of the actions of convulsant drugs should confirm our ideas of the role of excitatory and inhibitory mechanisms in regulating neuronal excitation.

The use of drugs which inhibit GABA synthesis should and does cause convulsions in experimental animals while drugs like aminooxyacetic acid and ethanolamine-*O*-sulphate (e.g. Baxter *et al.*, 1973) which inhibit GABA metabolism have anticonvulsant effects. Similarly, drugs like tetanus toxin which appear to act by inhibiting the release from nerve terminals of inhibitory amino acids like glycine produce co-ordinated 'spinal' seizures when given systemically. When injected locally into the cortex, tetanus toxin reduced the changes in cortical potentiation associated with recurrent collateral inhibition before the onset of gross convulsive activity (Brooks & Asanuma, 1965). However, in another

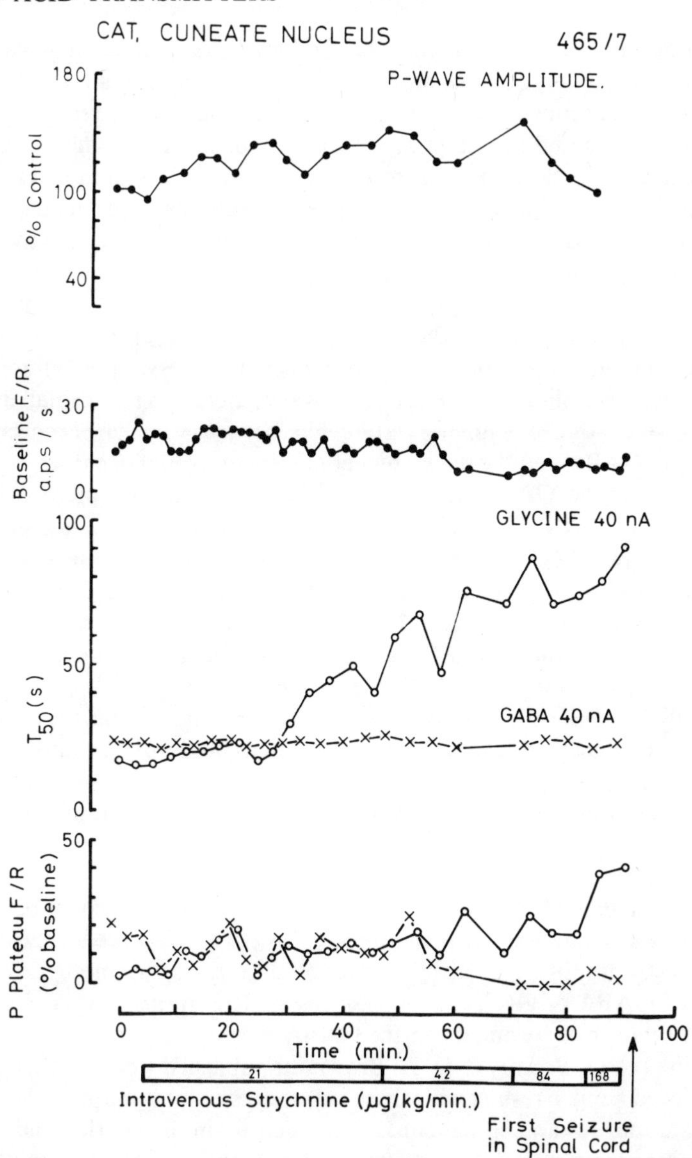

Fig. 2. *Data plotted from a cuneate neuron in an N_2O-halothane anaesthetized cat illustrating the effects of systemic strychnine on P-wave amplitude and GABA and glycine sensitivity*

The top record shows the amplitude of the surface positive (P) wave produced at regular intervals by stimulation of the ipsilateral radial nerve and recorded via silver ball macroelectrode from the surface of the cuneate nucleus. The second record from the bottom shows neuronal sensitivity to regular and alternate applications of glycine and GABA expressed as T_{50}, or the time for which the agonist application has to be maintained to produce a 50% depression of firing rate.

During the intravenous infusion of strychnine 21 µg/kg/min there was a progressive increase in P-wave amplitude to 50% above control values. This was followed after a short interval by a decreasing sensitivity to glycine (increasing T_{50}) but without any alteration in GABA sensitivity. A substantial change in baseline and plateau firing rates was not seen. Only after a substantial period of decreasing sensitivity to glycine was the first seizure in the spinal cord or cuneate seen, which is indicated by ↑ below the time axis.

Ordinate: Elapsed time in minutes.

This figure is reproduced with permission from Straughan (1974).

study, injection of tetanus toxin into the cerebral cortex caused no gross change in the inhibition of unit firing produced by epicortical stimulation (Krnjević *et al.*, 1966). Theoretically the block of postsynaptic inhibitory receptors should also cause seizures and this certainly seems to explain the convulsant action of systemic strychnine. Section experiments and regional electrographic recording show that the major site of action of systemic strychnine in producing co-ordinated convulsions is the spinal cord, where microelectrode studies have shown that it blocks glycine receptors and postsynaptic inhibition at similar times. Figure 2 illustrates that the block of glycine receptors in the cuneate and presumably the cord precedes the development of electrographic seizures by a substantial margin. Strychnine does not decrease presynaptic inhibition as judged from the amplitude of the surface P-wave; there is in fact a small increase.

It was hoped that a similar relationship between blocks of postsynaptic receptors for GABA and the development of seizures would occur when convulsants known to be GABA antagonists like bicuculline or picrotoxin were infused intravenously. Typically, no consistent reduction of sensitivity of cortical neurons to GABA or (as is illustrated in Fig. 3) of cuneate nucleus neurons to GABA or glycine could be detected prior to the development of electrographic seizures in cortex. The only consistent change seen with these antagonists was a reduction in P-wave amplitude suggesting a decrease in the effectiveness of presynaptic inhibition as has been shown for these drugs given systemically in numerous other studies. It might be argued that these drugs do reduce the response to endogenously liberated inhibitory amino acid, i.e. do block postsynaptic GABA receptors. The iontophoretic testing procedure used does not allow this effect to be seen but the strychnine results made this unlikely.

It is important to stress that almost identical results were achieved when experiments were made with slow intravenous infusion of leptazol or bemegride. These also caused a reduction in the cuneate P-wave which preceded the development of electrographic seizures by a substantial time. Furthermore, no evidence of block of GABA or glycine responses was evident in these experiments (Hill *et al.*, 1973*b*) or in experiments on the GABA receptor of the isolated ganglion preparation (N. G. Bowery & D. A. Brown, unpublished work). Other workers have achieved similar results with these substances on presynaptic inhibition in the cuneate nucleus and spinal cord. Thus picrotoxin, bemegride and leptazol block the increased excitability of cuneate presynaptic terminals and inhibition of lemniscal discharge produced by conditioning stimuli, as well as the amplitude of the associated surface positive P-wave (e.g. Boyd *et al.*, 1966; Banna & Hazbun, 1969; Banna & Jabbur, 1969, 1970). Also in the spinal cord, leptazol depressed presynaptic inhibition but was without effect on postsynaptic inhibition and the reduction of dorsal root potentials and presynaptic inhibition by picrotoxin precedes the development of convulsant activity (Schmidt & Willis, 1963). It is of interest that intravenous injections of penicillin also affect presynaptic inhibition and reduce dorsal root potentials, dorsal root reflexes and prolonged inhibition in cat spinal cord (Davidoff, 1972*b*). To what extent if at all this relates to block of pre- or postsynaptic GABA receptors is unclear. Unfortunately, although these studies suggest where picrotoxin and leptazol

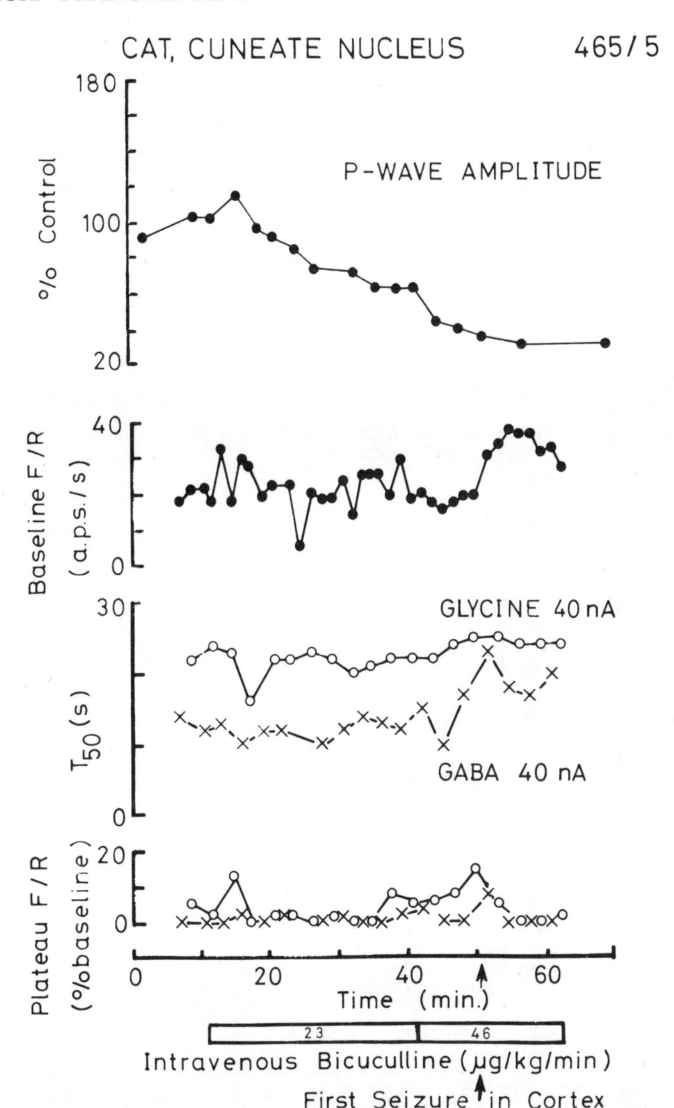

Fig. 3. *Data plotted from a cuneate neuron in an N_2O-halothane anaesthetized cat illustrating the effects of systemic bicucilline on P-wave amplitude and GABA and glycine sensitivity*

The top record shows the amplitude of the surface positive (P) wave corresponding to primary afferent depolarization and presynaptic inhibition. This was produced at regular intervals by stimulation of the ipsilateral radial nerve and recorded via a silver ball macroelectrode from the surface of the cuneate nucleus. Neuronal sensitivity to regular and alternate applications of glycine and GABA expressed as T_{50} was determined (second record from bottom). Regular estimates of baseline firing rate (second record from top) and plateau firing rate (bottom record) were taken.

During the intravenous infusion of bicuculline 23 µg/kg/min there was an early and progressively increasing decrease in P-wave amplitude which eventually reached about 40% of the control value. This fall in P-wave amplitude was not accompanied by any change in glycine or GABA sensitivity up until the time when the first electrographic seizure was recorded in the cerebral vortex (indicated by ↑ below the time axis). After this time there was apparently a small but sustained decrease in GABA sensitivity with an increasing base line firing rate.

Ordinate: Elapsed time in minutes.

This figure is reproduced with permission from Straughan (1974).

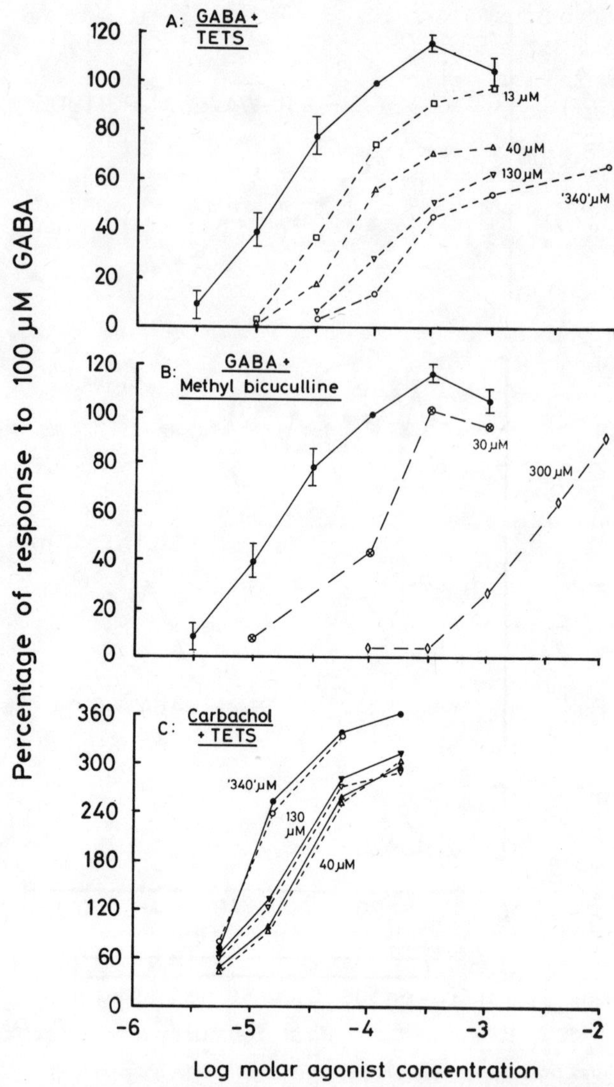

Fig. 4. *The effects of tetramethylenedisulphotetramine (TETS)* (A & C) *and methyl bicuculline* (B) *on ganglionic depolarization produced by γ-aminobutyric acid (GABA)* (A & B) *and carbachol* (C)

Ordinates in each case: peak depolarizations produced during 1 min applications of either GABA or carbachol, plotted as a percentage of the response to 100 μM GABA in that preparation (100 μM GABA = 100%). Abscissae: log molar concentrations of GABA (*A* & *B*) or carbachol (*C*). Solid symbols represent the control responsed to either GABA or carbachol; open symbols the responses obtained in the presence of different concentrations (μM) of TETS or methyl bicuculline as indicated on the figure. The mean control log-dose response curves to GABA (*A* & *B*) was calculated from results obtained in six separate ganglia. Vertical bars represent S.E. mean. GABA or carbachol was applied at not less than 15 min intervals. TETS or methyl bicuculine was in contact with the tissue at least 20 min before application of the agonists. One concentration of TETS or methyl bicuculline was examined on a single ganglion. Note that '340'μM refers to an incomplete solution of TETS.

This figure is reproduced with permission from Bowery *et al.* (1975).

act they do not show how they act. It is possible that the convulsant effects of picrotoxin are unrelated to block of pre- or postsynaptic GABA receptors and that picrotoxin and leptazol have some common but as yet undefined action.

Electrographic and section studies suggest that in producing seizures the convulsant drugs discussed act primarily in the brain and not the spinal cord, and a decrease in presynaptic inhibition allowing excessive excitatory bombardment of neurons may be the major mechanism through which seizures are produced. In this connection it may be of significance that a wide range of CNS depressants and sedative and non-sedative anticonvulsants, including pentobarbitone, diazepam, tridione, phenytoin, depress presynaptic inhibitory phenomena (see Miyahara *et al.*, 1966; Schmidt, 1971). Indeed, the potentiating effect of diazepam was particularly pronounced on primary afferent depolarizations and presynaptic inhibition but postsynaptic inhibition of motoneurons was not altered (Schmidt *et al.*, 1967; Stratten & Barnes, 1968).

Recent studies in our laboratory have been concerned with the pharmacology of tetramethylenedisulphotetramine (TETS). This resin was used by furniture manufacturers in Germany as a stiffening and antimould agent for fibres. It was discovered to be highly toxic and to produce convulsions through actions within the brain stem. Smythies (1974) suggested that it might be a GABA antagonist, and indeed in the isolated superior cervical ganglion of the rat TETS antagonized the depolarizing action of GABA (Bowery *et al.*, 1975). Although it was as potent as bicuculline, and discriminated between GABA and ACh depolarizations as effectively as bicuculline, TETS appeared to act in a non-competitive and reversible manner—this is illustrated in Fig. 4. In *in vivo* studies the sensitivity of a proportion of brainstem and cuneate nucleus neurons to both GABA and glycine was reduced during the systemic infusion of TETS in subconvulsant doses (Collins *et al.*, 1975; Dray, 1975). In view of the numbers of neurons not affected it is not clear whether this block of postsynaptic inhibitory amino acid responses does in fact cause the seizures. Although seizures could be produced in each animal, there was little temporal correlation between this and the changes in amino acid sensitivity. However, doses of TETS as low as 20 μg/kg effectively reduced the P-wave in rat cuneate nucleus—providing further evidence for the view that this is a very sensitive site and possibly a key site in the production of drug-induced seizures.

Acknowledgement

The support of the Medical Research Council for some of the research described in this review is gratefully acknowledged, as is the help of Mrs. L. Poidevin and Miss B. A. Streek in preparing the manuscript.

References

Ayala, G. F., Dichter, M., Gumnit, R. J., Matsumoto, H. & Spencer, W. A. (1973) *Brain Res.* **52**, 1

Banna, N. R. & Hazbun, J. (1969). *Experientia(Basel)*, **25**, 382

Banna, N. R. & Jabbur, S. J. (1969) *Int. J. Neuropharmacol.* **8**, 299

Banna, N. R. & Jabbur, S. J. (1970) *Int. J. Neuropharmacol.* **9**, 553

Barker, J. L. & Nicoll, R. A. (1973) *J. Physiol. (London)* **228**, 259.

Baxter, M. G., Fowler, L. J., Miller, A. A. & Walker, J. M. G. (1973) *Brit. J. Pharmacol.* **47**, 681P
Bell, J. A. & Anderson, E. G. (1972) *Brain Res.* **43**, 161
Biscoe, T. J., Duggan, A. W. & Lodge, D. (1972) *Comp. J. Gen. Pharmacol.* **3**, 423
Bowery, N. G. (1973) Ph.D. Thesis, University of London
Bowery, N. G. & Brown, D. A. (1974*a*) *Brit. J. Pharmacol.* **50**, 205
Bowery, N. G. & Brown, D. A. (1974*b*) *Brit. J. Pharmacol.* **52**, 436P
Bowery, N. G., Brown, D. A. & Collins, J. F. (1975) *Brit. J. Pharmacol.* **53**, 422
Boyd, E. S., Merritt, D. A. & Gardner, L. C. (1966) *J. Pharmacol. Exp. Ther.* **154**, 398
Brooks, V. B. & Asanuma, H. (1965) *Am. J. Physiol.* **208**, 674
Bruggencate, G. ten & Engberg, I. (1971) *Brain Res.* **25**, 43
Clark, R. & Collins, G. G. S. (1975) *J. Physiol. (London)* **246**, 16P
Clarke, G. (1975) Ph.D. Thesis, University of London
Clarke, G., Forrester, P. A. & Straughan, D. W. (1974) *Neuropharmacology* **13**, 1047
Collins, G. G. S. (1974) *Brain Res.* **76**, 447
Collins, J. F., Hill, R. G. & Roberts, F. (1975) *Brit. J. Pharmacol.* **54**, 239P
Crawford, I. L. & Connor, J. D. (1973) *Nature (London)* **244**, 442
Crowshaw, K., Jessup, S. J. & Ramwell, P. W. (1967) *Biochem. J.* **103**, 79
Curtis, D. R., Duggan, A. W., Felix, D. & Johnston, G. A. R. (1971) *Brain Res.* **32**, 69
Curtis, D. R. & Felix, D. (1971) *Brain Res.* **34**, **301**
Curtis, D. R., Game, C. J. A., Johnston, G. A. R., McCulloch, R. & MacLachlan, R. M. (1972) *Brain Res.* **43**, 242
Curtis, D. R., Hosli, L., Johnston, G. A. R. & Johnston, I. H. (1968) *Exp. Brain Res.* **5**, 235
Curtis, D. R. & Johnston, G. A. R. (1974) *Ergeb. Physiol.* **69**, 97
Curtis, D. R., Johnston, G. A. R., Game, C. J. A. & McCulloch, R. M. (1973) *Brain Res.* **49**, 467
Curtis, D. R. & Watkins, J. C. (1963) *J. Physiol. (London)* **166**, 1.
Davidoff, R. A. (1972*a*) *Science* **175**, 331
Davidoff, R. A. (1972*b*) *Brain Res.* **45**, 638
Davidson, N. & Simpson H. K. L. (1975) *J. Physiol. (London)* **244**, 83P
Davidson, N. & Southwick, C. A. P. (1971)
Davies, J. & Watkins, J. C. (1973) *Brain Res.* **59**, 311
Dennis, M. J. & Miledi, R. (1974) *J. Physiol. (London)* **237**, 431
Dray, A. (1975) *Brit. J. Pharmacol.* In press
Dreifuss, J. J., Kelly, J. S. & Krnjević, K. (1969) *Exp. Brain Res.* **9**, 137
Duggan, A. W. (1974) *Exp. Brain Res.* **19**, 522
Fonnum, F., Storm-Mathisen, J. & Walberg, F. (1970) *Brain Res.* **20**, 259
Fonnum, F. & Walberg, F. (1973) *Brain Res.* **54**, 115
Gottesfeld, Z. & Elazar, Z. (1975) *Brain Res.* **84**, 346
Gloor, P. & Testa, G. (1974) *Electroencephalogr. Clin. Neurophysiol.* **36**, 499
Hayashi, T. (1954) *Keio J. Med.* **3**, 183
Hayashi, T. (1956) *Chemical Physiology of Nerve and Muscle* Nakayama–Shoten, Tokyo
Hill, R. G., Simmonds, M. A. & Straughan, D. W. (1972) *Brit. J. Pharmacol.* **45**, 176P
Hill, R. G., Simmonds, M. A. & Straughan, D. W. (1973*a*). *Brit. J. Pharmacol.* **49**, 37
Hill, R. G., Simmonds, M. A. & Straughan, D. W. (1973*b*). *J. Physiol. (London)* **234**, 83P
Hill, R. G., Simmonds, M. A. & Straughan, D. W. (1975) *Brit. J. Pharmacol.* In press.
Iversen, L. L., Mitchell, J. F. & Srinivasan, V. (1971) *J. Physiol. (London)* **212**, 519
Jasper, H. H., Khan, R. T. & Elliott, K. A. C. (1965) *Science* **147**, 1448
Jasper, H. H. & Koyama, I. (1969) *Can. J. Physiol. Pharmacol.* **47**, 889
Koyama, I. (1972). *Can. J. Physiol. Pharmacol.* **50**, 740
Krnjević, K. (174) *Physiol. Rev.* **54**, 418
Krnjević, K. & Phillis, J. W. (1963) *J. Physiol. (London)* **165**, 274
Krnjević, K., Randić, M. & Straughan, D. W. (1966) *J. Physiol. (London)* **184**, 78
Krnjević, K. & Schwartz, S. (1967) *Exp. Brain Res.* **3**, 320
Levy, R. A. (1974) *Brain Res.* **76**, 155
Lux, H. D. (1974) *Epilepsia* **15**, 375
Michaelis, E. K., Michaelis, M. L. & Boyarsky, L. L. (1974) *Biochim. Biophys. Acta* **367**, 338
Minchin, M. & Iversen, L. L. (1974). *J. Neurochem.* **23**, 533
Miyahara, J. T., Esplin, D. W. & Zablocka, B. (1966) *J. Pharmacol. Exp. Ther.* **154**, 118
Nishi, S., Minota, S. & Karczmar, A. G. (1974) *Neuropharmacol.* **13**, 215
Obata, K. & Takeda, K. (1969) *J. Neurochem.* **16**, 1043
Otsuka, M., Obata, K., Miyata, Y. & Tanaka, Y. (1971) *J. Neuropharmacol.* **13**, 215
Prince, D. A. (1968) *Exp. Neurol.* **21**, 307
Prince, D. A. & Farrell, D. (1969) *Neurology (Minneapolis)* **19**, 309

Roberts, P. J. (1974*a*) *Brain Res.* **67**, 419
Roberts, P. J. (1974*b*) *Nature (London)* **252**, 400
Schmidt, R. F. (1971) *Ergeb. Physiol.* **63**, 20
Schmidt, R. F., Vogel, M. E. & Zimmermann, M. (1967) *Arch. Pharmacol. Exp. Pathol.* **258**, 69.
Schmidt, R. F. & Willis, W. D. (1963) *J. Neurophysiol.* **26**, 44
Smythies, J. R. (1974) *Annu. Rev. Pharmacol.* **14**, 9
Storm-Mathisen, J. & Fonnum, F. (1971) *J. Neurochem.* **18**, 1105
Stratten, W. P. & Barnes, C. D. (1968) *Fed. Proc. Am. Soc. Exp. Biol.* **27**, 571
Straughan, D. W. (1974) *Neuropharmacology* **13**, 495
Tebecis, A. K. & Di Maria, A. (1972) *Brain Res.* **40**, 373
Van Gelder, N. M., Sherwin, A. M. & Rasmussen, T. (1972) *Brain Res.* **40**, 385
Young, A. B. & Snyder, S. H. (1973) *Proc. Nat. Acad. Sci. U.S.A.* **70**, 2832
Zukin, S. R., Young, A. B. & Snyder, S. H. (1974) *Proc. Nat. Acad. Sci. U.S.A.* **71**, 4802

Chapter 18

Carbohydrate and Energy Metabolism in Relation to Mechanisms of Epilepsy

By H. S. BACHELARD

Department of Biochemistry, Institute of Psychiatry, De Crespigny Park, Denmark Hill, London SE5 8AF, U.K.

The possibility of a defect in cerebral energy metabolism in the epilepsies has been recognized since Boyle's (1660) experiments on air exsuction when he observed convulsions in animals and birds placed in an atmosphere of lowered air pressure. Boyle commented that 'the Engine could but considerably rarefie the Air' and that 'at the end of them there remained in the Receiver no inconsiderable quantity [of air]'. The atmosphere was therefore analogous to hypoxia rather than to anoxia. This concept of a potential biochemical lesion seems largely to have been ignored until relatively recently, when revived by the classical studies of Sir Rudolph Peters on thiamine deficiency (Peters, 1924; Kinnersley & Peters, 1925; Meiklejohn *et al.*, 1932). As Tower has pointed out in his excellent monograph (Tower, 1960), the concept was promoted by neurologists such as Lennox some 50 years ago, before biochemists themselves began generally to take an active interest (Lennox & Cobb, 1928).

From the large and diffuse evidence of the subsequent literature in this field, it seems clear that there is unlikely to be one basic biochemical lesion common to all of the epilepsies, and also that there may not necessarily be one common to any particular type of epilepsy. However, while we know, in broad terms, how many of the convulsant agents act metabolically, we still do not understand, in molecular terms, how any metabolic defect can lead to any of the epilepsies.

It is intriguing that thiamine deficiency should be expected to cause an impairment of acetyl CoA formation since the major metabolic effect is on pyruvate dehydrogenase activity, yet normal concentrations of both acetyl CoA and acetylcholine have been found in the brains of animals with thiamine deficiency sufficient to produce severe symptoms (Reynolds & Blass, 1975).

Respiration from pyruvate is impaired and it seems possible that formation of acetylcholine from acetyl CoA may be protected in comparison with its oxidation in the Krebs cycle. Alternative points of involvement of thiamine (such as transketolase activity) may also be important in this context (McIlwain & Bachelard, 1971).

This is typical of the broad pattern of biochemical observations in the epilepsies, which seem to fall into two main areas of cerebral function: synaptic events (with emphasis on neurotransmitter metabolism described in other contributions to this volume) and some aspects of carbohydrate and energy metabolism including the major energy-dependent systems such as cation transport. For example, anticonvulsant drugs affect postsynaptic potentials

(Richards, 1972) and inhibit glucose utilization (Strang & Bachelard, 1973). Excitant drugs, such as the amphetamines, interfere with the transport and metabolism of catecholamines (Snyder, 1972) and stimulate cerebral glycolysis (Manning *et al.*, 1974). Convulsant agents can affect both areas in a variety of ways (Fig. 1). It seems impossible at our present state of knowledge to assess

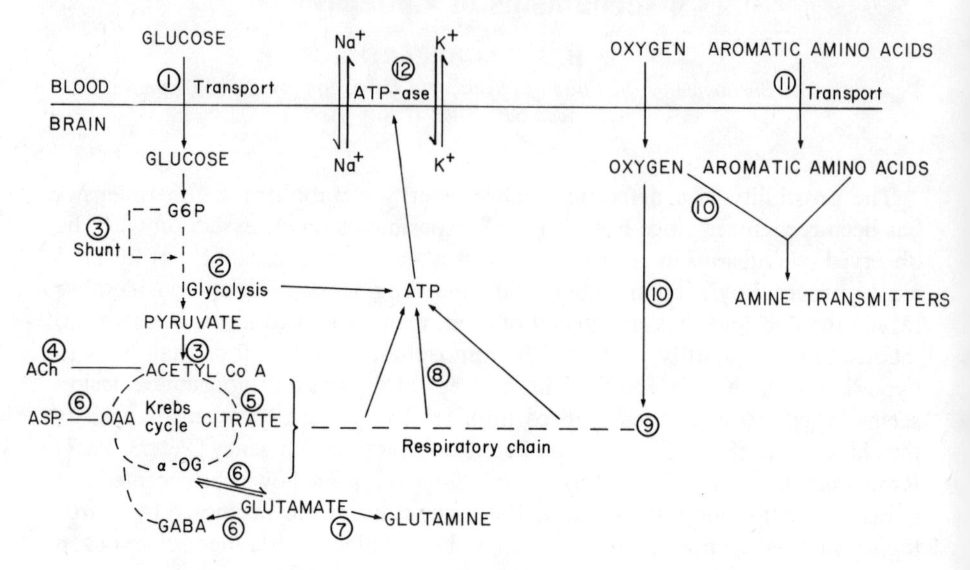

Fig. 1.

Key: (1) hypoglycaemia; (2) iodoacetate; (3) thiamine deficiency; (4) organophosphate poisons; (5) fluorocitrate; (6) pyridoxine—deficiency, antagonists; (7) hyperammonaemia?; (8) uncouplers (e.g. dinitrophenol); (9) cyanide; (10) hypoxia; (11) amino acid transport; genetic errors (e.g. phenylketonuria); (12) ouabain, metals.

which of these two areas is the more basic as a potential causative factor in any particular type of epilepsy. Both are intimately related, and while this contribution concentrates on aspects of carbohydrate and energy metabolism it cannot be discussed in isolation without reference to the functions of neurotransmitters.

It is generally clear that anything which interferes with the normal energy metabolism of the brain can cause convulsions (Fig. 1), but it must be emphasized that the convulsions which ensue need not necessarily be related to the epilepsies. They may not be identical both in the physical manifestations of the fit or in terms of the observed EEG. Even if the symptoms do seem to be identical to clinical cases of epilepsy, the experimental causative factor may produce a biochemical defect which is not necessarily related to the defects which occur naturally.

It is also worth noting that a convulsant agent (or the naturally occurring defect) may not cause hyperactivity directly but may act to inhibit or prevent processes which normally operate to end a transient period of activity, i.e. they may act to prevent normal endogenous inhibitory processes. This forms much of the rationale for the studies on inhibitory transmitter function, described in earlier chapters, and can apply equally well to regulatory factors which normally act to control metabolic rates. It follows that a metabolic defect is likely to be a subtle abnormality, i.e. some imbalance in metabolism which keeps the cells closer to the threshold of excitability; and hyperexcitation may then be triggered by a slight change in the metabolic balance. Is epilepsy a peculiar form of homeostasis, i.e. does the hyperexcitability help to return the cells to a more normal situation? Do the convulsions represent a response of the tissue attempting to regain its normal metabolic balance?

Table 1. *Energy metabolism and convulsions*

Type of study	Effect	Reference
1. Man		
Insulin	Hypoglycaemia	Tower, 1960
Low oxygen	Hypoxia	Boyle, 1660; Tower, 1960
2. Biopsy (man)		
Hexokinase	Elevated K_m(G)	Bachelard *et al.*, 1975
K^+ re-entry	Impaired	Tower, 1960, 1965
3. Animal models		
Audiogenic mice	Decreased ATPase	Hertz *et al.*, 1974
Alumina cream lesions	Decreased ATPase	Harmony *et al.*, 1968
Cobalt lesions	Decreased ATPase	Hunt & Craig, 1973
Freezing lesions	Increased ATPase	Lewin & McCrimmon, 1967
Ouabain	Inhibited ATPase	Barbeau & Donaldson, 1974
4. Animals, convulsants		
MSO	Increased G, glycogen	Folbergrova *et al.*, 1969
DOG	Hypoglycaemia	Landau & Lubs, 1958; Horton *et al.*, 1973
Insulin	Hypoglycaemia	Himwich & Fazekas, 1937
Various drugs* (Indoklon, CN, electroshock, high O_2)	Decreased G, G6P* ATP, CP Increased Na_i	Various (see McIlwain & Bachelard, 1971)
5. Animals, anticonvulsants		
Barbiturates	Increased G, glycogen	Mayman *et al.*, 1964
	Increased G uptake	Heaton & Bachelard, 1975
	Decreased glycolysis	Strang & Bachelard, 1973
	Decreased Na_i	Tower, 1960
6. Slices, *in vitro*		
Low glucose	Changed evoked potentials	C. D. Richards & H. S. Bachelard unpublished work

* Consequences of convulsions. The other effects preceded convulsions.

One of the problems in assessing the observations made so far on this topic is that most of the animal experiments have been on the metabolic consequences of convulsions. These studies have told us much about the capacity of the brain's metabolic machinery but little on the identity of possible defects which may lead to convulsions. For example, the increased consumption of glucose, glycogen, ATP and creatine phosphate, the increased production of lactate and the associated decreased in pH, are all subsequent events (Table 1). Oxygen consumption and rates of cerebral blood flow also increase in seizures, but have never been shown to be changed before the onset of seizures (Tower, 1960;

McIlwain & Bachelard, 1971). These metabolic consequences occur no matter what convulsant agent is employed (chemical, electrical or pharmacological) and so represent the effects of the hyperactivity itself.

Biochemical studies of human epileptic brain are few and are difficult to interpret, due to limitations in availability of samples, in problems of the history of drug treatment and in the type of experiment which can be employed. Recently the emphasis has been on the animal models discussed by Dr. Meldrum (this volume, Chapter 13) and on the use of slow-acting convulsant devices which allow for investigations of biochemical changes which occur before the onset of convulsions. Useful also is the study of the biochemical changes caused by anticonvulsant drugs (Table 1).

Energy-utilizing Systems

Active transport of Na^+ and K^+

The brain stores its readily utilizable energy as ATP and creatine phosphate; the major use of this energy is in active transport of Na^+ and K^+, through the action of the membrane-bound Na^+, K^+-ATPase activity. Studies on the overall ATPase activity of human brain showed no difference in epileptic tissue (Tower, 1960) but these measurements were performed before the presence of a specific Na^+, K^+-ATPase activity was recognized. The brain is one of the most active sources of this enzyme which is specifically inhibited by ouabain (Schwartz *et al.*, 1962).

Studies of the total concentrations of Na^+ and of K^+ in epileptic human tissue have also shown no difference (Tower, 1960) but changes in their transport have been detected. When slices of the cerebral cortex of man or experimental animals are prepared, potassium is lost from the cells and sodium enters. Incubation in oxygenated glucose-containing media promotes redistribution of the cations, with Na^+ extrusion and K^+ re-entry (McIlwain & Bachelard, 1971). Examination of such processes in human biopsy samples showed that the epileptic tissue had an impaired ability to pump K^+ back into the cells: the K^+ re-entry rate was calculated to be some 8 μequiv./g h compared to a normal rate of 28 μequiv./g h (Tower, 1965). Involvement specifically of Na^+ or K^+ transport has been suggested from the anticonvulsant action of Ca^{2+} in the cerebrospinal fluid, from the actions of anticonvulsant drugs and from the seizures produced by extracellular K^+ (Woodbury, 1955; Tower, 1960; Hillman, 1970; Isquierdo & Isquierdo, 1971; Zuckermann & Glaser, 1973).

It seems important to note that changes in distribution or transport of Na^+ and K^+ may not be due directly to effects on the energy state or on the ATPase activity, but may be a reflexion of a generalized effect on membrane permeability which could include changes in permeability also to Ca^{2+} and neurotransmitters. Indeed, the studies with animal models on the Na^+, K^+-ATPase have proved difficult to interpret (Table 1). Its activity is reported to be decreased in the brains of audiogenic strains of mice (Hertz *et al.*, 1974; see also Abood & Gerard, 1955) and in the focal lesions produced by cobalt and by alumina cream (Harmony *et al.*, 1968; Hunt & Craig, 1973). Yet in the freezing lesions produced by ethyl chloride, the activity is increased (Lewin & McCrimmon, 1967; Harmony *et al.*, 1968). Zinc produces convulsions similar to those produced by ouabain and is also thought to act by inhibiting Na^+, K^+-ATPase activity. Zinc is concentrated mainly in the hippocampus in normal brain, an area associated with some of the epilepsies. Further interest in zinc comes from observations of lower serum levels in treated epileptic patients (Barbeau & Donaldson, 1974).

Hexokinase

This enzyme, one of the control points in cerebral glycolysis, is not only an energy-utilizing stage but also is essential for energy production. Being the first enzymic stage of glycolysis it is very dependent upon the supply of glucose reaching the brain (Bachelard, 1967). Its maximum activity was found to be unchanged in biopsy specimens from patients with drug-resistant epilepsy but a significant change was observed in its K_m value for glucose (Table 2);

Table 2. *Hexokinase in biopsy samples: K_m values for glucose*

Specimen	No.		K_m (mM glucose)
Epileptic			
Temporal lobe	14		0.09 ± 0.017
Controls			
Temporal lobe	6	0.054	0.052 ± 0.008*
Frontal lobe	8	0.050	

* $P < 0.005$; V_{max} was not significantly altered. (Bachelard *et al.*, 1975.)

the hexokinase of the epileptic tissue showed less affinity for glucose than the enzyme in non-epileptic tissue. The possibility of artefacts due to the drugs given the patients was eliminated (Bachelard *et al.*, 1975). Such a change in enzymic property could be due to one of various factors: a genetic predisposition (Ounsted *et al.*, 1968; Falconer, 1971) or the glial scars associated with such forms of epilepsy. It could also be secondary to a change in the Ca^{2+} or Mg^{2+} of the tissue causing a change in the subunit structure or conformation of the enzyme (Bachelard *et al.*, 1975).

These studies on hexokinase form a bridge in the discussion between energy-utilizing systems and energy-producing systems, with particular reference to availability of glucose.

Energy-producing Systems

Limitations in availability of the two major substrates (glucose and oxygen) required for energy production in the brain can produce convulsions (without necessarily causing any detectable change in brain levels of ATP or creatine phosphate, below). A decrease in their rates of consumption, from whatever cause, can cause fits, and these rates are increased during convulsions. They are both needed for convulsions to proceed; a more severe lack produces coma. Yet, paradoxically, both hypoglycaemia and hypoxia can have the opposite effect, as both can abolish the effects of certain convulsant agents: for example, hypoxia can abolish the convulsions induced by metrazole and the convulsions induced by metrazole and the convulsions return if the animals are then given normal oxygen (Tower, 1960). This also may be an indication of the importance of the balance of the metabolic state.

Glucose

One of the fascinating features of the studies on convulsant drugs which interfere with the metabolism of glutamate and of γ-aminobutyrate (Fig. 1) is the accumulation of glucose which occurs before the onset of convulsions. Methionine sulphoximine, long known as a convulsant drug (Tower, 1960), is an inhibitor of glutamine synthesis, and produces the doubling of brain glucose shown in Table 3 (Folbergrova *et al.*, 1969). Since then, drugs which

Table 3. *Methionine sulphoximine and convulsions*

	Time after injection (h)				
	Preceding convulsions				During convulsions
	0	1	2	2.75	3.7–5.2
ATP	3.0 (100)	3.0 (100)	3.2 (107)	2.9 (97)	2.9 (97)
PC	4.4 (100)	4.3 (98)	4.7 (107)	4.5 (102)	4.6 (105)
Glucose	1.1 (100)	1.6 (146)	1.5 (136)*	2.2 (200)*	2.6 (236)*
Glycogen	2.1 (100)	2.6 (124)*	3.3 (157)*	3.4 (162)*	4.8 (229)*
Lactate	0.8 (100)	0.8 (100)	0.95 (119)	0.75 (94)	1.3 (162)*
NH_3	0.24 (100)				1.18 (490)*

The results (Folbergrova *et al.*, 1969) are μmol/g mouse brain with percentage of control values in parentheses.

* Statistically significant difference from control.

interfere with γ-aminobutyrate production, such as allylglycine and deoxypyridoxine, have been shown to produce similar but more pronounced increases in brain glucose (B. S. Meldrum & R. W. Horton, personal communication). It is tempting to consider that this may reflect an attempt to protect the tissue against convulsions. This was originally suggested by Tower (1960), who commented that raised glucose may be a protective device or an emergency defence mechanism, and is discussed below.

Hypoglycaemia

The evidence for the convulsions produced by hypoglycaemia came first from accidentally high doses of insulin in the treatment of diabetics and, especially, from the development of Sakel's insulin shock therapy in the treatment of psychoses. The seizures and the observed changes in EEG have been reviewed by Poire (1969). About 90% of the cases observed in a carefully designed study showed epileptic manifestations after insulin-induced hypoglycaemia, including some with seizures of 'grand mal' type. The variations in susceptibility to hypoglycaemia between individuals were taken as an indication of genetic variations but were concluded not to be due to the presence of glial scars.

Recovery from hypoglycaemia symptoms due to glucose injection can be remarkably quick. Intravenous injections in man act within 30 s (Poire, 1969) and in the cat, an intracardial injection of glucose produced recovery within 3 s

(Waltregny, 1969). Waltregny commented that it 'seems as though the lack of *immediately available* hexose in the neurons was directly responsible for the disturbance in cerebral function as shown by the almost immediate reversion of symptoms after [giving] small quantities of sugar'.

Intracellular concentrations of glucose in the normal brain are considered to be low (Bachelard, 1969) and may be limited by the capacity of the brain to take glucose. Glucose transport from the bloodstream to the brain is known to be a saturable stereospecific process of facilitated diffusion and follows Michaelis–Menten kinetics with a K_m (glucose) of 5 to 8 mM and a maximum velocity of 70 to 80 μmol/g h (Bachelard, 1975). At normal plasma glucose concentrations of 6 to 8 mM, the maximum capacity of the uptake process is about twice that required to maintain normal rates of glucose consumption, and uptake becomes limiting at concentrations below 2 mM, where hypoglycaemia symptoms begin. There seems to be a threshold effect: a slight theoretical limitation in glucose availability produces the changes in EEG and can lead to convulsions, yet the overall metabolism from glucose seems unimpaired. In the hypoglycaemia induced by insulin in man and various experimental animals, there need be no detectable change in respiration or in ATP levels, and no change in P_{O_2} or P_{CO_2}, in the time period before the onset of seizures (Himwich & Fazekas, 1937; Gibbs *et al.*, 1939; Della Porta *et al.*, 1964; Brierley *et al.*, 1971; Ferrendelli & Chang, 1973; Lewis *et al.*, 1974).

Similar results have been observed with hypoglycaemia produced using 2-deoxy-D-glucose: 'cellular' hypoglycaemia at normal blood glucose levels (Landau & Lubs, 1958; Horton *et al.*, 1973). Deoxyglucose exerts interesting effects on glucose metabolism in the brain (Fig. 2): it competes for, and inhibits,

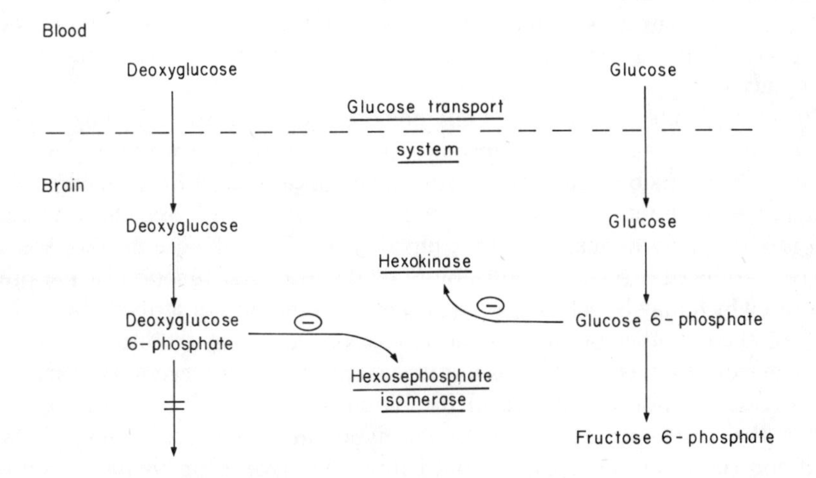

Fig. 2. *Glucose and deoxyglucose transport and phosphorylation*

glucose uptake and is phosphorylated by hexokinase. Yet the product, deoxyglucose 6-phosphate, has no effect on hexokinase (glucose 6-phosphate is a potent inhibitor) and is not further metabolized (Bachelard, 1972). It does not act as a substrate, unlike glucose 6-phosphate, for hexose phosphate isomerase, but inhibits it (Horton *et al.*, 1973).

Studies on the time-course of metabolic events following administration of deoxyglucose produced results which showed that the convulsions were produced by the hypoglycaemia itself, rather than by the subsequent inhibition of hexose phosphate isomerase, and therefore of glycolysis (Table 4). These

Table 4. *Deoxyglucose and convulsions*

Time (min)	Concentration in brain (% of control)					Behaviour	Metabolism
	G	G6P	Lactate	ATP	CP		
0	100	100	100	100	100	Changed EEG	Glucose uptake: 60% inhibited
5	62*	137	104	103	106	Tonic, clonic jerks	Glycolysis: 70% inhibited
10	137*	189*	69				

* Statistically significant (see Horton *et al.*, 1973).

results produced the same conclusion as those from the use of insulin: convulsions are produced by a simple and relatively slight deficiency in glucose with no discernible preceding interference with cerebral energy metabolism.

The threshold effects of subtle changes in the concentration of available glucose are also shown in Fig. 3. The prepiriform cortex preparation used in this experiment has been described by Dr. Richards (this volume, Chapter 15). The recording technique requires the slice to be floated upon the surface of the incubation fluid in the chamber. The tissue therefore needs about twice the concentration of glucose which would normally be required for a totally immersed slice. The appearance of the evoked potentials in 10 mM glucose was maintained in 6 mM glucose but electrical activity completely disappeared in 2 mM glucose.

However, in the presence of 4 mM glucose the observed recording slowly changed from normal to the abnormal type of discharge shown in Fig. 3, and proved to be reversible with time. The original appearance of the evoked potential returned without any change in the glucose concentration in the medium. This mild hypoglycaemia, roughly equivalent to 2 mM glucose *in vivo*, seems to have produced a reversible instability of the electrical function of the preparation which may be relevant to the *in vivo* observations described above.

All of these studies point to a subtle role of glucose in the brain which is different from its classical role as a nutrient. Specific glucoreceptors, sensitive to the concentration of glucose in the bloodstream, have been described in cells of the ventromedial nucleus of the hypothalamus (Cross, 1964). These could contribute to events *in vivo* but cannot be involved in the *in vitro* preparations. We may speculate therefore on the presence of glucoreceptors, or areas of special sensitivity to glucose, elsewhere in the brain. Could these react to changes in glucose concentrations to produce changes in permeability to other molecules or to modify some aspect of neurotransmitter function? Isolated preparations of synaptosomes have been shown to possess a glucose transport system which is quite distinct in its kinetic properties from the systems described elsewhere (Diamond & Fishman, 1973; Heaton & Bachelard, 1973), which may be an indication of a specialized function of glucose at the synapse.

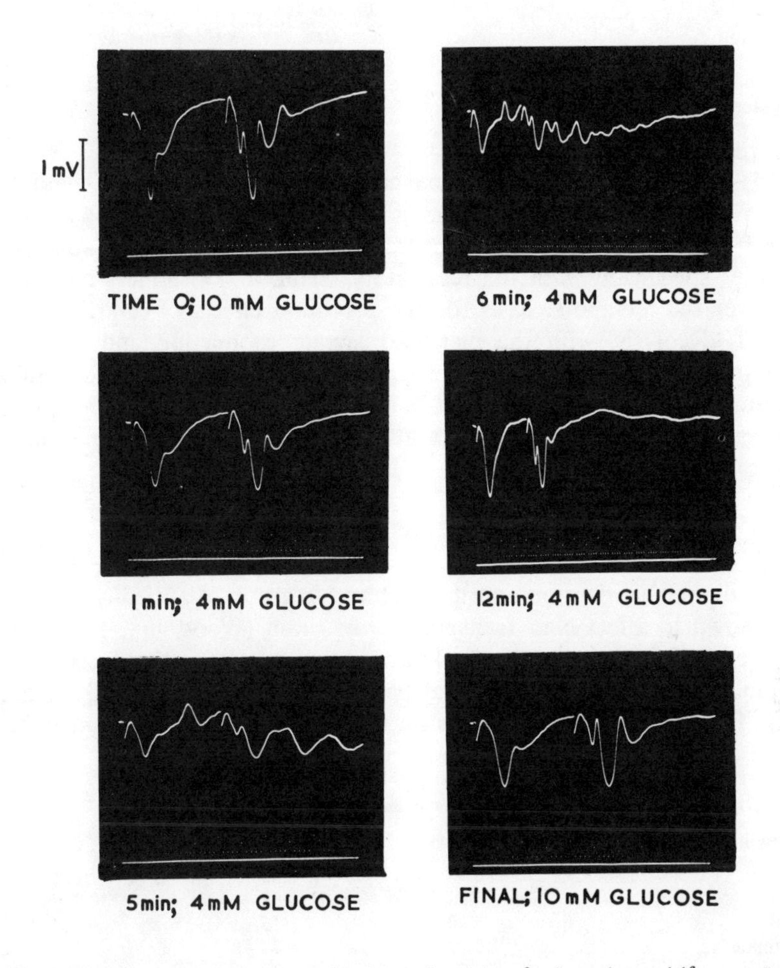

Fig. 3. *The effects of low glucose on the electrical activity of guinea-pig prepiriform cortex*

Oxygen

Studies using mild conditions of controlled arterial hypoxia in man and in experimental animals have shown a profound increase in glycolysis without any change in respiration (due to increased cerebral blood flow) or in the brain concentrations of ATP or of creatine phosphate (Cohen, 1971; Bachelard *et al.*, 1974). Is a lack of *respiratory* oxygen therefore involved? A comparison of the need for oxygen of relevant processes in the brain indicate that the respiratory chain is not the most sensitive (Table 5). The O_2-requiring hydroxylation reactions involved in the production of serotonin and the catecholamines

Table 5. *K_m values for oxygen of cerebral processes*

Process	K_m value (nM-O_2)	Reference
Respiration	0.1	Clark *et al.* 1975
Amine hydroxylation		
Tyrosine hydroxylase	500	Fisher & Kaufman, 1972
Tryptophan hydroxylase	1000	Kaufman, 1974

(Fig. 1) are far more sensitive to oxygen than is the respiratory chain and it seems possible therefore that the major or primary effect of hypoxia may be on the function of the biogenic amine transmitters. Nevertheless, cyanide, which interferes specifically with the respiratory chain, also produces convulsions. Moreover the metabolic basis for the profound effects of mild hypoxia on cerebral glycolysis, with the increased lactate production and decrease in pH, remains to be explained. Does it indicate some undiscovered regulatory mechanisim, e.g. some interaction of amines in the pathway specifically of glycolysis? Could it also reflect an attempt by the brain to redress a subtle metabolic imbalance?

Anticonvulsant Drugs

Both diphenylhydantoin and the barbiturate group of anticonvulsants have been shown to affect Na^+ transport in the brain (Woodbury, 1955; Tower, 1960), and diphenylhydantoin is known to exert profound effects on Ca^{2+} metabolism especially in peripheral tissues (Richens & Rowe, 1970; Frame, 1971). The majority of the studies on the effects of anticonvulsant drugs on

Table 6. *Phenobarbitone and cerebral metabolite levels*

Metabolite	Control	Phenobarbitone	% Change
Glucose	0.99	2.02	+102*
G6P	0.08	0.09	+ 16*
Lactate	2.30	1.55	− 33*
Glutamate	10.0	6.9	− 31*
Glycogen	1.2	1.3	+ 8

* Statistically significant; Strang and Bachelard, 1973.

the intermediary metabolism of the brain have been performed using barbiturates. These are of particular interest in view of the convulsions which can occur when barbiturate addicts are treated by withdrawal (Essig, 1967). Barbiturates decrease cerebral rates of glycolysis and respiration and produce increased concentrations of carbohydrates in the brain (Table 6) without necessarily affecting concentrations of high energy phosphates (Mayman *et al.*,

1964; Strang & Bachelard, 1973; Nilsson & Siesjö, 1974). The increase in brain glucose occurs with normal concentrations in the blood.

The use of [^{14}C]glucose *in vivo* has shown that the metabolic effect of phenobarbitone is specifically within the pathway of glycolysis and that the decreased respiration is a consequence of this: the lowered rate of metabolism in the tricarboxylic acid cycle is due to the inhibition of glycolysis (Table 7). The

Table 7. *Phenobarbitone and cerebral metabolism*

Precursor	→	Product	^{14}C rate constant (min^{+1}) Control	Phenobarbitone	% Change
Glucose		Lactate	0.185	0.068	−64
Glucose		Glutamate	0.225	0.073	−67
Lactate		Glutamate	0.43	0.35	−18

Strang & Bachelard, 1973.

accumulation of glucose in the brain could not be interpreted solely in terms of inhibition of its rate of consumption (Mayman *et al.*, 1964; Strang & Bachelard, 1973). Glucose transport may also be affected. Table 8 shows that 0.25 mM

Table 8. *Phenobarbitone and glucose transport*

K^+ (mM)	Phenobarbitone (0.25 mM)	K_m (glucose) mM	V_{max} (μmol/g h)
6.5	—	5	70
6.5	+	2	90
45.0	—	7	90
45.0	+	2	120

Heaton & Bachelard, 1975.

phenobarbitone lowers the K_m values, and slightly increases the V_{max}, from kinetic studies of hexose uptake *in vitro* (Heaton & Bachelard, 1975). Barbiturates also lower the K_m value for D-xylose uptake (Gilbert *et al.*, 1966). Calculations of theoretical rates of unidirectional influx from these kinetic constants (Fig. 4) showed the rate could be increased by 50% by phenobarbitone in the presence of normal 6.5 mM K^+ and by 100% with phenobarbitone and high 45 mM K^+ (which causes depolarization). High K^+ alone had no significant effect.

This effect on glucose transport could reflect a generalized change in membrane transport phenomena, caused by the drugs (see the comments by Dr. Richards—this volume, Chapter 15). It could also be a further indication of the function of glucose as a protective agent.

Comment

This chapter has concentrated on carbohydrate and energy metabolism in the epilepsies, an area which may be as important as are synaptic events. A general pattern which appears from these studies is that a very slight change in the availability of essential nutrients can trigger off convulsions with no noticeable

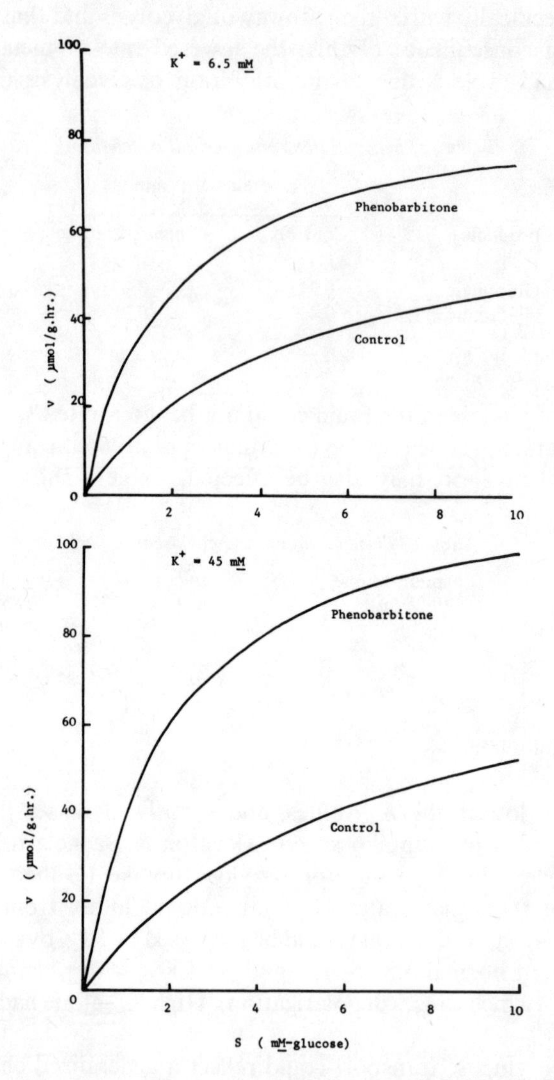

Fig. 4. *Phenobarbitone and glucose uptake to the brain* (Heaton & Bachelard, 1975).

impairment of energy metabolism. It is worth noting that a lack of change in levels of energy intermediates does not necessarily mean that their rates of turnover are unaltered; levels can be maintained by adjustment of the relative rates of energy-producing and energy-consuming metabolic events. Nevertheless, all the evidence suggests no change in the overall energy metabolism, from such as the unaltered rates of cerebral respiration observed *in vivo*.

The convulsant agents included in this description, and the anticonvulsant drugs, are known to cause changes also in membrane permeability or some aspect of synaptic function and the relationship between these and the availability of the essential nutrients is unclear. Following from the comments made in the text, some questions emerge.

(*a*) Are certain regions of the brain especially and specifically sensitive to variations in immediately available glucose, which fulfils a role quite distinct from its function as a metabolic energy substrate? Is this role a modulation of synaptic events? Is the synaptic area particularly sensitive to glucose as a nutrient?

(*b*) Is there present in the brain, an unsuspected group of regulatory mechanisms through which events at the synapse can exert some type of feedback control on rates of glucose transport and consumption?

There is little evidence (analogous to the sensitivity of transmitter metabolism to hypoxia) which bears on these questions at present, but they may indicate routes along which research could proceed.

Acknowledgements

The studies described in the chapter were supported by the Medical Research Council, the U.S. Public Health Service (NS 07918) and the Research Fund of the Bethlem Royal and Maudsley Hospitals, London.

References

Abood, L. G. & Gerard, R. W. (1955) in *Biochemistry of the Developing Nervous System* (Waelsch, H., ed.), pp. 467–472, Academic Press, New York

Bachelard, H. S. (1967) *Biochem. J.* **104**, 286–292

Bachelard, H. S. (1969) *Handb. Neurochem.* **1**, 25–31

Bachelard, H. S. (1972) *Biochem. J.* **127**, 83P

Bachelard, H. S. (1975) in *The Working Brain* (Ingvar, D. & Lassen, D., eds.), Munksgaard, Copenhagen

Bachelard, H. S., Lewis, L. D., Pontén, U. & Siesjö, B. K. (1974) *J. Neurochem.* **22**, 359–401

Bachelard, H. S., Polkey, C. E. & Thompson, M. F. (1975) *Epilepsia* (in press)

Barbeau, A. & Donaldson, J. (1974) *Arch. Neurol. Psychiat.* **30**, 52–58

Boyle, R. (1660) *New Physicomechanical Experiments Touching the Spirit of Air, and Its Effects,* Hall, Oxford

Brierley, J. B., Brown, A. W. & Meldrum, B. S. (1971) *Brain Res.* **25**, 483–499

Clark, J. B. (1975) *J. Neurochem.* **24**, 533–538

Cohen, P. J. (1971) in *Ion Homeostasis of the Brain* (Siesjö, B. K. & Sørensen, S. C., eds.), Munksgaard, Copenhagen

Cross, B. A. (1964) *Symp. Soc. Exp. Biol.* **18**, 157–193

Della Porta, P., Maiolo, A. T., Negri, V. U. & Rosella, E. (1964) *Metabolism* **13**, 131–140

Diamond, I. & Fishman, R. A. (1973) *J. Neurochem.* **20**, 1533–1542

Essig, C. F. (1967) *Epilepsia* **8**, 21–30
Falconer, M. A. (1971) *Epilepsia* **12**, 13–31
Ferrendelli, J. A. & Chang, M.-M. (1073) *Arch. Neurol. Psychiat.* **28**, 173–177
Fisher, D. B. & Kaufman, S. (1972) *J. Neurochem.* **19**, 1359–1365
Folbergrova, J., Passoneau, J. V., Lowry, O. H. & Schulz, D. W. (1969) *J. Neurochem.* **16**, 191–203
Frame, B. (1971) *Ann. Intern. Med.* **74**, 294–295
Gibbs, F. A., Gibbs, E. L. & Lennox, W. G. (1939) Influence of the blood sugar levels on the wave and spike formation in petit mal epilepsy. *Arch. Neurol. Psychiat.* **41**, 1111–1116
Gilbert, J. C., Ortiz, W. R. & Millichap, J. G. (1966) *J. Neurochem.* **13**, 247–255
Harmony, T., Urba-Holmgren, R., Urbay, C. M. & Szava, S. (1968) *Brain Res.* **11**, 672–680
Heaton, G. M. & Bachelard, H. S. (1973) *J. Neurochem.* **21**, 1099–1108
Heaton, G. M. & Bachelard, H. S. (1975) (in preparation)
Hertz, L., Schousboe, A., Formby, B. & Lennox-Buchthal, M. (1974) *Epilepsia* **15**, 619–631
Hillman, H. (1970) *Lancet* **ii**, 23–24
Himwich, H. E. & Fazekas, J. F. (1937). *Endocrinology* **21**, 800–805
Horton, R. W., Meldrum, B. S. & Bachelard, H. S. (1973) *J. Neurochem.* **21**, 507–520
Hunt, W. A. & Craig, C. R. (1973) *J. Neurochem.* **20**, 559–567
Isquierdo, I. & Isquierdo, J. A. (1971) *Annu. Rev. Pharmacol.* **11**, 188–208
Kaufman, S. (1974) in *Aromatic Amino Acids in the Brain,* CIBA Foundation Symposium no. 22, pp. 85–108, Elsevier, Amsterdam
Kinnersley, H. W. & Peters, R. A. (1925) *Biochem. J.* **19**, 820–826
Landau, B. R. & Lubs, H. A. (1958) *Proc. Soc. Exp. Biol. Med.* **99**, 124–127
Lennox, W. G. & Cobb, S. (1928) *Epilepsy,* Williams and Wilkins, Baltimore
Lewin, E. & McCrimmon, A. (1967) *Arch. Neurol. Psychiat.* **16**, 321–325
Lewis, L. D., Ljunggren, B., Ratcheson, R. A. & Siesjö, B. K. (1974) *J. Neurochem.* **23**, 673–680
McIlwain, H. & Bachelard, H. S. (1971) *Biochemistry and the Central Nervous System,* 4th edn, Churchill, London
Manning, D. H., Strang, R. H. C. & Bachelard, H. S. (1974) *Biochem. Pharmacol.* **23**, 1205–1209
Mayman, C. I., Gatfield, P. D. & Breckenridge, B. McL. (1964) *J. Neurochem.* **11**, 483–487
Meiklejohn, A. P., Passmore, R. & Peters, R. A. (1932) *Proc. Roy. Soc. B.* **111**, 391–395
Nilsson, L. & Siesjö, B. K. (1974) *J. Neurochem.* **23**, 29–36
Ounstead, C., Linday, J. & Norman, R. (1968) *Clinics in Developmental Medicine* **22**, Heinemann, London
Peters, R. A. (1924) *Biochem. J.* **18**, 858–865
Poire, R. (1969) in *The Physiopathogenesis of the Epilepsies* (Gastaut H., Jasper, H., Bancaud, J. & Waltregny, A., eds.), pp. 75–110, Thomas, Springfield, Ill.
Reynolds, S. F. & Blass, J. P. (1975) *J. Neurochem.* **24**, 185–186
Richards, C. D. (1972) *J. Physiol. (London)* **227**, 749–767
Richens, A. & Rowe, D. J. F. (1970) *Brit. Med. J.* **iv**, 73–75
Schwartz, A., Bachelard, H. S. and McIlwain H. (1962) *Biochem. J.* **84**, 626–637
Snyder, S. H. (1972) *Arch. Gen. Psychiat.* **27**, 169–179
Strang, R. H. C. & Bachelard, H. S. (1973) *J. Neurochem.* **10**, 987–996
Tower, D. B. (1960) *Neurochemistry of Epilepsy,* Thomas, Springfield, Ill.
Tower, D. B. (1965) *Epilepsia* **6**, 183–197
Waltregny, A. (1969) in *The Physiopathogenesis of the Epilepsies* (Gastaut, H., Jasper, H., Bancaud, J. & Waltregny, A., eds.), Thomas, Springfield, Ill.
Woodbury, D. A. (1955) *J. Pharmacol. Exp. Ther.* **115**, 74–95
Zuckermann, E. G. & Glaser, G. H. (1973) Anticonvulsant action of increased calcium concentration in cerebrospinal fluid. *Arch. Neurol. Psychiat.* **29**, 245–252

Discussion Paper

Folate and Epilepsy

By E. H. REYNOLDS

University Department of Neurology, Institute of Psychiatry and King's College Hospital, Denmark Hill, London SE5 8AF; and Medical Research Council Clinical Research Centre, Northwick Park, Harrow, Middlesex, U.K.

First I will summarize the effects of the major anticonvulsant drugs on folate metabolism in the clinical situation, and then, with reference to experimental studies, discuss the possible role of the vitamin and its derivatives in the antiepileptic action of the drugs, in seizure mechanisms, and in cerebral metabolism.

Clinical Observations

I have reviewed these in detail elsewhere (Reynolds, 1972). Megaloblastic anaemia, due to folic acid deficiency, was first reported as a rare complication of diphenylhydantoin (DPH), phenobarbitone or primidone therapy in 1952, but it was not until the mid 1960s that a less severe form of the vitamin deficiency was recognized in the majority of *non-anaemic* epileptic patients treated with these drugs. Several studies revealed subnormal serum folate levels in

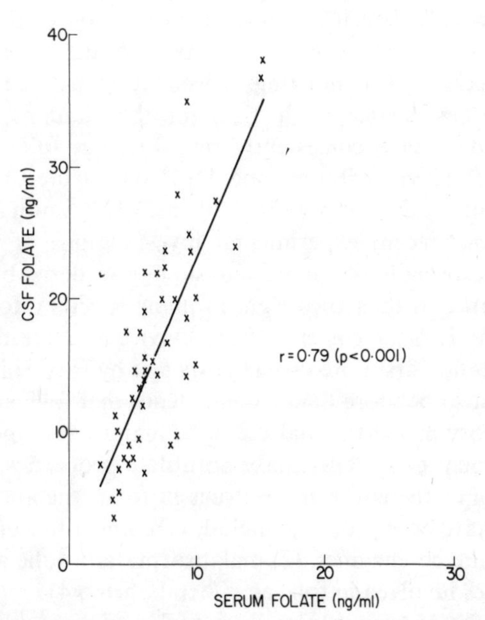

Fig. 1. *Relationship between CSF and serum folate in drug-treated epileptics*
From Reynolds *et al.* (1972).

between 27% and 91% (mean 55%) of such patients. The fall in serum folate is accompanied by a corresponding rise in mean red cell volume (macrocytosis) and fall in red cell folate, especially 'free' red cell folate (i.e. glutamate chain length 1–3) (Chanarin *et al.*, 1975).

Of particular importance to the present discussion is the fact that cerebrospinal fluid (CSF) folate is normally approximately three times greater than serum folate and that the drugs result in a fall in CSF folate which is closely correlated with that in serum (Reynolds *et al.*, 1972) (Fig. 1). There is a negative correlation between drug and folate levels in both serum and CSF.

There appears to be an efficient blood–brain barrier mechanism for the vitamin as illustrated by the failure to elevate CSF folate levels even after three months of treatment with large doses of the vitamin and despite a considerable rise of serum folate levels (Spaans, 1970). Chanarin *et al.* (1974) have shown that labelled 5-methyltetrahydrofolate is detectable in the CSF of neurological patients at least within four hours of parenteral administration, but rapidly declines; and so it seem there is considerable circulation of folate between serum and CSF, but presumably the vitamin is either rapidly taken up by brain cells or returned to the circulation.

Relationship Between Antiepileptic and Antifolate Mechanisms

This is reviewed by Reynolds (1973). In view of the fall in serum and CSF folate levels produced by all three major anticonvulsants in so many patients, and the apparent deterioration in seizure control in some following the administration of folic acid, Reynolds *et al.* (1966) suggested a possible relationship, at least in part, between the antiepileptic action of the drugs and their antifolate properties. Subsequent clinical reports of the effect of the vitamin on seizure control have been conflicting. This may in part be explained by the efficient blood–brain barrier mechanism for the vitamin. However, several experimental studies have consistently revealed that folic acid partially reverses the effect of phenobarbitone and DPH on seizure threshold (de Wolff *et al.*, 1971; Woodbury & Kemp, 1971; Obbens, 1973; Smith & Racusen, 1973). Furthermore, other recent experimental investigations have confirmed that folic acid, and especially its more physiological derivatives, have potent excitatory properties in their own right in many species (Hommes & Obbens, 1972; Spector, 1972; Baxter *et al.*, 1973; Davies & Watkins, 1973; Obbens, 1973). This was in fact first noted some years ago by Hayashi (1959) in dogs.

It would appear to be more than a coincidence that folic acid and its derivatives have excitatory properties and the three major anticonvulsant drugs long used in the treatment of epilepsy, have antifolate properties.

The mechanism of the effect of the drugs in folate metabolism is unknown. Hypotheses that have been proposed include (1) competitive interaction between the drugs and folate co-enzymes, (2) malabsorption of folic acid, (3) induction of hepatic enzymes involved in folic acid metabolism, (4) increased demand for folic acid either for anticonvulsant drug hydroxylation or for other hepatic enzymes induced by the drugs. The conflicting evidence for these hypotheses is reviewed by Reynolds (1972). It is of interest that Spector (1971) has reported that methotrexate, a folate antagonist used in the treatment of malignant disease, has anticonvulsant properties when administered parenterally in the rat.

It has also been suggested that folic acid may lead to deterioration in seizure control by increasing the metabolism of DPH, leading to a fall in the blood level of this drug (Baylis *et al.*, 1971), but evidence for this hypothesis has also been conflicting (Reynolds, 1972) and further studies are indicated.

Folic Acid and Seizure Mechanisms

The demonstration of the excitatory properties of folic acid and its derivatives has naturally raised questions about the mechanism of this effect and the possible role of the vitamin in seizure mechanisms. Obbens (1973) has shown that a cobalt or heat lesion to rat cerebral cortex will increase the penetration of the blood–brain barrier in the region of the focus by parenterally administered folic acid, and this is associated with increased susceptibility to folate-induced seizures. She quotes Mayersdorf *et al.*'s (1971) brief report of elevated folate content in the region of cobalt foci, to support her suggestion that local accumulation of folate due to damage to the blood–brain barrier mechanism for the vitamin, as a result of a focal brain lesion, may contribute to clinical focal epilepsy.

The functions of folic acid in brain are largely unknown but one amongst several possibilities is that the vitamin is involved in synaptic events, especially in view of its convulsant properties, and also the report that much of it is localized in synaptosomal fractions, at least in the mouse (Bridgers & Maclain, 1972). Spector (1972) has even suggested that the vitamin may have a transmitter role. The fact that the folate molecule includes one or several glutamate moeties has naturally focused attention on the possibility that it is the glutamate, with its well-known excitatory properties, that confers the convulsant properties on folic acid (whether mono- or polyglutamate). Obbens (1973) thought that this was unlikely as she found that following intracisternal injection Na-folate was, molecule for molecule, 100 times more convulsant than Na-glutamate. She suggests that folate (in the form of polyglutamate) is the glutamate carrier postulated by Krnjević (1970) for the action of glutamic acid and Na-K-ATPase in neuronal transmission. Davies & Watkins (1973) reported that iontophoretic application of folate and folinate to single neurons in the cat pericruciate cortex increased the firing of spontaneously active and glutamate-activated cells, but were only weakly excitatory to quiescent neurons. They also found evidence that the vitamin could block by up to 80% the inhibitory effects of γ-aminobutyric acid (GABA). Roberts (1974) found that in the rat dorsal root ganglion folate could inhibit the high affinity glial uptake of [^{14}C]glutamate, but he thought this was unlikely to have physiological significance. It has also been suggested that folic acid, like picrotoxin, may block presynaptic inhibitory receptors in the rat cuneate nucleus (Hill & Miller, 1974) but in this situation no GABA antagonism was observed (Hill *et al.*, 1974).

Other Cerebral Metabolic Functions of Folic Acid

The functions of folic acid in brain need not be and possibly are not confined to synaptic events. From our knowledge of the functions of the vitamin in other tissues it may be presumed, for example, that it is involved in cerebral nucleoprotein and methionine synthesis, and there are complex but ill-understood relationships with vitamin B_{12} metabolism (Chanarin, 1969).

Another area of recent interest has been a possible role in monoamine metabolism, and this too has implications for neurotransmission. Tyrosine and tryptophan hydroxylation require pteridine co-factors but whether or not these are derived from folate still seems uncertain. It has been suggested, on the basis of *in vitro* studies, that folate may be a physiological methyl donor in brain. Laduron (1972) and Leysen & Laduron (1974*a*) have recently reported the existence in several species of a brain methyltransferase requiring 5-methyltetrahydrofolate as methyl donor and capable of *N*-methylation of nearly all primary and secondary catacholeamines and indoleamines *in vitro*. Korevaar *et al.* (1973), who found a close correlation between the regional distribution of 5-methyltetrahydrofolate and indoleamine (but not catacholeamine) pathways in rat brain, suggest that *N*-methylation may be an inactivation step in indoleamine metabolism. Banerjee & Snyder (1974) have even suggested that 5-methyltetrahydrofolate may be the physiological donor in *O*- as well as *N*-methylation of indoleamines. It should be stressed, however, that *S*-adenosylmethionine

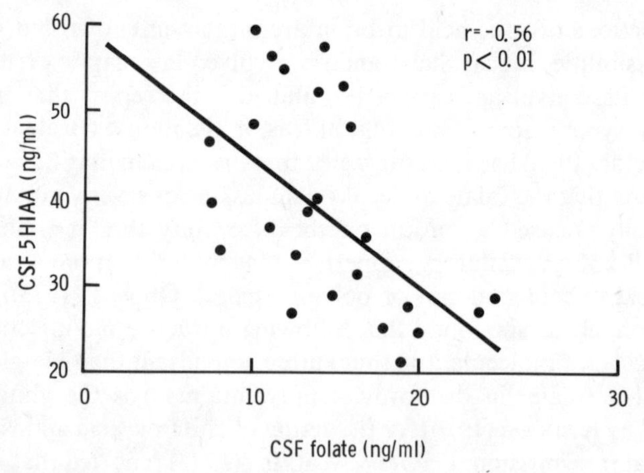

Fig. 2. *Relationship between CSF folate and CSF 5-hydroxyindoleacetic acid in drug-treated epileptics* From Reynolds *et al.* (1975).

is still widely regarded as the physiological methyl donor in brain and the possible physiological significance of the above studies is uncertain at the present time, not least because the postulated *N*-methylated amines have not been identified, even in *vitro* (Leyson & Laduron 1974*b*; Mandel *et al.*, 1974; Meller *et al.*, 1975).

It is of interest, in relation to the above discussion, that anticonvulsants also affect monoamine metabolism. Thus Chadwick *et al.* (1975) have recently shown for the first time in man that 'therapeutic' blood levels of phenobarbitone and DPH are associated with a significant elevation of CSF 5-hydroxyindoleacetic acid (5HIAA) and possibly also homovanillic acid (HVA). This confirms earlier experimental studies demonstrating an influence of these and other anticonvulsants on serotonin (5HT) synthesis and 5HIAA exit from CSF (Bonnycastle *et al.*, 1957; Chase *et al.*, 1969). We have examined the relationship between CSF folate and CSF 5HIAA in our drug-treated epileptic patients (Reynolds *et al.*, 1975) and found a significant negative correlation (Fig. 2). This correlation is not seen in untreated epileptics. It may be that the negative correlation observed in Fig. 2, and which was also seen to a less significant extent between CSF folate and HVA, is no more than coincidental, but in view of the *in vitro* studies referred to above it would seem to repay further investigation. Indeed, the largely unknown functions of folic acid in brain would appear to be a profitable area for further basic studies, with definite clinical implications.

References

Banerjee, S. P. & Snyder, S. H. (1974) *Advan. Psychopharmacol.* **11**, 85

Baxter, M. G., Miller, A. A. & Webster, R. A. (1973) *Brit. J. Pharmacol.* **48**, 350P

Baylis, E. M., Crowley, J. M., Preece, J. M., Sylvester, P. E. & Marks, V. (1971) *Lancet* **i**, 62

Bonnycastle, D. D., Giarman, N. J. & Paasonen, M. K. (1957) *Brit. J. Pharmacol.* **12**, 228

Bridgers, W. F. & Maclain, L. D. (1972) *Advan. Psychopharmacol.* **4**, 81

Chadwick, D., Jenner, P. & Reynolds, E. H. (1973) *Lancet* **i**, 473

Chanarin, I. (1969) *The Megaloblastic Anaemias,* Blackwell, Oxford

Chanarin, I., Perry, J. & Reynolds, E. H. (1974) *Clin. Sci.* **46**, 369

Chanarin, I., Perry, J., Lumb, M., Laundy, M., Chadwick, D. & Reynolds, E. H. (1975) In preparation.

Chase, T. N., Katz, R. I. & Kopin, I. J. (1969) *Trans. Am. Neurol. Ass.* **94**, 236

Davies, J. and Watkins, J. C. (1973) *Biochem. Pharmacol.* **22**, 1667

de Wolff, F. A., Hillen, F. C., Sprangers, W. J. J. M., Suijkerbuijk-van Beek, M. M. A. & Noach, E. L. (1971) *Arch. Int. Pharmacodyn. Ther.* **194**, 316

Hayashi, T. (1959) *Neurophysiology and Neurochemistry of Convulsion,* Dainihan-Tosho, Tokyo

Hill, R. G. & Miller, A. A. (1974) *Brit. J. Pharmacol.* **50**, 425

Hill, R. G., Miller, A. A., Straughan, D. W. & Webster, R. A. (1974) in *Epilepsy. Proceedings of the Hans Berger Centenary Symposium* (Harris, P. & Mawdsley, C., eds.), Churchill Livingstone, Edinburgh

Hommes, O. R. & Obbens, E. A. M. T. (1972) *J. Neurol. Sci.* **16**, 271

Korevaar, W. C., Geryer, M. A., Knapp, S., Hsu, L. L. & Mandell, A. J. (1973) *Nature (London)* **245**, 244

Krnjević, K. (1970) *Nature (London)* **228**, 119

Laduron, P. (1972) *Nature (London)* **238**, 212

Leysen, J. & Laduron, P. (1974*a*) *Advan. Psychopharmacol.* **11**, 65

Leyson, J. & Laduron, P. (1974*b*) *FEBS Lett.* **47**, 299

Mandel, L. R., Rosegay, A., Walker, R. W., Van den Heuvel, W. J. A. & Rokach, J. (1974) *Science* **186**, 741

Mayersdorf, A., Streiff, R. R., Wilder, B. J. & Hammer, R. H. (1971) *Neurology (Minneapolis)* **21**, 418

Meller, E., Rosengarten, H., Friedhoff, A. J., Stebbins, R. D. & Silber, R. (1975) *Science* **187**, 171

Obbens, E. A. M. T. (1973) *Experimental Epilepsy Induced by Folate Derivatives,* Thesis, Nijmegan

Reynolds, E. H. (1972) in *Antiepileptic Drugs* (Woodbury, D. M., Penry, J. K. & Schmidt, R. P., eds.), p. 247, Raven Press, New York

Reynolds, E. H. (1973) *Lancet* **i**, 1376

Reynolds, E. H., Chanarin, I., Milner, G. & Matthews, D. M. (1966) *Epilepsia* **7**, 261

Reynolds, E. H., Gallagher, B. B., Wattson, R. H., Bowers, M. B. & Johnson, A. L. (1972) *Nature (London)* **240**, 155

Reynolds, E. H., Chadwick, D., Jenner, P. & Chanarin, I. (1975) To be published
Roberts, P. J. (1974) *Nature* (*London*) **250**, 429–430
Smith, D. B. & Racusen, L. C. (1973) *Arch. Neurol.* **28**, 18
Spaans, R. (1970) *Epilepsia* **11**, 403
Spector, R. G. (1971) *Biochem. Pharmacol.* **20**, 1730
Spector, R. G. (1972) *Nature* (*London*) **240**, 247
Woodbury, D. M. & Kemp, J. W. (1971) *Psychiat. Neurol. Neurochir.* **74**, 91

Discussion Paper

A General View of Possible Causative Factors in Epilepsy and Their Investigation by Means of Pharmacological Agents

By J. C. WATKINS

Department of Pharmacology, The Medical School, University Walk, Bristol BS8 1TD, U.K.

Reduced to the simplest terms, epilepsy is probably the consequence of some specific malfunction of conduction or transmission within the central nervous system such that one or more foci produce an inadequately controlled excitatory output. Although in experimental epilepsies in animals these impaired mechanisms within the primary foci might well be detectable at the single cell level by the use of appropriate recording techniques, it is possible that more subtle malfunctions are involved in human epilepsy and that these manifest as convulsive activity only as a result of the summation of subliminal effects with a complex neuronal system. It may be useful to list some hypothetical ways in which conduction and transmission may be impaired, and to consider ways in which actual pathological conditions might be investigated. In the classification below, mechanisms 1–3 relate to conduction and 4–6 to transmission.

Conduction of electrical impulses along fibres may be impaired by:

(1) Structural defects in conducting membrane, that is, in the membrane through which the ion movements associated with action potentials take place. This could include defects in the organic ligands responsible for the binding of Ca^{2+} to the conducting membrane, which is of critical functional importance (Shanes, 1958). The fault may lie either in the synthesis of the membrane macromolecules or in the enzymes responsible for providing the unit molecules (amino acids, fatty acids, glycerides, amino sugars etc.) from which the macromolecules are assembled.

(2) Malfunction of ionic pumps situated in the cytoplasmic membrane of the neurons, and responsible for maintaining optimal electrochemical gradients across the membrane for action potential generation and conduction. Such malfunction may be due to defects in enzyme structure (such as in Na^{+}- and K^{+}-activated membrane ATPase) or, conceivably, by way of a defective association of the enzyme with a structural membrane matrix. Impaired function could also be due to inadequate ATP synthesis arising from a deficient carbohydrate transport or metabolism, or to malfunction of the enzymic processes, such as electron transport, involved in oxidative phosphorylation.

(3) Malfunction of glial-situated ion pumps responsible for 'corrective maintenance' of extracellular concentrations of ions, e.g. in the active uptake of KCl into glial cells in cases of abnormally high extracellular concentration

of K^+. Impairment of function could be due to similar causes as detailed under (2).

Transmission may be impaired by:

(4) Too little transmitter available for release from presynaptic endings on the arrival of an action potential in the terminal This could be due to (*a*) deficient transmitter synthesis, which in turn might be due to low levels of the synthesizing enzyme, co-factor or precursor. Alternatively (*b*), re-uptake into terminals of released transmitter, or of transmitter metabolites might be inadequate. The fault here could lie in defective structure of the membrane components involved in uptake or in a defective energy supply—since uptake against a concentration gradient is an active process—or in a defective coupling of the uptake components to the energy supply.

(5) Defective transmitter release mechanism, which could result in too little or too much transmitter being released per impulse in the presynaptic fibre. Structural defects may be present in the membrane components involved in release, or the processes involved in the 'mobilization' of transmitter, prior to release, during repetitive firing of the presynaptic fibre may be inadequate.

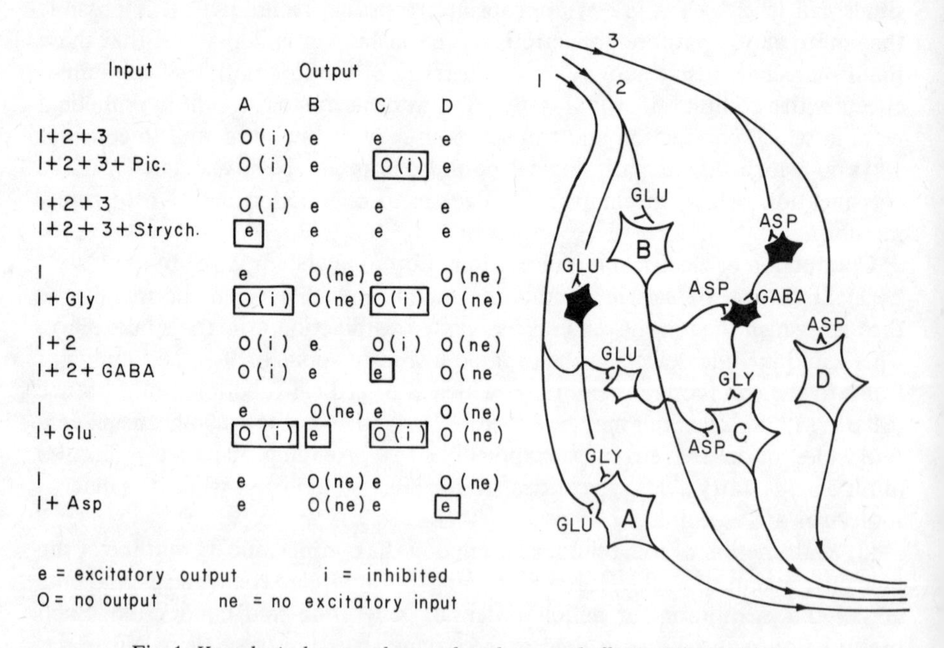

Fig. 1. *Hypothetical neuronal network and expected effects of chemical lesions*

Fibres 1, 2 and 3 constitute the excitatory input, and cells A, B, C and D the output. Three inhibitory interneurons (black) and one excitatory interneuron complete the network. Four different transmitters are involved, glutamate and aspartate as excitatory transmitters and GABA and glycine as inhibitory transmitters. The effect of substances present in the extracellular fluid (ECF) would depend on the number of fibres of the excitatory input which happen to be active. A few examples are given, neglecting effects of overdepolarization (leading to depression) of cells A–D which could arise from an excess of glutamate or aspartate present in the ECF. Changes in the expected output *v.* input relationship, due to the presence of particular substance in the ECF, are indicated by rectangles.

Figure 1 illustrates the difficulties of elucidating the exact mechanism underlying any neural dysfunction which results in an abnormal output from a focus. In the simple neuronal network depicted, it can be seen that the effect of an added agent, or of an excitatory or inhibitory substance present in extracellular fluid due to leakage from cellular compartments, depends on the excitatory input. In the examples given, the presence of γ-aminobutyric acid (GABA) in the extracellular fluid could cause an increased excitatory output, whereas the presence of glutamate or picrotoxin could reduce the excitatory

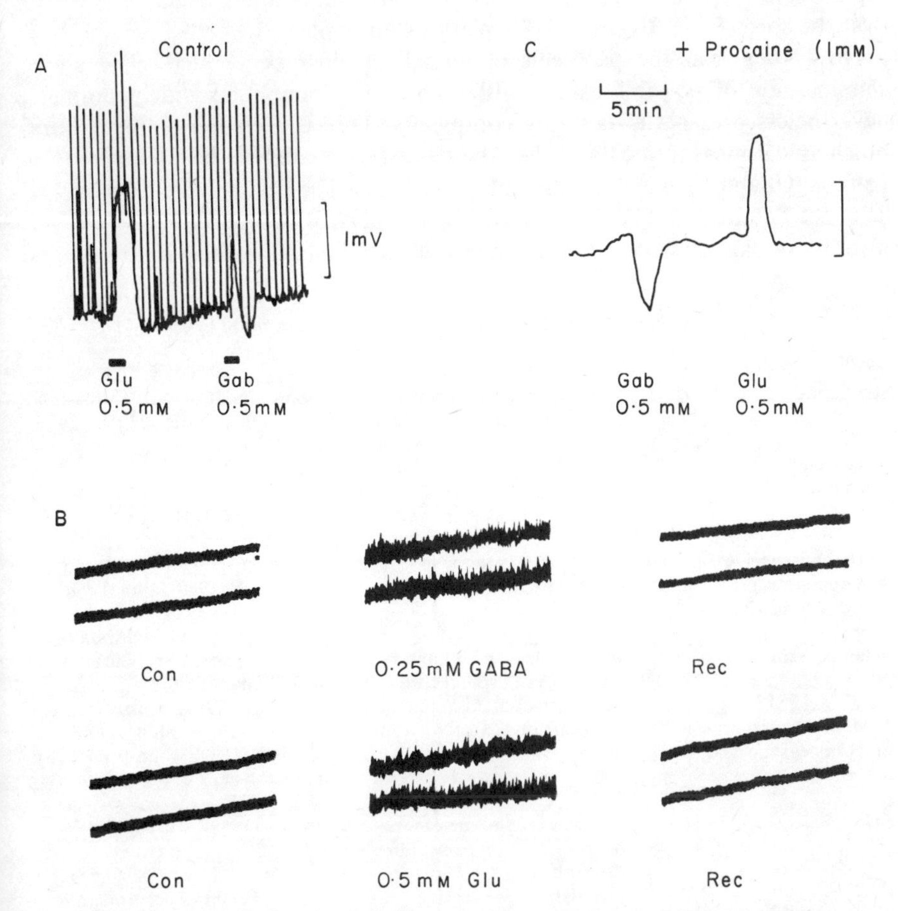

Fig. 2. *Excitatory action of* L-*glutamate and GABA on the isolated hemicord of the frog*

Responses were measured in the ventral root. In *A* (slow recording) the dorsal root was stimulated at 30 s intervals, causing a transient depolarization of motoneurons due to release to excitatory transmitter from terminals of primary afferents and interneurons. The presence of glutamate in the perfusion medium caused a depolarization of motoneurons during the period over which the glutamate was in contact with the cord. The presence of GABA in the perfusion medium caused an initial depolarization, followed by hyperpolarization. Fast recordings made from another preparation during the depolarizing phases of the action of glutamate and GABA (*B*) both showed repetitive discharge of action potentials in the ventral root. The GABA-induced depolarization was not present when interneuronal activity was blocked by procaine (*C*) and is considered to be due to inhibition by GABA of the activity of inhibitory neurons synapsing with motoneurons, i.e. to a GABA-induced 'disinhibition' of motoneurons (R. H. Evans & J. C. Watkins, unpublished work).

output, depending on what input fibres were active at any given time. It is self-evident that if a lesion is present in a neural focus, then whether or not this gives rise to an epileptiform discharge in fibres leaving the focus could well be dependent on the activity of particular fibres entering the focus. The fact of a hyperexcitatory output from the focus under these conditions might well obscure an essentially *inhibitory* lesion (in terms of effects on individual cells) within the focus. On the other hand, the lesion could equally well be an excitatory one at the single cell level. Figure 2 shows repetitive discharges in the ventral root of the isolated amphibian cord produced by both GABA and glutamate perfused over the cord (R. H. Evans & J. C. Watkins, unpublished work).

How, then, can the problems of human epilepsy be tackled? The above classification of possible causes, although grossly simplified and incomplete, nevertheless presents a daunting complexity. Direct electrophysiological and biochemical measurements of the type described in animals by Meldrum (this volume, Chapter 13) are not possible in humans. The best possibility may well lie in the detailed analysis of cerebrospinal fluid (CSF) composition. However prior to, or parallel with any such study of human CSF, an extensive programme

Table 1. *Pharmacological agents with defined mechanisms of action*

Agents	Action	References
Strychnine[1]	Blocks postsynaptic spinal inhibition and the depressant actions of glycine, β-alanine	Bradley *et al.* (1953); Curtis *et al.* (1968*a, b*); Curtis & Johnston (1974)
Bicuculline[1] Picrotoxin[1]	Block GABA-mediated pre- and postsynaptic inhibition	Curtis *et al.* (1971*a, b*); Schmidt (1971); Meldrum & Horton (1971); Curtis & Johnston (1974)
Thiosemicarbazide[1] 3-Mercaptopropionic acid[1] 4-Deoxypyridoxine[1]	Inhibit the synthesis of GABA in presynaptic endings	Killam & Bain (1957); Baxter & Roberts (1960); Wood *et al.* (1966); Baxter (1970); Lamar (1970); Horton & Meldrum (1973)
Tetanus toxin[1]	Blocks release of inhibitory transmitter in spinal cord	Brooks *et al.* (1957); Curtis & de Groat (1968)
Kainic acid[1,2] *N*-Methyl-DL-aspartic acid[1,2] DL-Homocysteic acid[1,2]	Stimulate excitant amino acid receptors on central neurons	Crawford (1963); Curtis & Watkins (1963); Shinozaki & Konishi (1970); Johnston (1973); Johnston *et al.* (1974); Biscoe *et al.* (1975)
EDTA[2]	Reduces level of membrane bound Ca^{2+}, increases fibre excitability	Shanes (1958); Curtis *et al.* (1960)
K^+ ions[2]	Depolarizes conducting membrane, increases fibre excitability	Shanes (1958); Zuckermann & Glaser (1968)
Ouabain[2]	Inhibits Na^+- and K^+-activated membrane ATPase, alters ion distribution, increases fibre excitability	Donaldson *et al.* (1972)
2-Deoxyglucose[1] Fluoroacetate[1] Fluorocitrate[1]	Interfere with energy metabolism, probably causing altered ion distribution and increased fibre excitability	Peters (1957, 1963); Tower (1963); Horton *et al.* (1973)

[1] Intravenous or intraperitoneal administration.
[2] Intracerebral injection or perfusion.

Table 2. *CSF composition as a means of correlating biochemical and electrical changes in brain*

Metabolites	Rationale
Amino acids[a]	1–5
Fatty acids[a]	1, 4, 5
Metals (e.g. Na, K, Ca, Mg, Zn, Fe)[a]	1, 4, 5
Phosphate[a]	1, 4, 5
Amino sugars[a]	1, 4
Glycerol[a] and glycerides	1, 4
Proteins and peptides	1, 3 (?)[b]
Carbohydrate metabolites (e.g. glucose, pyruvate, lactate, malate)	5 (3)[c]
Adenine metabolites	5 (2, 3?)[d]
Transmitter metabolites	2, 3
Co-factor metabolites (e.g. of B_1, B_6, nicotinamide, folic acid)	3, 4, 5

1. Involvement in membrane structure and maintenance.
2. Transmitter function.
3. Transmitter synthesis/turnover
4. Intermediary metabolism, general.
5. Energy metabolism.

[a] These should be assayed in both unmodified and hydrolysed CSF samples so that free and bound amounts can be determined.

[b] Substance P is currently under consideration as a transmitter candidate (Konishi & Otsuka, 1974; Takahashi & Otsuka, 1975).

[c] The synthesis of putative amino acid transmitters, glutamate, aspartate and GABA proceeds through carbohydrate metabolites.

[d] Adenosine has been postulated to play a role in the control of electrical excitability (McIlwain, 1972; see also Phillis *et al.*, 1975)

should be conducted on animals such that the changes in both tissue and CSF composition can be correlated with electrophysiological manifestations in experimental epilepsies caused by as large a number as possible of clearly defined mechanisms. Key metabolites could be selected so as to enable the monitoring of a wide range of enzymic and other systems which could be involved. A table of some of the most useful pharmacological agents (Table 1) is given below, and also a general outline of the range of CSF analyses required (Table 2). It should be noted that Table 1 excludes epileptogenic agents such as cobalt metal and other similar agents for which the mechanism of convulsive action has still to be determined, although these agents may also yield useful information relating to convulsive phenomena in general (Emson, this volume, Chapter 14) and the mechanism may soon be elucidated (e.g. Bradford, this volume, Chapter 16).

Any such exhaustive study of CSF composition must include controls (*a*) for separating cause and effects of metabolic dysfunction, and (*b*) for the drug therapy being undergone by most epileptic patients. In relation to the problem of cause and effect, changes in brain tissue and CSF composition resulting from epileptogenic treatments must be monitored both before and after seizures become apparent. To examine the effects of anticonvulsant drug therapy, control animals and animals undergoing epileptogenic treatment must be subjected to a similar chronic drug administration both prior to and during the induction of convulsive states.

As a parallel approach to CSF analysis, an intensive search could be undertaken for specific antidotes to seizures caused by convulsive drugs, and other treatments, which produce their actions by way of defined mechanisms. Once developed, and their mechanism of action clearly defined, the new drugs could be subjected to clinical trials for anticonvulsant activity against different types of epilepsy. If effective, any treatment would, by corollary, elucidate the underlying dysfunction and thus pave the way for improved therapy. All new drugs must of course be able to cross the blood–brain barrier. For picrotoxin- (and bicuculline-)induced convulsive states, as well as for seizures resulting from glutamic acid decarboxylase (GAD) inhibition, a range of substances incorporating structural similarities to GABA might be investigated. Unfortunately, one series of compounds offering promise of such activity, namely, *β-p*-chlorphenyl-GABA and related substances, seem to exert depressant effects on single neurons of the mammalian CNS more by way of their phenylethylamine moiety than through any interaction with GABA receptors (Curtis *et al*., 1974; Davies & Watkins, 1974). However, the scope for further synthesis along these lines is wide. Antistrychnine agents might likewise be based on the structures of glycine, *β*-alanine and taurine. Additionally, effort might fruitfully be directed towards the development of potent inhibitors of the uptake of synaptically released GABA or of the other putative inhibitory transmitters (Iversen & Johnston, 1971; Balcar & Johnston, 1973). Inhibition of the enzymic breakdown of GABA following cellular re-uptake, such as that produced by amino-oxyacetic acid and di-*n*-propylacetate (Bradford, this volume, Chapter 16), may also prove effective, although the inhibition of the breakdown enzymes seems to occur mainly in compartments other than nerve endings. Such inhibition may, however, indirectly lead to higher than normal extracellular levels of the transmitter, perhaps by reducing uptake. Unlike the case of inhibitory amino acids, agents which specifically antagonize the excitatory action of glutamate and/or aspartate have not yet been discovered, although an active search for such compounds is being pursued in a number of laboratories. Their development may well open a new chapter of anticonvulsive therapy. Other prospective avenues that suggest themselves are molecules with local anaesthetic properties to counteract the excitability increases likely to occur in conducting membranes due to defective Ca^{2+} binding. However, the expected action of such local anaesthetic-like agents on peripheral conduction might constitute a difficult problem.

References

Balcar, V. J. & Johnston, G. A. R. (1973) *J. Neurochem.* **20**, 529–539
Baxter, C. F. (1970) *Handb. Neurochem.* **3**, 289–353
Baxter, C. F. & Roberts, E. (1960) *Proc. Soc. Exp. Biol. Med.* **104**, 426–427
Biscoe, T. J., Evans, R. H., Headley, P. M., Martin, M. & Watkins, J. C. (1975) *Nature* (*London*) **255**, 166–167
Bradley, K., Easton, D. M. & Eccles, J. C. (1953) *J. Physiol.* (*London*) **122**, 474–488
Brooks, V. B., Curtis, D. R. & Eccles, J. C. (1957) *J. Physiol.* (*London*) **135**, 655–672
Crawford, J. M. (1963) *Biochem. Pharmacol.* **12**, 1443–1444
Curtis, D. R. & de Groat, W. C. (1968) *Brain Res.* **10**, 208–212
Curtis, D. R., Duggan, A. W., Felix, D. & Johnston, G. A. R. (1971*a*) *Brain Res.* **32**, 69–96

Curtis, D. R., Duggan, A. W., Felix, D., Johnston, G. A. R. & McLennan, H. (1971*b*) *Brain Res.* **33**, 57–73
Curtis, D. R., Game, C. J. A., Johnston, G. A. R. & McCulloch, R. M. (1974) *Brain. Res.* **70**, 493–499
Curtis, D. R., Hösli, L. & Johnston, G. A. R. (1968*a*) *Exp. Brain Res.* **6**, 1–18
Curtis, D. R., Hösli, L., Johnston, G. A. R. & Johnston, I. H. (1968*b*) *Exp. Brain Res.* **5**, 235–258
Curtis, D. R. & Johnston, G. A. R. (1974) *Ergeb. Physiol.* **69**, 97–188
Curtis, D. R., Perrin, D. D. & Watkins, J. C. (1960) *J. Neurochem.* **6**, 1–20
Curtis, D. R. & Watkins, J. C. (1963) *J. Physiol. (London)* **166**, 1–14
Davies, J. & Watkins, J. C. (1974) *Brain Res.* **70**, 501–505
Donaldson, J., Minick, J. L. & Barbeau, A. (1972) *Can. J. Biochem.* **50**, 888–896
Horton, R. W. & Meldrum, B. S. (1973) *Brit. J. Pharmacol.* **49**, 52–63
Horton, R. W., Meldrum, B. S. & Bachelard, H. S. (1973) *J. Neurochem.* **21**, 507–520
Iversen, L. L. & Johnston G. A. R. (1971) *J. Neurochem.* **18**, 1939–1950
Johnston, G. A. R. (1973) *Biochem. Pharmacol.* **22**, 137–140
Johnston, G. A. R., Curtis, D. R., Davies, J. & McCulloch, R. M. (1974) *Nature* (*London*) **252**, 804–805.
Killam, K. F. & Bain, S. A. (1957) *J. Pharmacol. Exp. Ther.* **119**, 255–262
Konishi, S. & Otsuka, M. (1974) *Nature* (*London*) **252**, 734–735
Lamar, C. (1970) *J. Neurochem.* **17**, 165–170
McIlwain, H. (1972) *Biochem. Soc. Symp.* **36**, 69–85
Meldrum, B. S. & Horton, R. W. (1971) *Brain Res.* **35**, 419–436
Peters, R. A. (1957) *Advan. Enzymol.* **18**, 113–159
Peters, R. A. (1963) *Biochemical Lesions and Lethal Synthesis*, pp. 88–130, Pergamon, Oxford
Phillis, J. W., Kostopoulos, G. K. & Limacher, J. J. (1975) *Eur. J. Pharmacol.* **30**, 125–129
Schmidt, R. F. (1971) *Egeb. Physiol.* **63**, 20–101
Shanes, A. M. (1958) *Pharmacol. Rev.* **10**, 59–273
Shinozaki, H. & Konishi, S. (1970) *Brain Res.* **24**, 368–371
Tower, D. B. (1963) *Am. J. Clin. Nutr.* **12**, 308–320
Takahashi, T. & Otsuka, M. (1975) *Brain Res.* **87**, 1–11
Wood, J. D., Watson, W. J. & Stacey, N. E. (1966) *J. Neurochem.* **13**, 361–370
Zuckermann, E. G. & Glaser, G. H. (1968) *Exp. Neurol.* **22**, 96–117

General Discussion on Epilepsy

H. F. Bradford: Those of us who are proponents of an excitatory role for glutamate in epilepsy have noted that folic acid contains a pteroyl unit linked to several series-linked glutamic acid moieties (1 to 7) and is even called polyglutamate. This raises the question of whether its analogue relationship to glutamic acid is important in understanding the excitation it produces. Thus it might work as a glutamate agonist at glutamate receptors or it might simply block the re-uptake of glutamate released at the synapse or elsewhere and thereby raise extracellular glutamate to convulsive levels [H. F. Bradford & P. R. Dodd (1975) in *Biochemistry and Neurological Diseases* (A. N. Davison, ed.) chap. 3, Blackwell, Oxford]. Roberts has recently examined the capacity of folate to block glutamate uptake and finds that it is effective at low glutamate concentrations (2 μM) with a K_i of 500 μM [Roberts (1974) *Nature* (*London*) **250**, 429–430]. The much greater potency of folate over glutamate when introduced intracerebrally (1000-fold) mentioned by Dr. Reynolds could be due to the rapid uptake of glutamate which would be ensured by high affinity transport processes thus rapidly reducing its extracellular concentration to zero. No such uptake system for folate has yet been demonstrated and it would maintain its extracellular concentration.

A. N. Davison: One central problem which concerns us all is what is the most appropriate experimental model of epilepsy. All of them seem to have something to offer but none of them appear to have the complete answer. For the biochemist the cobalt-metal implant model appeals because it is easily produced and fairly reproducible.

B. S. Meldrum: As you say, there are many different models in use. In fact, no perfect model can be conceived since epilepsy encompasses such a wide range of clinical conditions. You must simply choose the model which is appropriate for the type of tests which you are going to conduct. I feel that the data presented this morning on the cobalt model do not distinguish between factors which may be involved in causing epilepsy and those which are its consequences. It is not clear, for instance, whether changes in amino acids such as glutamate and GABA are due to the loss of neurons, and proliferations of astrocytes and microglia due to the abnormal activity, or whether the changes are actually causing the epilepsy.

H. H. Hillman: All models involving focal lesions have the same problem. They can themselves cause local biochemical changes which can be mistaken for the causes or effects of the epileptic state. Unless you use systemically an agent like picrotoxin which causes a generalized seizure, any agent you implant on the brain will generate the same problem.

M. H. Joseph: With reference to Dr. Bradford's paper, I should like to say that we regard the amino acid changes as a consequence rather than a cause of epilepsy. Taking together the enzyme and the histological changes and those in aspartate, glutamate and GABA we believe that these are changes secondary to neuronal damage and degeneration. The change in glycine and glutamine are likely to be secondary to the extensive gliosis which occurs. Strong supporting evidence for this conclusion is that there are no amino acid changes in the secondary focus. These conclusions probably also apply to the changes observed in excised human cortical foci, and the release of glutamate to superfusates by cobalt foci is probably due to neuronal breakdown. The critical experiments would be those performed on the secondary foci. Work on the cobalt foci must now begin to study turnover rates. Since this model shows a delay in the onset of epilepsy following cobalt implantation, there *is* a chance of detecting changes which occur before the spiking is detected and which may therefore be causal. On this point I must disagree with Dr. Meldrum. By manipulating the rates of spiking it may be possible to study the effects of activity on specific biochemical parameters and learn something of factors controlling epileptogenicity.

D. W. Straughan: I should like to make two points. Firstly, large changes in amino acid levels in fact may have no functional significance at all with regard to epilepsy since, as I showed this morning, synaptic function of neurons in the foci appear normal despite all these changes. Secondly, regarding Dr. Bradford's paper showing glutamate release from the focus, there does seem to be an interesting correlation here. Thus the only certain electrical abnormality of the epileptic focus is the generation of abnormal large depolarizations which are believed to involve the transmitter at these neurons and this might well be glutamate or aspartate. Equally the changes could be due to changed metabolism and other effects, but the correlation between release of glutamate and these paroxysmal depolarizations which are observed in the focus remains very striking.

J. C. Watkins: I should like to endorse this point. The correlation might also highlight the role of glutamate in excitatory synaptic transmission. This apparently specific release of glutamic acid closely parallels the development of the hyperactivity and presents a striking coincidence of effects.

C. J. van den Berg: There is about 10μmol glutamate g wet wt brain. 1% to 10% of this could be acting as a neurotransmitter, the rest has other functions. If the glutamate pool is decreased by 30–50% how do we know whether the pool concerned with transmission is changed rather than that concerned with other functions? We cannot say that the effect on glutamate is due to a change in its transmitter role.

H. F. Bradford: We don't know whether the amino acid changes are causally connected with the epileptic activity or not. The fact remains that if you are looking for a mechanism you need to find some kind of hint of it in the epileptogenic tissue. It is not acceptable simply to find a series of agents which, say, reduce the levels of GABA in normal brain and in parallel cause convulsions, and conclude that an aberration in the GABA system must be important in generating human epilepsy. It must simply rank with the many other possibilities until the appropriate aberrations in the GABA system are consistently found in human epilepsy. As far as the involvement of glutamate is concerned, similar considerations must apply, though diminished glutamate and GABA pools and changes in their enzymes in focal tissue may well prove to be more difficult to ascribe to cause than to effect. However, a strong circumstantial case for glutamate as an excitatory factor may be made. A central observation which bridges between the observed glutamate release and the precipitation of epileptic states, is the convulsive effect of glutamate introduce either intracerebrally in small amounts or systematically in larger amounts, and this has been observed by many researchers sometimes inadvertently when introducing glutamate during metabolic experiments (see page 205). Aspartate is also a convulsant at higher doses but other amino acids at the same dosage have no effect. we do not know whether the loss of amino acids is due to lowered production, leakage, or loss of a specific compartment. Any specific compartment lost would have to be large to account for all of the change (30–50%) and could not, for instance, be nerve endings. Also, it must be remembered that the change is often reversible. In cobalt epilepsy alone levels slowly return to normal when the epileptic condition ceases. This could be due to gliosis, with glial cells replacing lost neurons and making an equivalent amino acid contribution to the tissue, the net change in composition being zero. Equally, it might be a reversible change within the neurons themselves. At any event there remains the positive observation on the potent convulsant properties of glutamate itself when present extracellularly.

A. N. Davison: Can we link together epilepsy induced by hypoglycaemia and these other forms of epilepsy? For instance, do you see the same pattern of amino acid change during fits induced by hypoglycaemia?

H. S. Bachelard: Experiments on mild hypoglycaemia, sufficient to produce EEG changes in the period leading to onset of convulsions have shown no significant changes in amino acid levels. But, as emphasized by others today, since only a small proportion of the total pool of a transmitter amino acid is likely to change, these changes might be hard to detect.

H. S. Bachelard: Professor Straughan commented on the possible involvement of an excitatory transmitter (glutamate) and Dr. Bradford emphasized the importance of studying human epileptic tissue samples. Why is acetylcholine (ACh) not equally considered? It is the original excitatory neurotransmitter, and was shown by Tower to be changed in human epileptic biopsy samples [D. B. Tower (1960) *The Neurochemistry of Epilepsy*, Thomas, Springfield, Ill.] The total ACh was not changed but the ratio of free to bound was affected. Why cannot ACh be the important excitant in the clinical epilepsies, even if it is not in the animal models?

D. W. Straughan: I think there is a simple answer to this, namely, that in those neurons in epileptic foci which have been examined the electrophyisological changes seen are not typically those which acetycholine would produce. Thus acetylcholine would increase membrane conductance whilst glutamate application would decrease it. The observed changes show a decreased membrane conductance.

C. D. Richards: I would like to make three points in relation to the possible involvement of acetycholine (ACh). Firstly, the widely distributed acetylcholine-esterase would rapidly destroy any released acetylcholine. Secondly, acetylcholine excitation is of slow onset in many regions of the brain, and thus, unless a sustained and massive release of ACh were taking place, it would be surprising if it could propagate a seizure. Thirdly, if we take the observation that barbiturates depress the glutamate excitation of neurons as one possible mechanism of action of anticonvulsants, then it is important to note that barbiturates do not suppress the acetylcholine-induced firing of cortical neurons, indicating a clear difference between ACh receptors and glutamate receptors. This adds strength to the idea that a leakage of glutamate could be involved in the primary event in initiating a seizure, although all the reservations expressed must be clearly borne in mind.

D. A. Brown: I was struck by Professor Straughan's last slide showing a neuron which appeared electrophysiologically quite normal. The depolarization looked just like that expected to follow a massive release onto the neuron of glutamate, either by a sudden synchronous excitatory input or even due to disruption of a neighbouring cell. The depolarization did not produce sustained spiking because it obviously accommodated to the large depolarization (the spike height was clearly reduced). I was also struck by the concurrence between the time course of the subsequent hyperpolarization and the glial depolarization. The neuronal depolarization might cause a leakage of K^+ which would then depolarize the glial cell, and this might then release GABA directly from the glial cell onto the neuron to hyperpolarize it and prevent spiking.

D. W. Straughan: It has been shown that synaptosomes from epileptic foci have an impaired potassium uptake capacity, e.g. those from freeze foci, [A. G. Escueta & E. L. Reilly, (1971) *Neurology* (*Minneapolis*) **21**, 418]. This ties in with the observation of changed Na-K^+-ATPase (sodium pump) activity observed in experimental foci, and would tie in also with the abnormally high extracellular levels of potassium which are observed. It is reasonable to imagine that a slow extracellular accumulation of potassium occurs during the initial spiking and is responsible for the progression to a major seizure. I think biochemists should concentrate on changes in these ion pumps as they will prove to be a key factor in epilepsy.

H. W. Reading: Taking up Professor Straughan's point, I wish to point to the existence of a possibly relevant situation in recurrent endogenous depression which is a genetic disorder. Here intracellular erythrocyte sodium concentration is increased in parallel with a specific reduction in sodium pump activity. Patients do not show any hyperactivity or seizure, in fact quite the opposite, so that some caution must be exercised when considering faulty sodium pump activity as underlying excitatory mechanisms in epilepsy. On the other hand, Dr. Davidson from Edinburgh has evidence from a small group of drug-free epileptics that these patients showed a reduction in the rate of Na^+ efflux from their erythrocytes.

A. N. Davison: This data is on sodium pump activity in red cells rather than on brain tissue isn't it?

H. W. Reading: This is right but since it is a genetic disorder we can expect similar changes in the brain.

D. Davidson: As Dr. Reading said, these findings were with a small group of carefully selected patients, all with idiopathic epilepsy and not on drug treatment. We found a reduced rate of $^{24}Na^+$ efflux from red cells of patients compared with controls suggesting an impaired sodium pump. This was a brief study and further investigation is required.

B. S. Meldrum: Can Professor Straughan give us the evidence that the transition from spiking to sustained seizure activity is caused by a sudden increase in extracellular potassium? Studies have shown a rise to 9–10 mM during brief bursts of spikes and a fall to a slightly lower plateau level (8 mM) during prolonged seizure activity [D. A. Prince, H. D. Lux & E. Neher (1973) *Brain Res.* **50**, 489–495; H. D. Lux (1974) *Epilepsia* **215**, 375–393; G. W. Sypert & A. A. Ward (1974) *Exp. Neurol.* **45**, 19–41].

D. W. Straughan: It may be that reduced inhibitory hyperpolarization reduces the output of inhibitory neurotransmitter at the terminal.

A. N. Davison: Could I introduce some new topics and ask if there is any increased receptor sensitivity in primary or secondary focal areas, whether there are reported changes in glutamate decarboxylase or any other transmitter synthesizing enzyme in focal tissue from experimental models or excised human foci?

C. D. Richards: There is evidence from Ward that deafferentation occurs in aluminum-induced foci and it has been suggested that hypersensitivity results as it does following muscle denervation. The difficulty is working out what the input is onto these foci, if they are causing them to discharge.

R. Balazs: What we really need is a quantitative estimate of change in receptor sensitivity and in change in circuitry. It is clear that major changes in neuronal circuitry will follow if there is massive loss of neurons, and dendritic arborization.

H. F. Bradford: I brought along these slides from the paper from Ward's group [L. E. Westrum *et al.* (1965) *J. Neurosurg.* **21**, 1033–1046] which show the massive loss of dendritic arborization, smoothing of dendritic surface and loss of neurons which can be seen in the chronic alumina focus of monkeys, following Golgi–Cox staining. Very similar patterns are seen in excised human temporal lobe samples [M. E. Scheibel & A. B. Scheibel (1973) in *Epilepsy: Its Phenomenon in Man* (M. A. B. Brazier, ed.), pp. 315–335, Academic Press, New York; W. Jann Brown (1973) ibid, pp. 431–387]. As Dr. Richards said, this apparent deafferentation raises the question of whether receptor supersensitivity also develops due to increased receptor formation on the postsynaptic membrane as occurs following denervation of muscle. There have been several studies on this question using cortex deafferented by undercutting as a model [K. Krnjević (1973) in *Basic Mechanisms of the Epilepsies* (H. M. Jasper, A. A. Ward & A. Pope, eds.), pp. 159–165, Churchill Livingstone, Edinburgh; S. K. Sharples (1973) ibid, pp. 329–348; L. T. Rutledge (1973) ibid, pp. 349–356 and earlier work referred to therein]. Krnjević examined the excitability of chronically isolated cortical slabs (3–8 weeks) with glutamate or ACh released by iontophoresis from microelectrodes and the inhibitory effects of GABA and cortical shock on this evoked activity. Little difference from the response of normal cortex was detected. Sharples, Rutledge and many earlier workers, on the other hand, have found that chronically isolated cortex is able to sustain epileptiform after discharges five- to tenfold longer than normal tissue, which is evidence of supersensitivity. There is in parallel a 25% fall in acetylcholinesterase content of the tissue which might explain a cholinergic supersensitivity. Daily electrical stimulation prevented the onset of supersensitivity and appeared to reduce or prevent deafferentation, suggesting a correlation between the two events.

W. R. D. Smith: With regard to Professor Davison's questions about observed changes in glutamate decarboxylase in epileptic tissue, we have results from experiments on offspring of vitamin B_6 deficient rats which display all the features of B_6 deficiency (i.e. convulsions, ataxia, atonia, etc.). One injection of pyridoxine results in a remarkable and immediate recovery. We felt that if GABA synthesis was involved in this recovery we should be able to measure the increase. In fact we found that brain glutamic acid decarboxylase (GAD) is activated immediately, but no increase in GABA is detectable for about 20 minutes. It may well be synthesized before that time, of course. We feel that the important step is the reactivation of GAD and the synthesis of GABA at the 'right' sites. Total levels of brain GABA do not seem to be so important. Thus it seems important to establish whether there is impaired GAD activity in experimental epileptic foci.

M. H. Joseph: Dr. Emson and I have examined GAD activity in cobalt foci [P. C. Emson & M. H. Joseph (1975) *Brain Res.* in press] and find it lowered. We link this change to the degenerative changes occurring in the foci. There is also a small but significant fall in GAD activity in the secondary focus. However, amino acid levels in the secondary focus were normal, and this underlines the importance of measurements of amino acid turnover in the foci, so that some distinction can be made between different pools of amino acids.

B. S. Meldrum: In answer to Professor Davison's earlier question, hypoglycaemia does not provide a satisfactory model for the study of epilepsy. In a minority of patients (especially in children with petit mal) the usual type of seizures can be provoked by hypoglycaemia but this is rare. In non-epileptic patients subjected to hypoglycaemia by Poire [R. Poire (1969) in *The Physiopathogenesis of the Epilepsies* (H. Gastaut, H. H. Jasper, J. Bancaud & A. Waltregny, eds.), pp. 75–110, Thomas, Springfield, Ill.] more than 90% showed focal or generalized myoclonus, but only 3% showed tonic–clonic seizures. The myoclonus occurs after behavioural and EEG signs of hypoglycaemia are severe. At this stage other forms of epilepsy are often inhibited (such as photically induced epilepsy in baboons [R. Naquet, B. S. Meldrum, E. Balzan & J. P. Chanier (1970) *Brain Res.* **18**, 503–512]. Hypoglycaemia can also block seizure activity induced by pentylenetetrazole or bicuculline. Myoclonus induced by hypoglycaemia is an unsuitable model for most studies in experimental epilepsy and especially for tests involving anticonvulsant drugs.

H. S. Bachelard: Yes, it is true that hypoglycaemia can protect against convulsions under certain conditions just as it is also true for hypoxia. For example, hypoxia can protect against metrozole convulsions which return at normal P_{O_2}. This comes back to the argument I made about metabolic balance and metabolic stability. These points about hypoglycaemia reinforce this rather than argue against it. I argued this morning (see p. 238) that induced hypoglycaemia in animals is a good, simple model, not for the epilepsies as a whole, but for hypoglycaemic-induced convulsions in man.

Author Index

Numbers in italics are pages where References are listed at the end of each article.

A

Abood, L. G., 236, *245*
Abraham, J., 166, *172*
Abrahams, D. E., 146, *161*
Adams, R. C. F., 207, *211*
Adembri, G., 167, *172*
Adey, W. R., 209, *211*
Aghajanian, G. J., 58, 60, 63, 64, *70*
Aghajanian, G. K., 73, *76*, 149, 150, *160*
Agid, Y., 16, *20*
Ahtee, L., 61, 66, 67, *70*
Albano, J. D. M., 36, *44*
Alberici, M., 198, 200, *210*
Alberici de Canal, M., 198, 199, 200, 204, 208, *211*
Albert, M. L., 50, *55*
Ambrose, J. A., 129, *140*
American College of Neuropsychopharmacology, 49, *54*
Anagnoste, B., 63, 64, *70, 71*
Anden, N. E., 27, 28, 32, *33, 34,* 41, *44,* 58, 61, 63, 65, 68, *70, 71*
Anderson, E. G., 147, *160,* 222, *230*
Angst, J., 41, *44*
Anlezark, G. M., 67, *71,* 147, 149, 151, 152, 153, 159, *160, 161*
Apter, N. S., 146, 147, *161*
Aquilonius, S. M., 98, *101,* 108, *109*
Arbuthnott, G. W., 9, *11,* 28, 29, 30, 31, 33, 98, *101*
Arias, D., 49, *55*
Arias, L. P., 144, 158, *160*
Asakura, T. 147, *161*
Asanuma, H., 224, *230*
Ashcroft, G. W., 58, *71,* 138, *140,* 171, *172*
Asselman, P., 50, *55*
Atack, C., 60, 63, *71*
Axelrod, J., 84, *92*
Ayala, G. F., 224, *229*

B

Bachelard, H. S., 233, 234, 235, 236, 237, 239, 240, 241, 242, 243, *245, 246,* 256, *259*
Bain. J. A., 198, 199, 204, 211
Bain, S. A., 256, *259*
Bak, I. J., 99, *101*
Baker, A. B., 51, *54*
Baker, J., 73, *76*
Balazs, R., 203, *211*
Balcar, V. J., 205, *211,* 258, *258*
Baldessarini, R. J., 8, 10, *11,* 47, 48, 50, 51, 53, *54*
Baldwin, M., 180, *183*
Baldy-Moulinier, M., 144, 158, *160*
Ballinger, B. R., 73, *77*
Balzamo, E., 145, 147, 148, 149, 150, 151, 152, 153, *161*
Balzer, H., 146, *160*
Bamji, A. N., 21, *26*
Banerjee, S. P., 250, *251*
Banna, N. R., 226, *229*
Barbeau, A., 6, *11,* 18, *20,* 21, *26,* 235, 237, *245,* 256, *259*
Barker, J. L., 222, *229*
Barnes, C. D., 192, *192,* 229, *231*
Barrera, S. E., 163, *173*
Bartholini, G., 32, *34,* 63, 64, 68, *71*
Bartolini, A., 167, *172*
Bartolini, R., 167, *172*
Bass, N. H., 165, *173*
Bassendine, M. F., 28, *33*
Baxter, C. F., 256, *258*
Baxter, M. G., 224, *230,* 248, *251*
Baylis, E. M., 249, *251*
Bedard, P., 41, *44,* 64, *71*
Bell, J. A., 147, *160,* 222, *230*
Benakis, A., 192, *193*
Ben Ari Y., 63, 64, *71*
Bender, H. J., 49, *54*
Bente, D., 41, *44*
Bergheim, O., 151, *160*
Bergmann, S., 15, *20*
Berl, S., 197, 202, 203, 206, 207, *210*
Berner, P., 41, *44*
Bernheimer, H., 13, 15, 16, 17, *20,* 93, *101*
Berridge, T. L., 68, *71*
Bhagaran, H. N., 205, *210*
Bhalerao, V. R., 52, *55*
Bieger, D., 19, *20,* 76, *76*

Binsack, K. F., 49, *54*
Bird, E. D., 6, *11,* 13, 17, 18, *20,* 83, 84, *92,* 93, 97, 98, *101,* 103, 104, 105, 106, 107, *109*
Birkmayer, W., 13, 15, 16, 17, *20,* 93, *101*
Biscoe, T. J., 221, *230,* 256, *258*
Black, R. G., 166, *172*
Blackwood, W., 52, *54*
Blanshard, K. C., 207, *210, 211*
Blass, J. P., 233, *246*
Bliss, T. V. P., 187, *192*
Bloom, F. E., 28, *33,* 35, *44,* 104, *109*
Boakes, A. J., 24, *26*
Boek, U., 39, *44*
Bohm, M., 50, *55*
Bonham-Carter, S., 15, *20*
Bonnycastle, D. D., 251, *251*
Bordelean, J-M., 48, *55*
Borg, S., 75, *76*
Bouchard, R., 29, *34*
Boucher, R., 29, *34*
Bouchet, 181, *183*
Bouillon, D., 19, *20*
Boukma, S. J., 35, *45*
Bovill, K. T., 50, *55*
Bowers, M. B., 150, *161,* 247, 248, 249, *251*
Bowers, M. C., *173*
Bowery, N. G., 215, 221, 223, 228, 229, *230*
Bownds, M. D., 83, *92*
Boyarsky, L. L., 208, *212,* 220, *230*
Boyd, A. E., 87, *92*
Boyd, E. S., 226, *230*
Boyle, R., 233, 235, *245*
Bradford, H. F., 166, *172, 173,* 191, *193,* 202, 203, 204, 205, 207, 208, *210, 211, 212*
Bradley, K., 256, *258*
Bradley, W., 186, *193*
Braestron, C., 58, *71*
Brandon, S., 49, *54*
Bratton, A. C., 151, *160*
Bratz, I., 181, *183*
Breckenridge, B. McL., 235, 242, 243, *246*
Bridgers, W. F., 249, *251*
Brierley, J. B., 157, 158, 159, *160, 161,* 239, *245*
Briones, R. V., 50, *54*
Brodie, K. H., 52, *54*
Brody, J. A., 15, *20*
Brooks, V. B., 224, *230,* 256, *258*
Brooks, W. H., 208, *212*
Brown, A. W., 157, 159, *161,* 239, *245*
Brown, B. L., 36, *44*
Brown, D. A., 215, 221, 228, 229, *230*
Brown, D. J., 208, *210*
Brown, J. H., 32, *34,* 35, 38, 39, *44*
Browning, R. A., 30, *34*
Bruggencate, G. ten., 220, *230*
Bruton, C. J., 182, *184*
Bruyn, G. W., 6, *11,* 93, *101*
Buchwald, N. A., 30, *34*
Buffaloe, W. J., 52, *54*
Bunney, B. S., 58, 60, 63, 64, *70,* 73, *76*
Burgen, A. S. V., 41, *44,* 103, 104, 108, *109*
Burehardt, C. R., 38, *45*
Burkard, W. P., 58, 68, *71*
Burks, J., 50, *54*
Burns, B. D., 179, *183,* 189, *193*
Butcher, L. L., 19, *20,* 28, *34*
Butcher, S. G., 28, 32, *33, 34,* 61, *70*
Byers, R. K., 86, *92*

C

Cabreels, F. J. M., 167, *173*
Calne, D. B., 9, *11,* 21, 23, 24, *26,* 28, *33, 34,* 50, *54*
Calvin, W. H., 164, *172,* 189, *193*
Campbell, B. A., 16, *20*
Campbell, D. B., 37, *44*
Cannon, J. G., 38, *44*
Cannon, W. B., 27, *33*
Carl, J. L., 201, *211*
Carlsson, A., 58, 60, 63, 67, *71*
Caro, A. J., 83, *92*
Castellion, A. W., 152, *161*
Caveness, W. F., 159, *160*
Cazauviehl, 181, *183*
Cerletti, A., 74, *77*
Chadwick, D., 150, *160,* 248, 250, *251, 252*
Chadwick, D. W., 98, *101*
Champlain, J., de., 28, 29, *33*
Chanarin, I., 248, 250, *251, 252*
Chang, M-M., 239, *246*
Charamza, O., 198, 200, *211*
Chase, T. N., 6, *11,* 15, *20,* 21, 22, *26, 33,* 52, 53, *54,* 75, *76,* 251, *251*
Chen, G., 151, *160*
Cheney, D. L., 68, *71*
Chien, C. P., 17, *20*
Christensen, E., 52, *54,* 75, *76*
Christiansen, A. V., 39, *44*
Christie, J., 28, *33*
Chusid, J. G., 151, *160,* 165, *172*
Ciesieisli, L., 204, *212*

Clark, J. B., 242, *245*
Clark, J. E., 133, *140,* 204, *211*
Clark, R., 217, *230*
Clarke, G., 218, 219, *230*
Claveria, L. E., 21, *26,* 28, *33,* 50, *54*
Clayton, P. R., 166, *172*
Clement-Cormier, Y. C., 35, 38, 39, *44,* 60, *71*
Clyde, D. J., 41, *44*
Cobb, S., 233, *246*
Cohen, D. J., 150, *161, 173*
Cohen, P. J., 241, *245*
Colasanti, B. K., 202, *210*
Colburn, R. W., 15, *20*
Cole, J., 17, *20*
Cole, J. O., 41, *44,* 49, *55*
Coleman, M., 19, *20*
Collins, G. G. S., 198, 199, 204, *210,* 214, 217, *230*
Collins, J. F., 228, 229, *230*
Colstanti, B. K., 165, *172*
Connor, J. D., 28, *33,* 217, *230*
Cooper, I., 183, *183*
Cooper, I. S., 139, *140*
Corrodi, H., 21, *26,* 28, 32, *33, 34,* 61, 63, *70, 71,* 150, *160*
Corsellis, J. A. N., 151, *160,* 181, 182, *184*
Costa, E., 28, *33,* 35, *45,* 63, 68, *71,* 104, *109*
Costall, B., 15, *20,* 37, *44,* 95, *101*
Cotzias, G. C., 21, 24, *26*
Coursin, D. B., 205, *210*
Courtois, A., 203, *212*
Craig, C. R., 165, 166, *172, 173,* 202, *210,* 235, 236, *246*
Crandall, P. H., 159, *161,* 166, *173*
Crane, G. E., 19, *20,* 49, 51, *54*
Crapper, D. R., 166, *172*
Crawford, I. L., 217, *230*
Crawford, J. M., 191, *193,* 256, *258*
Creasy, W. A., 52, *54*
Creese, I., 32, *34*
Cross, B. A., 240, *245*
Crossett, P., 10, *11,* 16, *20*
Crossland, J., 205, *210*
Crossman, A. R., 94, *101*
Crow, T. J., 28, *33,* 42, *44*
Crowley, J. M., 249, *251*
Crowshaw, K., *230*
Csillik, B., 146, *160*
Cuatrecasas, P., 42, *44*
Cuello, A. C., 35, *44,* 84, *92*
Cull-Candy, S. G., 210, *211*
Cummings, J. N., 52, *54*
Currie, S., 180, *183*
Curry, S. H., 73, *76, 77*
Curtis, D. R., 96, 99, *101,* 145, 149, 150, *160,* 189, 191, *193,* 195, 198, 205, *211,* 213, 218, 220, 221, 222, 223, *230,* 256, 258, *258, 259*
Curto, E. M., 190, *193*
Curzon, G., 4, *11,* 13, 15, 17, 19, *20,* 74, *76*

D

Dahl, E., 131, *141*
Dana, N., 10, *11,* 16, *20*
Davidoff, R. A., 222, 223, 226, *230*
Davidson, D., 186, *193*
Davidson, L., 15, 16, 18, *20,* 21, *26*
Davidson, N., 222, *230*
Davies, J., 218, *230,* 248, 249, *251,* 256, 258, *259*
Davis, J. M., 74, *77*
Davis, M., 19, *20*
Davis, R., 149, 150, *160*
Davison, A. N., *173,* 204, 205, *211*
Declaration of Helsinki, 140, *140*
De Feudis, V., 199, 208, *211*
Degos, C. F., 19, *20,* 150, *160*
Degwitz, R., 49, *54*
De Jaramillo, G. A. V., 74, *76*
Della Porta, P., 239, *245*
De Meyer, R., 67, *71*
De Moor, J., 159, *160*
Dennis, M. J., 215, *230*
Derkach, P., 76, *76*
De Ropp, R. S., 199, *211*
Dharmapal, N., 137, *141*
Dhasmana, K. M., 24, *26*
Diamond, I., 241, *245*
Dichter, M., 224, *229*
Di Maria, A., 221, *231*
Di Maschio, A., 41, *45,* 49, *54*
Dodd, P. D., 208, *210*
Dodd, P. R., 205, 207, 208, *210, 211*
Dominic, J. A., 32, *33*
Donaldson, J., 235, 237, *245,* 256, *259*
Donato, D., 73, *77*
Donnelly, E. S., 52, *54*
Doose, H., 169, *173*
Dougan, D., 22, *26*
Dow, R. C., 138, *140,* 159, *160,* 164, 165, 171, *172*
Drawbaugh, R. B., 17, *20*
Dray, A., 229, *230*
Dreifuss, J. J., 220, *230*
Dresse, A., 60, 67, *71*

Driver, M. V., 182, *184*
Duby, S. E., 21, *26*
Duffy, K. M., 15, *20*
Duggan, A. W., 96, *101,* 220, 221, 222, *230,* 256, *258, 259*
Dunitz, J. D., 40, *44*
Dunn, A., 201, 203, *211*
Dunn, B., 52, *54*
Dynes, J. B., 47, *54*

E

Eadie, M. J., 152, *160*
Earl, C. J., 51, *54*
Earle, K. M., 180, *183*
Eastman, R., 21, *26,* 50, *54*
Easton, D. M., 256, *258*
Eccles, J. C., 189, 192, *193,* 256, *259*
Eccleston, D., 32, *33*
Edwards, H., 49, *54*
Eggleston, L. V., 208, *212*
Ehringer, H., 67, *71,* 83, *92,* 103, *109*
Eichenberger, E., 60, 65, *71*
Ekins, R. D., 36, *44*
Elazar, Z., 197, *211,* 223, *230*
Elizan, T. S., 50, *54*
Ellaway, P. H., 150, *160*
Elliott, K. A. C., 199, 201, 202, *211, 212,* 215, *230*
Elmasry, A. M., 204, *211*
Emson, P. C., 138, *140,* 163, 164, 165, 166, 167, 170, *172,* 197, *211*
Enberg, I., 220, *230*
Engel, D. J. C., 38, *44*
Engel, G., 19, *20*
Ensor, C. R., 151, *160*
Erba, G., 150, *161*
Ericsson, A. D., 15, *20*
Ernst, A. M., 27, *33,* 65, *71*
Escutea, A. V., 186, *193*
Eser, H., 40, *44*
Esplin, 192, *193*
Esplin, D. W., 190, *193,* 229, *230*
Essig, C. F., 242, *246*
Essman, W. B., 203, *211*
Estrada Villaneuva, F., 167, *173*
Etherington, L. G., 192, *192*
Evans, R. H., 256, *258*
Everetti C. M., 151, *160*
Everitt, B. J., 16, *20*
Ey, H., 48, *54*

F

Falconer, M. A., 127, 136, *140,* 159, *160,* 180, 182, 183, *184,* 237, *246*
Falstein, E. I., 86, *92*
Farnebo, L. O., 28, *33*
Farr, A. L., 104, 107, *109*
Farrell, D., 223, *230*
Faurbye, A., 19, *20,* 48, 50, 52, 53, *54, 55,* 75, *76*
Faure, H., 48, *54*
Fazekas, J. F., 235, 239, *246*
Felix, D., 96, *101,* 221, 222, *230,* 256, *258, 259*
Fellman, J. H., 15, *20*
Feltz, E. E., 164, 167, *172, 173*
Feltz, P., 28, 29, *33, 34*
Ferrendelli, J. A., 239, *246*
Feser, Prof., 66, *71*
Fibiger, H. C., 16, 18, *20,* 32, *34,* 42, *44,* 93, 97, 98, 99, *101, 102,* 103, 104, *109*
Field, P. M., 16, *20*
Fiftkova, E., 208, 209, *212*
Filotto, J., 48, *55*
Findley, L. J., 24
Fink, G. B., 152, *161*
Finkelstein, R. A., 42, *44*
Fischer-Williams, M., 127, *140*
Fisher, D. B., 242, *246*
Fishman, R. A., 241, *245*
Fjallan, B., 39, *44*
Foerster, O., 181, *184*
Fogel, B. J., 49, *55*
Folbergrova, J., 235, 238, *246*
Fonnum. F., 81, *92,* 214, *230, 231*
Forfar, J. C., 164, 165, *172*
Formby, B., 235, 236, *246*
Forno, L. S., 87, *92*
Forrest, F. M., 74, *76*
Forrester, P. A., 218, 219, *230*
Fowler, L. J., 224, *230*
Frame, B., 242, *246*
Frane, K. F., 39, *44*
Friedhoff, A. J., 250, *251*
Friedman, E., 63, 64, *70*
Friedrl, J., 15, *20*
Frigyesi, T. L., 28, *34*
Fry, J. P., 10, *11,* 17, *20*
Fukunchi, Y., 208, *211*
Fung, C., 86, *92*
Fuxe, K., 16, 19, 20, 21, *26,* 27, 28, 32, *33, 34,* 61, 63, 65, *70, 71,* 150, *160*
Fyrö, B., 75, *76*

G

Gadae-Ciria, M., 32, *34,* 68, *71*
Gadea, M., 145, 147, *161*
Gallagher, B. B., 247, 248, 249, *251*

Gallant, D., 49, *55*
Gamberger, B., 28, *33*
Game, C. J. A., 218, 223, *230,* 258, *259*
Gardner, A., 49, *55*
Gardner, E. L., 35, *44*
Gardner, L. C., 226, *230*
Garelis, E., 13, *20*
Garris, P., 138, *140*
Gastaut, H., 127, 129, *140,* 182, 183, *184*
Gatfield, P. D., 235, 242, 243, *246*
Gazengel, J., 98, *102*
Geiger, P. F., 148, *160*
George, C. F., 24, *26*
Gerard, R. W., 236, *245*
Geryer, M. A., 250, *251*
Gey, K. F., 58, *71*
Gianutsos, G., 17, *20*
Giarman, N. J., 251, *251*
Gibbs, F. A., 239, *246*
Gibbs, E. L., 239, *246*
Gibson, M., 205, *211*
Gilbert, J. C., 243, *246*
Gilkes, M. J., 52, *55*
Gill, M., 187, *193*
Gillbe, C., 42, *44*
Gilles, F. H., 86, *92*
Gillo-Joffroy, L., 21, *26*
Gilman, A., 50, *54*
Ginos, J. Z., 21, 24, *26*
Giotti, A., 167, *172*
Girado, M., 197, 202, 206, *210*
Glaser, G. H., 144, 156, *160, 161,* 236, *246,* 256, *259*
Glasson, B., 192, *193*
Glaubiger, G. A., 21, *26*
Globus, A., 159, *160*
Gloor, P., 223, *230*
Giötzner, F. L., 144, *160,* 164, *172*
Glowinski, J., 16, *20,* 36, *44*
Godwin-Austen, R. B., 15, *20*
Goetz, C., 19, *20*
Goldberg, L. I., 35, 38, *44*
Goldstein, M., 63, 64, *70, 71*
Gollnitz, P., 205, *212*
Gonzales-Monteagudo, O., 203, *210*
Goodman, L. S., 50, *54,* 152, *161*
Gordon, E. K., 53, *54,* 75, *76*
Gordon, R. J., 98, 99, *102*
Gottsfeld, Z., 197, *211,* 223, *230*
Grabowska, M., 16, 17, *20*
Grahame-Smith, D. G., 17, *20*
Green, A. R., 17, *20*
Greenacre, J. K., 21, *26,* 50, *54*
Greenberg, D., 41, *45,* 107, 108, *109*
Greenblatt, D. L., 49, *54*
Greengard, P., 23, 24, *26,* 32, *34,* 35, 36, 37, 38, 39, 43, *44,* 60, *71*
Greenhouse, A. M., 51, *54*
Greiner, A. C., 74, *76*
Grewaal, D. S., 32, *34*
Groat, W. C., de, 256, *258*
Grob, D., 145, *160*
Grossman, R. G., 144, *160*
Guiddita, A., 201, 203, *211*
Guidotti, A., 63, *71*
Guldberg, H. C., 75, *76*
Gumnit, R. J., 224, *229*
Gumpert, J., 17, *20*
Guth, P. S., 74, *76*
Guttnick, M. J., 171, *172*

H

Haber, C., 198, 199, *211*
Haefely, W., 22, *26,* 68, *71*
Haigler, H. J., 149, 150, *160*
Haljamae, H., 144, *160*
Hall, Z. W., 83, *92*
Halliday, A. M., 7, *11*
Hamberger, A., 144, *160*
Hamilton, M., 50, *54*
Hammer, R. H., 249, *251*
Hammerschlag, R., 195, 208, *212*
Hanlon, T. E., 50, *55*
Hanna, G. R., 165, *173*
Hansen, S., 18, *20,* 86, 90, *92,* 93, 97, *102,* 196, 204, *211*
Harmony, T., 235, 236, 237, *246*
Harris, A. B., 163, 166, *172*
Harris, P., 171, *172,* 185, *193*
Harrison, M. J. G., 7, *11*
Hartman, E. R., 165, 166, *172*
Hartwig, G., 186, *193*
Harvey, P. K. P., *173,* 204, 205, *211*
Hassler, R., 99, *102*
Hattori, T., 88, *101*
Hawk, P. B., 151, *160*
Hawkins, J. E., Jr, 198, *211*
Hayashi, T., 205, *211,* 213, *230,* 248, *251*
Hazbun, J., 226, *229*
Headley, P. M., 256, *258*
Healey, A. F., 73, *76*
Heathfield, K. W. G., 180, *183*
Heaton, G. M., 235, 241, 243, *246*
Heiman, H., 41, *44*
Heinrich, K., 49, *54*
Heller, A., 30, *34*
Helmchen, H., 41, *44*
Henn, F. A., 144, *160*
Hennecke, A., 205, *211*

Henson, R. A., 180, *183*
Herkert, H., 49, *54*
Hertz, L., 235, 236, *246*
Herz, A., 28, *34*
Hicks, G., 207, *211*
Hiley, C. R., 6, 9, *11*, 41, *44*, 103, 104, 105, 106, 107, 108, *109*
Hiley, G. R., 18, *20*
Hill, A. G., 159, *160*, 167, 171, *172*, *173*
Hill, D. F., 22, *26*
Hill, R. G., 221, 223, 226, 229, *230*, 249, *251*
Hillen, F. C., 248, *251*
Hillman, H., 236, *246*
Himwich, H. E., 235, 239, *246*
Hippius, H., 41, 42, *44*, *45*
Hippius, V. H., 51, *54*
Hoehn, M., 90, *92*
Hoffer, B. J., 31, 32, *34*, 35, *44*
Hokfelt, T., 16, *20*, 21, *26*, 27, *33*, *34*, 65, *70*, 150, *160*
Holden, E., 15, *20*
Holtz, P., 146, *160*
Hommes, O. R., 248, *251*
Horn, A. S., 15, *20*, 23, *26*, 35, 37, 38, 39, 40, *44*, 84, *92*
Hornykiewicz, O., 4, *11*, 13, 15, 16, 17, 18, 19, *20*, 21, 22, *26*, 60, 67, *71*, 76, *76*, 83, *92*, 93, *101*, 103, *109*
Horsley, 181, *184*
Horton R. W., 94, 95, *101*, 146, 147, 151, 152, 153, 155, 157, 158, 159, *160*, *161*, 235, 239, 240, *246*, 256, *259*
Hosli, L., 198, *211*, 220, *230*, 256, *259*
Howard, R. O., 52, *54*
Hrebicek. J., 198, 200, *211*
Hsu, L. L., 250, *251*
Hull, C. D., 30, *34*
Hungen, K., 22, *26*
Hunt, W. A., 166, *173*, 235, 236, *246*
Hunter, K. R., 15, *20*, 24, *26*
Hunter, R., 51, 52, *54*
Hynes, M. D., 17, *20*

I

Ilahi, M. M., 22, *26*
Ingleby, J., 138, *140*, 171, *172*
Isquierdo, I., 236, *246*
Isquierdo, J. A., 236, *246*
Iversen, L. L., 10, *11*, 13, 15, 17, 18, *20*, 23, *26*, 32, *34*, 35, 36, 37, 38, 39, 40, *44*, 60, *71*, 84, *92*, 93, 97, 98, *101*, 103, 104, 106, 107, *109*, 204, 205, *212*, 215, *230*, 258, *259*
Iversen S. D., 27, 32, *34*

J

Jabbur, S. J., 226, *229*
Jalfre, M., 68, *71*
Janssen, P. A. J., 60, 65, *71*
Jasper, H., 128, *141*
Jasper, H. H., 205, *211*, 215, *230*
Javoy, F., 16, *20*
Jeavons, P. M., 129, 133, *140*, 204, *211*
Jellinger, K., 13, 15, 16, 17, *20*, 93, *101*
Jenner, P., 37, *44*, 66, *71*, 150, *160*, 250, *251*, *252*
Jennett, W. B., 159, *160*
Jenney, E. H., 146, 147, *161*
Jessup, S., *230*
Jobe, P. C., 148, *160*
Johansson, B., 17, *20*
Johnson, A. L., 247, 248, 249, *251*
Johnson, A. W., 52, *54*
Johnson, G. A., 35, *45*, 148, *161*
Johnston, G. A. R., 96, 99, *101*, 145, *160*, 198, *211*, 213, 218, 220, 221, 222, 223, *230*, 256, 258, *258*, *259*
Johnston, I. H., 220, *230*, 256, *259*
Johnstone, G. A. R., 195, 198, 205, *211*
Jones, B. J., 61, *71*
Jones, D. A., 203, *211*
Jose, C., 87, *92*
Joseph, M. H., 138, *140*, 163, 164, 165, 166, 167, 170, *172*, 197, *211*
Julian, T., 203, *211*
Jung, I., 84, 85, 86, *92*
Jüül-Jensen, P., 131, *140*

K

Kääriänen, I., 61, *70*
Kaczmarek, L. K., 209, *211*
Kaiser, C., 40, *44*, *45*
Kalyanarman, S., 137, *141*
Kanazawa, L., 18, *20*
Kanda, T., 208, *211*
Karczmar, A. G., 222, *230*
Karobath, M., 38, 39, *44*
Kartzinel, R., 23
Katz, R. I., 251, *251*
Katzman, R., 35, *44*
Kaufman, S., 242, *246*
Kazamatsuri, H., 17, *20*
Kebabian, J. W., 23, 24, *26*, *32*, *34*, 35, 36, 37, 38, 43, *44*, *60*, *71*
Kehr, W., 60, 63, *71*
Keller, H. H., 64, 68, *71*
Kelly, J. S., 63, 64, *71*, 220, *230*
Kelly, P., 42, *44*
Kelly, P. H., 37, *44*
Kemp, J. W., 248, *252*

Kennard, O., 40, *44*
Kennedy, J., 204, *211*
Kennedy, W. A., 183, *184*
ter Keurs, W. J., 187, *193*
Khan, R. T., 215, *230*
Killam, E. K., 153, *160*
Killam, K. F., 146, 147, 153, *160*, *161*, 198, 199, 204, *211*, 256, *259*
Kim, J. S., 99, *101*
King, L. J., 201, *211*
Kinnersley, H. W., 233, *246*
Kiørboe, E., 131, *140*
Kiss, J., 146, *160*
Klawans, H. L., 16, 18, 19, *20*, 22, *26*, 83, *92*
Klawans, H. L., Jr, 4, 9, 10, *11*, 51, 52, 53, *54*, 93, 98, *101*
Kleijn, E., van der, 167, *173*
Kline, N. S., 49, *54*
Klippel, R. A., 94, 100, *101*
Kloster, B., 18, *20*
Kloster, M., 86, *92*, 93, 97, *102*
Knapp, S., 250, *251*
Knyihar, E., 146, *160*
Kolousek, J., 198, 200, *211*
König, J. F. R., 94, 100, *101*
Konishi, S., 256, 257, *259*
Kopeloff, L. M., 151, *160*, 163, 165, *172*, *173*
Kopeloff, N., 151, *160*, 163, *173*
Kopia, I. J., 24, *26*
Kopin, I. J., 15, *20*, 58, *71*, 251, *251*
Korevaar, W. C., 250, *251*
Kostopoulos, G. K., 257, *259*
Koyama, I., 197, 205, 206, *211*, 215, 223, *230*
Kravitz, E. A., 83, *92*
Krebs, H. A., 208, *212*
Krnjevic, K., 145, *160*, 195, 205, *211*, 213, 214, 218, 220, 226, *230*, 249, *251*
Krynauw, R. A., 183, *184*
Kuczenski, R., 63, *71*
Kuffler, S. W., 144, *160*
Kuhar, M. J., 107, 108, *109*
Kurland, A. A., 50, *55*
Kusske, J. A., 164, *172*

L

Lader, M., 47, 48, 49, 52, *55*
Lader, M. H., 73, *76*, *77*
Lader, S., 47, 48, 49, 52, *55*
Laduron, P., 250, *251*
Lahti, R. A., 68, *71*
Lal, H., 17, *20*
Lal, S., 13, *20*
Lamar, C., 198, *211*, 256, *259*
Landau, B. R., 235, 239, *246*
Lanoir, J., 151, *161*
Lao, L., 201, *211*
Larochelle, L., 19, *20*, 64, *71*, 76, *76*
Larrabee, M. G., 187, *193*
Larson, M. D., 192, *193*
Lauener, H., 60, 65, *71*
Laundy, M., 248, *251*
Laurence, D. R., 24, *26*
Laverty, R., 58, 61, 68, 69, *71*
Lawrence, W. H., 21, *26*
Leao, A. A. P., 208, *211*
Lebovitz, H. E., 87, *92*
Lechner, K., 67, *71*
Lee, T., 62, 64, *71*
Lee, L. W., 202, *211*
Lehman, A., 204, *212*
Lehmann, A., 147, *160*
Leigh, P. N., 21, *26*
Leitich, H., 38, 39, *44*
Lenaerts, F. M., 60, *71*
Lennox, M. A., 180, *184*
Lennox, W. G., 180, *184*, 233, 239, *246*
Lennox-Buchtal, M., 140, *140*, 235, 236, *246*
Lennox-Buchtal, M. A., 159, *160*
Leonard, B. E., 201, *211*
Leopold, N., 52, *55*, 87, *92*
Levine, M. S., 30, *34*
Levy, R. A., 222, *230*
Lewin, E., 235, 236, *246*
Lewis, J., 90, *92*
Lewis, L. D., 239, 241, *245*, *246*
Lewis, P. J., 24
Leysen, J., 250, *251*
Hermitte, F., 19, *20*, 98, *102*, 150, *160*
Liao, G. L., 202, *211*
Lidbrink, P., 21, *26*, 28, *34*, 150, *160*
Limacher, J. J., 257, *259*
Linday, J., 237, *246*
Lindsay, J., 182, *184*
Lindqvist, M., 58, 60, 63, 67, *71*
Lindquist, N. G., 74, *76*
Lipman, F., 203, *211*
Lippmann, W., 40, *44*
Ljungberg, T., 28, 31, 32, *33*, *34*
Ljunggren, B., 239, *246*
Lloyd, K., 64, *71*
Lloyd, K. E., 32, *34*
Lloyd, K. G., 15, 16, 18, *20*, 21, *26*, 63, 68, *71*
Lodge, D., 221, *230*
Logeman, G., 51, *54*

Lorez Arnaiz, R. G., 198, 199, 200, 204, 208, 210, *211*
Losey, E. G., 68, *71*
Lowry, O. H., 104, 107, *109*. 235, 238, *246*
Lubs, H. A., 235, 239, *246*
Lumb, M., 248, *251*
Lux, H. D., 224, *230*
Lyons, L. E., 74, *76*

M

McAfee, D. A., 35, *44*
Maccario, M., 156, *160*
McClelland, H. A., 49, *54*
McCrimmon, A., 235, 236, *246*
McCulloch, R., 218, 223, *230*
McCulloch, R. M., 256, 258, *259*
McDonald, C. J., 52, *54*
McDowell, F. H., 21, *26*, 28, *34*
McGeer, E. C., 16, 18, *20*, 42, *44*, 84, 85, 86, *92*, 93, 97, 98, 99, *101*, *102*. 103, 104, *109*
McGeer, P. L., 16, 18, *20*, 84, 85, 86, *92*, 93, 97, 98, 99, *101*, *102*, 103, 104, *109*
McGinness, J., 74, *76*
Machiyama, Y., 203, *211*
McIlwain, H., 186, *193*, 203, 205, 208, *210*, *211*, 233, 235, 236, *246*, 257, *259*
Mackay, A. V. P., 73, *76*, 84, *92*, 93, 97, *101*, 103, *109*
McKendall, R. R., 51, 52, *54*
McKenzie, G. M., 76, *76*, 98, 99, *102*, 159, *160*, *161*
McKey, R., 49, *55*
Mackie, J. C., 74, *76*
MacLachlan, R. M., 223, *230*
Maclain, L. D., 249, *251*
McLennan, H., 28, *34*, 53, *54*, 104, *109*, 256, *259*
McNay, J. L., 35, 38, *44*
McQueen, J. K., 138, *140*, 159, *160*, 164, 165, 171, *172*
Mabry, P. D., 16, *20*
Magnusson, T., 60, 63, *71*
Mahapatra, S. B., 50, *54*
Maiolo, A. T., 239, *245*
Maitre, M., 204, *212*
Major, M. A., 192, *193*
Makman, M. H., 32, *34*, 35, 38, 39, *44*
Mandel, L. R., 250, *251*
Mandel, P., 204, *212*
Mandell, A. J., 250, *251*
Manning, D. H., 234, *246*
Marcus, E. M., 138, *141*, 167, *173*
Margerison, J. H., 181, *184*
Marian, A. A., 74, *77*
Marks, V., 249, *251*
Marsden, C. D., 3, 7, 8, 9, 10, *11*, 15, *20*, 47, 48, 50, 51, 53, *54*, *55*, 66, *71*, 74, *76*, *77*, 93, 94, 98, 100, *101*, *102*
Marshall, M. H., 51, *54*
Marshall, W. H., 146, 147, *161*
Marteau, R., 19, *20*, 98, *102*, 150, *160*
Martin, M., 256, *258*
Marzuli, F. N., 74, *77*
Masland, R. L., 129, *141*
Massieu, H. G., 204, *212*
Matsumoto, H., 224, *229*
Matsuzaki, M., 153, *160*
Matthews, D. M., 248, *251*
Mayanagi, Y., 138, *141*
Mayersdorf, A., 249, *251*
Mayman, C. I., 235, 242, 243, *246*
Maynert, E. W., 148, *160*
Mearrick, P., 22, *26*
Medina, M. A., 146, *160*
Meek, J. L., 67, *71*
Meire-Ruge, W., 74, *77*
Meiklejohn, A. P., 233, *246*
Meldrum, B. S., 67, *71*, 94, 100, *101*, *102*, 144, 145, 146, 147, 148, 149, 150, 151, 152, 153, 155, 156, 157, 158, 159, *160*, *161*, 165, *173*, 225, 239, 240, *245*, *246*, 256, *259*
Meller, E., 250, *251*
Mena, J., 21, *26*
Merker, J., 40, *44*
Merritt, D. A., 226, *230*
Merritt, H. H., 151, *161*
Messina, F. S., 75, *77*
Meyer, J. S., 208, *211*
Michaelis, E. K., 220, *230*
Michaelis, M. L., 220, *230*
Michaluk, J., 16, 17, *20*
Miledi, R., 215, *230*
Miller, A. A., 224, *230*, 248, 249, *251*
Miller, R. J., 9, 10, *11*, 14, 15, *20*, 23, *26*, 35, 37, 38, 39, 40, 41, 42, *44*, 60, *71*
Millichap, J. G., 151, 152, *161*, 243, *246*
Milner, G., 248, *251*
Minard, F. N., 84, *92*
Minchin, M., 215, *230*
Minick, J. L., 256, *259*
Minota, S., 222, *230*
Mishra, R. K., 32, *34*, 35, *44*
Mistrorgio de Pacheo, M., 200, 204, 208, *211*
Mitchell, J. F., 68, *71*, 215, *230*
Miyahara, J. T., 229, *230*

Miyahata, Y., 192, *193,* 214, *230*
Miyata, Y., 18, *20*
Molina-Negro, P., 150, *161*
Molinoff, P. B., 84, *92*
Moller, J. E., 52, *54,* 75, *76*
Moller, Nielsen, I., 39, 40, *44*
Monachon, M-A., 68, *71*
Mones, R. J., 14, *20,* 50, *54*
Moore, K. D., 32, *33*
Moore, K. E., 17, *20,* 32, *34*
Morrell, F., 186, *193*
Mould, G. P., 73, *76, 77*
Muir, R. B., 189, *193*
Murphy, G. F., 16, *20,* 61, *71*
Musgrave, F. S., 186, *193*
Mushahwar, I. K., 84, *92*

N

Nahorsky, S. R., 199, 201, *211*
Naquet, R., 145, 147, 149, 150, 151, 153, *161*
National Institute of Mental Health, 47, *54*
Naylor, R. J., 15, *20,* 37, *44,* 95, 98, *101, 102*
Neff, N., H., 67, *71*
Negri, V. U., 239, *245*
Neumeyer, J. L., 37, *44*
Ng, K. Y., 15, *20*
Nicholls, J. C., 144, *160*
Nicholson, G. A., 74, *76*
Nicoll, R. A., 192, *193,* 222, *229*
Nielson, M., 58, *71*
Neimegeers, C. J. E., 60, 65, *71*
Nilsson, L., 243, *246*
Nishi, S., 222, *230*
Noach, E. L., 248, *251*
Norman, R., 237, *246*
Norman, R. M., 182, *184*
Northfields, D. W., 181, *184*
Nyback, H., 58, *71*
Nymark, M., 39, *44*
Nys, G. G., 38, *44*
Nystrom, B., 14, *20*

O

Obata, K., 214, 217, *230*
Obbens, E. A. M. T., 248, 249, *251*
Ochs, S., 209, *211*
Ohye, C., 29, *34*
Ojemann, G. A., 164, *172*
Okada, Y., 99, *101*
Okamoto, K., 205, *211*
Okata, Y., 99, *102*
O'Keeffe, R., 62, 64, *71*
Oliver, A. P., 35, *44*
Olley, J. E., 95, 98, *101, 102*
Olson, L., 14, *20*
Olvera, A., 167, *173*
O'Malley, K, 73, *77*
Orrego, F., 203, *211*
Ortiz, W. R., 243, *246*
Otsuka, M., 18, *20, 214, 230,* 257, *259*
Ott, J. E., 50, *54*
Ounstead, C., 182, *184,* 237, *246*

P

Paasonen, M. K., 251, *251*
Pagluicca, N., 203, *211*
Palfryman, M. G., 201, *211*
Pallis, C., 24, *26*
Palm, D., 146, *160*
Papavasilious, P. S., 21, *26*
Papeschi, R., 150, *161*
Pappins, H., 202, *211*
Parkes, J. D., 3, 4, *11,* 15, *20,* 50, *55,* 75, *77*
Parks, L. C., 24, *26*
Passmore, R., 233, *246*
Passoneau, J. V., 235, 238, *246*
Passouant, P., 144, 158, *160*
Paulson, G. W., 6, *11,* 18, *20,* 50, *55*
Payan, H., 183, *184*
Pearce, J., 50, *55*
Pedersen, V., 39, *44*
Peesker, S. J., 198, 204, *212*
Penfield, W., 128, 135, *141,* 180, 181, *183, 184*
Penry, J. K., 138, *141*
Perez de la Mora, M., 204, *212*
Peringel, E., 66, *71*
Perrin, D. D., 256, *259*
Perry, J., 248, *251*
Perry, T. L., 18, *20,* 86, 90, *92,* 93, 97, *102,* 196, 204, *211*
Persson, T., 75, *77*
Peterfalvi, M., 98, *102*
Peters, R. A., 233, *246,* 256, *259*
Petersen, P. V., 40, *44*
Petrie, A., 21, *26,* 28, *33,* 50, *54*
Petzold, G. L., 23, *26,* 32, *34,* 35, 36, 37, 38, 39, 43, *44,* 60, *71*
Pfeiffer, C. C., 146, 147, *161*
Pfeiffer, J. B., 87, *92*
Phillips, G., 127, 128, *141*
Phillis, J. W., 191, *193,* 205, 209, *211,* 214, 218, *230,* 257, *259*
Phipps, J. A., 50. *55*
Pieri, L., 22, *26*
Pieri, M., 22, *26*

Pijnenburg, A. J. I., 42, *44*
Pilling, J. B., 83, *92*
Pinchard, A., 60, *71*
Pind, K., 19, *20,* 53, *55*
Pinder, R. M., 35, 37, 38, *44*
Pinsky, C., 189, *193*
Pletscher, A., 58, 64, 68, *71*
Plum, C. N., 208, *211*
Podolsky, S., 87, *92*
Poire, R., 238, *246*
Poirier, F., 183, *184*
Poirier, L. J., 16, *20,* 29, *34*
Polkey, C. E., 235, 237, *245*
Pollen, D. A., 144, *161*
Ponten, U., 241, *245*
Porter, R., 189, *193*
Portig, P. J., 68, *71*
Portman, R., 151, *160*
Post, M. L., 40, *44*
Posternak, J. M., 187, *193*
Precht, W., 94, 99, *102*
Preece, J. M., 249, *251*
Prince, D. A., 171, *172,* 223, 224, *230*
Pritchard, M. J., 207, *211*
Proctor, P., 74, *76*
Protheroe, C., 49, *54*
Ptashne, M., 186, *193*
Pugsley, T., 40, *44*
Purpura, D. P., 28, *34,* 138, *141,* 186, *193,* 197, 202, 203, 206, 207, *210*
Putnam, T. J., 151, *161*
Pycock, C., 94, 100, *102*
Pycock, C. J., 6, *11*

Q

Quastel, J. H., 205, *211*

R

Racagni, G., 68, *71*
Racusen, L. C., 248, *252*
Raisman, G., 16, *20*
Rajan, K. S., 74, *77*
Ramamurthi, B., 137, *141*
Ramwell, P. W., *230*
Randall, R. J., 104, 107, *109*
Randic, M., 226, *230*
Randrianariosa, H., 204, *212*
Rappard, P., 48, *54*
Rasmussen, T., 134, *141, 166, 173,* 180, *184,* 196, 203, *212,* 223, *231*
Ratcheson, R. A., 239. *246*
Rayner, A. N., 84, *92*
Rayner, C. N., 93, 97, *101,* 103, *109*
Reid, J. L., 21, 24, *26,* 28, *34*
Reilly, E., 186, *193*
Reilly, R. H., 146, 147, *161*
Rekker, R. F., 38, *44*
Reynolds, E. H., 98, *101,* 150, *160,* 247, 248, 249, 250, *251, 252*
Reynolds, S. F., 233, *246*
Richards, C. D., 187, 188, 189, 192, *192, 193,* 234, *246*
Richards, R. K., 151, *160*
Richens, A., 242, *246*
Ricklan, M., 139, *140*
Ringel, S. P., 18, *20*
Roa, P. D., 198, 199, 200, *211*
Robertis, E. de, 198, 199, 200, 204, 208, *210, 211*
Roberts, D. J., 61, *71,* 199, 201, *211*
Roberts, E., 256, *258*
Roberts, F., 229, *230*
Roberts, P. J., 217, 220, *230, 231,* 249, *252*
Roberts, S., 22, *26*
Robinson, D., 61, *71*
Robiolo, B., 200, 204, 208, *211*
Robson, R. D., 24, *26*
Roffler-Tarlov, S., 14, *20,* 58, *71*
Rokach, J., 250, *251*
Roos, B. E., 17, *20,* 58, *70, 71,* 75, *77*
Rose, P., 50, *55*
Rosebrough, N. J., 104, 107, *109*
Rosegay, A., 250, *251*
Rosella, E., 239, *245*
Rosenbleuth, A., 27, *33*
Rosengarten, H., 250, *251*
Rosman, J. L., 144, *160*
Rowe, D. J. F., 242, *246*
Rubenson, A., 27, *33,* 65, *70*
Rubovits, R., 18, *20*
Rudzik, A. D., 148, *161*
Rumarck, B., 50, *54*
Russell, W. R., 128, *141*
Rylander, G., 51, *55*

S

Saad, S. F., 204, *211*
Sacktor, B., 201, *211*
Sadof, M., 76, *76*
Sahakian, B. J., 42, *44*
Saidel, L., 198, 199, *211*
Saito, S., 199, *211*
Sakalis, G., 73, *76,* 77
Salamon, G., 183, *184*
Salmoiraghi, G. C., 28, *33,* 104, *109,* 179, *183*
Sandler, M., 15, *20*
Sano, K., 137, *141,* 171, *173*
Sarett, L. H., 198, *211*

Sawaya, M. C. B., 146, *161*
Sax, D. S., 52, *55,* 87, *92*
Sayeed, Z. A., 137, *141*
Schaefer, J. P., 40, *44*
Schaper, W. K. A., 60, *71*
Scheibel, A. B., 159, *160, 161,* 166, *173*
Scheibel, M. E., 159, *161,* 166, *173*
Schelkunov, E. L., 41, *44*
Schellekens, K. H. L., 60, 65, *71*
Schiele, B. C., 49, *55*
Schmidt, R., 192, *193*
Schmidt, R. F., 148, *161,* 222, 226, 229, *231,* 256, *259*
Schmidt, R. P., 127, *141*
Schnur, J. A., 53, *54,* 75, *76*
Schobben, F., 167, *173*
Scholz, W., 157, *161*
Schousboe, A., 235, 236, *246*
Schriver, C. R., 202, *211*
Schulz, D. W., 235, 238, *246*
Schwartz, A., 236, *246*
Schwartz, S., 220, *230*
Scott, D. F., 180, *183*
Scott, P. M., 204, *211*
Sears, E., 90, *92*
Sedvall, G., 39, *44,* 58, *71,* 75, *76*
Seeman, P., 62, 64, *71*
Seiger, A., 14, *20*
Seisjo, B. K., 241, *245*
Seitelberger, F., 13, 15, 16, 17, *20,* 93, *101*
Sellinger, O. Z., 198, *211*
Serafetinides, E. A., 182, *184*
Sercombe, R., 187, *193*
Serdaru, M., 98, *102*
Sethy, V. H., 19, *20*
Seymour, C. A., 21, *26,* 28, *34*
Shader, R. I., 41, *45,* 49, *54*
Shanes, A. M., 253, 256, *259*
Sharman, D. F., 10, *11,* 14, 16, 17, *20,* 58, 61, 62, 64, 68, 69, *71*
Sharpe, D., 17, *20*
Sharpless, S. K., 27, *34*
Shaywitz, B. A., 150, *161, 173*
Shenker, D., 22, *26*
Shenker, D. M., 19, *20*
Shepherd, M., 47, 48, 49, 52, *55*
Sheppard, H., 38, *45*
Sherwin, A. L., 166, *173,* 196, 203, *212*
Sherwin, A. M., 223, *231*
Shinohara, Y., 208, *211*
Shinozaki, H., 256, *259*
Shirron, C., 63, 64, *70, 71*
Shoulson, I., 28, *33*
Siegel, G. J., 50, *54*
Siesjö, B. K., 239, 243, *246*
Siggins, G. R., 31, 32, *34,* 35, *44*
Silber, R., 250, *251*
Simler, S., 204, *212*
Simmonds, M. A., 221, 223, 226, *230*
Simpson, G., 49, *55*
Simpson, G. M., 49, *55*
Simpson, H. K. L., 222, *230*
Sjöstrom, R., 98, *101,* 108, *109*
Skripkus, A., 74, *77*
Smaje, J. C., 192, *193*
Smith, D. B., 248, *252*
Smith, H. C., 140, *141*
Smith, M. C., 52, *54*
Smythies, J. R., 229, *231*
Snedeker, E. H., 199, *211*
Snider, R. S., 139, *140*
Snodgrass, S. R., 204, 205, *212*
Snyder, S. H., 38, 41, *44, 45,* 103, 107, 108, *109,* 221, *231,* 134, *246,* 250, *251*
Sokoi, M., 196, *211*
Somjen, G. G., 187, *193*
Sommer, W., 181, *181, 184*
Sonneville, P. F., 35, 38, *44*
Soroko, F. E., 159, *160, 161*
Soudijn, W., 40, *45*
Sourkes, T. L., 13, 16, *20,* 22, *26,* 150, *161*
Southwick, C. A. P., 222, *230*
Spaans, R., 248, *252*
Spector, R. G., 248, 249, *252*
Spencer, W. A., 224, *229*
Spiegel, E. A., 136, *141*
Spielmeyer, W., 157, *161*
Spilker, B. A., 24, *26*
Spirtes, M. A., 74, *76*
Sprangers, W. J. J. M., 248, *251*
Srinivasan V., 215, *230*
Stacey, N. E., 256, *259*
Stadler, H., 32, *34,* 64, 68, *71*
Stauder, H. K., 181, *184*
Stavraky, G. W., 27, *34*
Stebbins, R. D., 250, *251*
Stein, L., 89, *92*
Stern, G. M., 15, *20,* 24, *26*
Stevenson, I. H., 73, *77*
Stewart, C. N., 205, *210*
Stewart, G. G., 199. 201, *211*
Stille, G., 42, *45,* 60, 65, *71*
Stone, T. T., 86, *92*
Stone, W. E., 198, 199, 200, 208, *210, 211, 212*
Storm-Mathisen, J., 99, *102,* 214, *230, 231*
Stotsky, B. A., 49, *54*
Strang, R. H. C., 234, 235, 242, 243, *246*

Stratten, W. P., 229, *230*
Straughan, D. W., 218, 219, 221, 223, 225, 226, 227, *230, 231,* 249, *251*
Strauss, M. B., 140, *141*
Streiff, R. R., 249, *251*
Strickler, P., 40, *44*
Stull, R. E., 148, *160*
Suijkerbuijk-van Beek, M. M. A., 248, *251*
Svensmark, O., 192, *193*
Sweet, R., 28, *34*
Sweet, R. D., 21, *26*
Swinyard, E. A., 151, 152, *161*
Sylvester, P. E., 249, *251*
Symonds, C. P., 128, 129, 141
Sypert, G. W., 164, *173*
Szava, S., 235, 236, 237, *246*

T

Takagaki, G., 202, 207, *210*
Takahashi, T., 257, *259*
Takeda, K., 217, *230*
Tamer, A., 49, *55*
Tanaka, Y., 214, *230*
Tang, L. C., 24, *26*
Tapia, R., 204, *212*
Tarsy, D., 8, 10, *11,* 47, 48, 50, 51, 52, 53, *54, 55,* 94, 100, *102*
Tausig, M. D., 146, 147, *161*
Taylor, A. R., 37, *44*
Taylor, D. C., 136, *140,* 182, *184*
Taylor, E., 179, *184*
Tebecis, A. K., 145, *161,* 221, *231*
Tegerdine, P., 14, *20,* 58, *71*
Terenius, L., 104, 105, *109*
Terner, C., 208, *212*
Testa, G., 223, *230*
Tetreault, L., 48, *55*
Tews, J. K., 198, 199, 200, *211*
Teychenne, P. F., 21, 24, *26,* 28, *33,* 50, *54*
Thomas, A. J., 208, *210*
Thomas, L. B., 127, *141*
Thompson, M. F., 235, 237, *245*
Thornburg, J. E., 17, *20,* 32, *34*
Thornicroft, S., 51, *54*
Tiekert, C. G., 201, *211*
Toga, M., 183, *184*
Tokunaga, Y., 199, *211*
Toseland, P., 152, 153, 155, *161*
Tower, D. B., 145, *161,* 166, *173,* 200, 202, 205, *212,* 233, 235, 236, 238, 242, *246,* 256, *259*
Tower, D. M., 138, *141*
Townsend, H. R. A., 167, 171, *172, 173*
Toyokura, Y., 18, *20*
Trabucchi, M., 68, *71*
Trachtenberg, M. C., 144, *161*
Trimble, M., 67, *71,* 147, 149, 151, 152, 153, 159, *161*
Trott, J. R., 150, *160*
Trurek, I., 50, *55*
Turano, P., 73, *77*
Turnbull, M. J., 73, *77,* 205, *210*
Turner, W. J., 73, *77*
Tyrer, J. H., 152, *160*

U

Udvarhelyi, G. B., 138, *141*
Uhrbrand, L., 48, 50, *55*
Ullberg, S., 74, *76*
Ungerstedt, U., 14, 16, *20,* 21, *26,* 27, 28, 31, 32, *33, 34,* 61, 63, *70, 71,* 100, *102,* 150, *160*
Urba-Holmeren, R., 235, 236, 237, *246*
Urbay, C. M., 235, 236, 237, *246*
Urquhart, N., 90, *92,* 204, *211*
Usherwood, P. N. R., 210, *211*

V

Vaernet, K., 137, *141*
Vakil, S. D., 21, 24, *26,* 28, *34*
Vanden Heuvel, W. J. A., 250, *251*
Van Gelder, N. M., 165, 166, 167, *173,* 196, 197, 203, *212,* 223, *231*
Van Harreveld, A., 208, 209, *212*
Van Nueten, J. M., 60, *71*
Van Rossum, J. M., 42, *44*
Van Woert, M. H., 19, *20*
Velasco, F., 167, *173*
Velasco, M., 167, *173*
Verbruggen, F. J., 60, *71*
Vigouroux, M., 183, *184*
Vigouroux, R. A., 157, 158, *161*
Viik, K., 98, 99, *102*
Vogel, M. E., 229, *231*
Vogt, M., 62, 64, 68, *71*
Voigtlander, P. F., 35, *45*
Votzke, E., 167, *173*
Vuillon-Cacciuttolo, G., 148, 149, 150, *161*

W

Wada, G. H., 15, *20*
Wada, J. A., 84, 85, 86, *92,* 147, 148, 149, *161*
Wade, D., 22, *26*
Wade, J. A., 196, *211*
Waelsch, H., 197, 202, 203, 206, *210*
Wakenzie, J. S., 28, *34*

Walberg, F., 214, *230*
Walker, A. E., 127, 138, *141*
Walker, J., 50, *54*
Walker, J. E., 90, *92*
Walker, J. M. G., 224, *230*
Walker, R., 138, *141*
Walker, R. J., 38, *45,* 94, *101*
Walker, R. W., 250, *251*
Waltregny, A., 239, *246*
Ward, A. A., 159, *161,* 189, *193*
Ward, A. A., Jr, 127, *141,* 163, 164, 165, 166, 167, *172, 173*
Warren, R. J., 40, *44*
Wars, H., 21, *26*
Wasterlain, C. G., 21, *26*
Watanabe, A. M., 24, *26* 52, *54*
Watkins, J. C., 191, *193,* 218, *230,* 248, 249, *251,* 256, 258, *258, 259*
Watson, W. J., 199, *212,* 256, *259*
Wattson, R. H., 247, 248, 249, *251*
Weakly, J. N., 187, *193*
Webster, R. A., 248, 249, *251*
Wedege, E., 203, *212*
Wederman, M., 198, 200, *211*
Wegerer, I., 49, *54*
Weichert, G., 205, *211*
Weichert, P., 205, *212*
Weiler, P., 198, *211*
Weiner, W. J., 19, *20*
Weinreich, D., 195, 208, *212*
Weinshilboum, R., 84, *92*
Weiss, R., 35, *45*
Werdinius, B., 58, *70*
Wertman, B. G., 15, *20*
Westmoreland, B. F., 165, *173*
Westrum, L. E., 159, *161,* 166, *173*
Westruman, L. E., 166, *173*
Wheeler, D. D., 208, *212*
Wheeler, R. H., 52, *55*
Whelan, D. T., 202, *211*
White, L. E., 159, *161,* 166, *173*
Whitty, C. W. M., 128, *141*
Wickson, V., 42, *44*
Wijngaarden, I., van, 40, *45*
Wilder, B. J., 249, *251*
Williamsen, R., 131, *141*
Willis, W. D., 192, *193,* 226, *231*
Wilson, J. E., 201, *211*
Wise, C. D., 89, *92*
Wode-Helgodt, B., 75, *76*
Wolff, F. A., de, 248, *251*
Wood, J. D., 146, *161,* 198, 199, 204, *212,* 256, *259*
Woodbury, D. A., 236, 242, *246*
Woodbury, D. M., 133, *141,* 248, *252*
Woodruff, G. N., 38, *45,* 94, *101*
Woods, A. C., 21, *26*
Worley, L., 49, *55*
Wycis, H. T., 136, *141*
Wyler, A. R., 164, 167, *173*

X

Yahr, M. D., 24, *26*
Yamadori, A., 50, *55*
Yamamoto, C., 186, *193*
Yamamura, H., 41, *45*
Yamamura, H. I., 107, 108, *109*
Yates, C. M., 75, *76*
Yatsu, F. M., 202, *211*
York, D. H., 28, *34,* 53, *54,* 104, *109*
Yoshida, M., 94, 99, *102*
Yoshino, Y., 201, *212*
Young, A. B., 103, *109,* 221, *231*
Young, J. M., 41, *44,* 103, 105, 108, *109*
Young, S. N., 13, *20*

Z

Zablicka, 192, *193*
Zablocka, B., 229, *230*
Zieglgansberger, W., 28, *34*
Zilleti, L., 167, *172*
Zimmermann, M., 229, *231*
Zirkle, C. L., 40, *44, 45*
Zivkovic, B., 63, *71*
Zuckermann, E. G., 156, *161,* 236, *246,* 256, *259*
Zukin, S. R., 103, *109,* 221, *231*

Subject Index

A

Acaperone (butyrophenone), 67
Acetylcholine,
epileptogenic, 127, 145
in Huntingdon's chorea, 18
question of involvement of, in epilepsy, 263–264
selective antagonists of, 218, 219
striatal, 4, 32, 68, 111
Acetylcholinesterase, 145, 265
Acromegaly, effect of bromocriptine on growth hormone in, 79
Adenosine monophosphate, cyclic (cAMP), 35
distribution of, in brain, 35
in dopamine receptor response, 14, 23, 25, 32
formation of, *see* adenylate cyclase
stimulates dopamine synthesis in striatal slices, 64
S-Adenosylmethionine, methyl donor in brain, 250
Adenylate cyclase,
activated by cholera enterotoxin, 42–43
activated by dopamine antagonists, 14, 36–38, 113
dopamine-sensitive, in striatum, 23, 32, 43, 60
inhibited by dopamine antagonists, 14, 38–40
noradrenaline-sensitive, in cerebral cortex, 32
not affected by *O*-methyldopamine, 15
Adrenergic stimulants, α-methyltyrosine increases response to, 32
ADTN, dopamine analogue affecting both excitatory and inhibitory dopamine receptors of invertebrates, 113
Akathisia, induced by antipsychotic drugs, 47, 48, 118
Akinesia, 113, 114
Alanine,
in epileptogenic foci, 196, 197
release of, from cerebral cortex *in vivo*, 216, 217
Albino animals, chlorpromazine in, 74
Alcohol
chorea on withdrawal of, from addicts, 6
epilepsy induced by excess of, 176
Allyl glycine, convulsant, inhibitor of GAD, 95, 146, 198, 203, 204
amino acids in whole brain after, 199, 200
decreases release of inhibitory transmitters, 223
increases glucose in brain, 238
and responses in photosensitive baboons, 147, 153–155
Alumina cream foci in frontal cortex, induce epilepsy, 151, 159, 163, 182
compared with human epileptogenic foci, 165, 166
effects of, compared with human cortical epilepsy, 164
loss of dendritic arboization in, 265
Amantadine, and dyskinesias, 50, 51
Amines, transmitter, *see* monoamines
Amino acids, *see also under individual amino acids*
convulsants, and concentrations of, in whole brain, 199, 200, 261–262
convulsants, and processes inhibited by, 224–229
in epileptogenic foci, 196–197, 222, 262
experiments on release of, from CNS, 214–217
metabolic implications of changes in concentrations of, 201–204
in neurotransmission, 195–196, 213–214
postsynaptic action by (excitatory), 218–220
postsynaptic action by (inhibitory), 220–221
presynaptic inhibition by, 221–222
synaptic mechanisms mediated by, and seizure foci, 222–224
γ-Aminobutyric acid (GABA), inhibitory transmitter, 83, 86, 195, 204, 213, 214, 221, 254
apomorphine and metabolism of, 70
attempts to increase brain concentration of, 90
in brain excitation and inhibition, 204–205
in CSF, 257
differential sensitivity of neurons to glycine and, 220
drugs blocking action or inhibiting synthesis of, are convulsants, 145, 221, 224

γ-Aminobutyric acid (GABA), *continued*
drugs inhibiting metabolism of, are anticonvulsants, 68, 148, 224
in epileptogenic foci, 196, 197, 199, 200, 202, 261
excitatory effect of, in frog hemichord, 255, 256
extra-cellular, potently anticonvulsant, 205
folate blocks inhibitory effects of, 249
in Huntingdon's chorea, 6, 18, 84, 93, 111, 120, 121
increases in postmortem brain, 84
interactions of, with metabolism of other transmitters, 68–69, 120
involved in metabolism as well as in neurotransmission, 204
manipulation of content of, in substantia nigra, 99–100
metabolism of disturbed in brain trauma, 127
in Parkinsonism, 16, 18
penicillin and, 223;
receptors for, blocked by bicucullin, 96, and blocked by picrotoxin, 94–95
reduces sensitivity of neurons to acetylcholine and glutamate, 218, 219;
release of, from cerebral vortex *in vivo*, 215, 216, 217
vortex *in vivo*, 215, 216, 217
in spreading depression, 209
temporarily abolishes drug-induced myoclonus, 95–96, 99
transmitter for presynaptic inhibition, 221–222
in whole brain after convulsants, 199, 200
γ-Aminobutyric acid-glutamate transaminase (GABA-T), 84, 90, 145
effects of inhibition of, 204
inhibitors of, 70, 146, 169, 198
γ-aminobutyric acid shunt pathway, 200, 201, 202, 203
blocked by *n*-dipropylacetate, 205
Amino-oxyacetic acid, anticonvulsant, 90, 148
and ECOG in cobalt-implanted rat, 168, 169
inhibits GABA-T, 68, 70, 258
toxic to humans, 204
Ammonia, increased in brain during convulsions, 238
Amphetamines, dopamine agonists
and cAMP formation, 36
and "circling" in rats, 98
dyskinesia induced by excess of, 6, 50, 51
6-hydroxydopamine and response to, 32, 100
"innervation supersensitivity" induced by (guinea pig), 11, 16
stimulate cerebral glycolysis, 234
Amygdala,
dopamine receptors in, 63
muscarinic acetylcholine receptors in, 107, 108
Anaemia, megaloplastic (induced by folate deficiency), as a complication of therapy with some neuroleptics, 247
Anoxia, cerebral,
action myoclonus after, 7, 19
epilepsy after, 180
in status epilepticus, 182
Anticholinergic activity of some neuroleptics, 9–10, 118
ratio of, to antidopaminergic activity, 41–42
Anticholinergic drugs,
abolish dystonia, 48
given with antipsychotic drugs, 51–52
in epilepsy, 145
in myasthenia gravis, 23
in tardive dyskinesia, 10
without effect on drug-induced myoclonus, 99
Anticonvulsants, *see also individual drugs*
affect sodium transport, 242
and brain energy metabolism, 235
EEG in epilepsy not affected by, 176
and epilepsy, 127, 132–133
inhibit utilization of glucose, 234
long-term effects of, 122
membrane function and, 144–145
and monoamines in brain, 150, 250–251
suggested for dyskinesia, 122
suggested for prophylaxis, 133
tests of, on animals, 143–144, 151–156
Anti-epileptic drugs, *see also individual drugs*
sites of action of, 186–192
Antihistamines, as convulsants, 151
Apomorphine, dopamine agonist, 16, 17, 19
abolishes dystonia, 48
adenylate cyclase and, 37
antagonism of neuroleptics to, 66, 68
and "circling" in rats, 27, 28, 97–98, 100

effects of, on pigs, 66
effects of, on seizures in different animals, 159
and firing rate in lesioned and normal sides of striatum, 29–31, 32
metabolism of GABA and, 70
stereotyped behaviour induced by, in mice, 65
and responses of photosensitive baboons, 149
striatal injection of, 99
supersensitivity to, 32
in treatment of Huntingdon's chorea, 112
Aromatic amino acid decarboxylase, 15, 19, 24
depleted in brain in Parkinsonism, 21, 22
inhibitors of, 24–25, 28
Arteriovenous malformations of brain, epilepsy caused by, 180
Asparagine, in epileptogenic foci, 202
Aspartate, excitatory transmitter, 195, 204, 254, 263
differential sensitivity of neurons to glutamate and, 220
in epileptogenic foci, 196, 197, 222
release of, from cerebral cortex *in vivo*, 216, 217
in spinal cord, 214
in whole brain after convulsants, 199, 200
Astrocytes in epileptogenic brain lesions, 144, 158, 181, 261
ATP (adenosine triphosphate),
brain content of, preceding and during convulsions, 238, 240
GABA as substrate for brain synthesis of, 202
ATPase (Na, K),
in brain membranes, 253
in epileptogenic foci, 165, 236–237, 264
involved in epilepsy? 144–145, 236
Atropine, anticholinergic, 41
blocks responses to acetylcholine, 218, 219
HVA in striatum and, 64, 68

B

Baboons,
drug-induced Parkinsonism in, 116
experimental epilepsy in, 138
natural syndrome of photosensitive epilepsy in, 153
photosensitive, drugs enhancing and preventing seizures in, 147–150, 153–155
response of, to neuroleptics, 67, 115–116
Baclofen (Lioresal), tried for Huntingdon's chorea, 90, 91
Barbiturates, *see also pentobarbitone, phenobarbitone*
abolish dystonia, 48, 122
affect sodium transport, 242
analysis of anti-epileptic activity of, 186–192
convulsions in addicts after withdrawal of, 242
depress glutamate excitation of neurons, 264
Baroceptor reflexes, L-dopa and, 24
Basal ganglia,
balance between cholinergic and dopaminergic factors in, 80–81
contain highest concentrations in CNS of acetylcholine, dopamine, and GABA, 84
damage to, in dyskinesias, 3, 111
discharge preparatory to voluntary movement (monkey), 80
effects of damage to, in animals, 6
in Huntingdon's chorea, 18, 83, 85–86, 106
phenothiazines increase dopamine and manganese in (monkey), 76
respond to dopamine by increase of cAMP (homogenates), 35
Bemigride, blocks presynaptic inhibitory transmitter, 223, 226
Benzodiazepines, anticonvulsants, 122
and cerebral content and turnover of monoamines, 150
protect against enhancement of responses to convulsants, 153, 156
relieve prolonged epileptic seizures, 156
Benztropine, anticholinergic, 41
abolishes drug-induced dystonia in baboon, 116
exacerbates Huntingdon's chorea, 18
Bicuculline, convulsant,
blocks inhibitory effects of GABA, 198, 221, 223, 226, 227, 228, 229, 156
epileptic seizures induced by, 148, 157–158
hypoglycaemia may block convulsant effects of, 266
weak myoclonus induced by, 96, 99

Biopsy of brain, 90, 138
Birth injury,
epilepsy caused by, 158, 159, 180
torsion dystonia caused by, 6
Blepharospasm, induced by antipsychotic drugs, 48
Blood, *see also red blood cells*
circulation of, in prolonged seizures, 157, 158
transport of glucose to brain from, 239
Blood platelets,
metabolism of glutamate in, in Huntingdon's chorea, 117
Blood pressure,
effect on, of L-dopa combined with MAO inhibitor, 24
Brain
from autopsies, determinations on, 13
choreic, collection of, 84, 90
homeostasis in, 128, 235
in Huntingdon's chorea, 5, 85
oedema of, accompanying convulsions, 208
respiratory rate of, after barbiturates, 242, 295
transport from blood to, of folate, 248, and of glucose, 239
Bromocriptine,
dopamine receptors and, 8
enhances dyskineasis induced by L-dopa, 8, 50
and growth hormone in blood, 79
in Parkinsonism, 21–22, 23, 28
2-Bromlysergic acid diethylamide, and responses in photosensitive baboons, 150
Bulbocapnine, dopamine antagonist, 37
Butaclamol, neuroleptic, inhibits dopamine stimulation of adenylate cyclase, 40
Butyrophenones,
block dopamine receptors, 17, 22
for control of chorea, 5
dyskinesias induced by, 17, 19, 22, 50
inhibitors of dopamine stimulation of adenylate cyclase, 40
may induce Parkinsonism, 4, 22, 48

C

Caffeine, phosphodiesterase inhibitor, 23, 25–26
Calcium,
anticonvulsant effects of, 156, 236
in conduction of electrical impulses, 253, 256, 258
diphenylhydantoin and metabolism of, 242
and glutamate excitation, 218, 220
and glutamate release, 217
Carbachol, 99, 228
Carbamazepine (Tegretol), anticonvulsant, 133, 152
side effects of, 143, 156
Carbidopa, inhibitor of extracerebral aromatic amino acid decarboxylase, 24
Carbon monoxide poisoning, convulsions in, 125
Carnitine, sometimes abolishes drug-induced myoclonus, 96
Catalepsy,
dose-response curves for, in mice after pimozide, 61, 62, 63
methods of estimating, 61
potency of neuroleptics in producing, 60–61, 65, 66
Catecholamine terminals in brain, histological location of, 14
Catecholamines, *see also dopamine, noradrenaline*
amphetamines interfere with transport and metabolism of, 234
decarboxylase inhibitors prevent formation of, 24
and drug-induced myoclonus, 98
drugs decreasing concentration of, for Huntingdon's chorea, 19
effects of, mediated by cAMP, 35
production of, sensitive to hypoxia, 242
Caudate nucleus,
effects of interference with GABA activity in, 95, 97, 99, 100
in Huntingdon's chorea, 17, 85–86, 88
muscarinic acetylcholine receptors in, 103–104, 106
two populations of neurons in, with respect to dopamine, 53
Cerebellum,
changes in, after anoxia or seizures, 158
lacks dopamine-containing nerve terminals and dopamine-sensitive adenylate cyclase, 35
muscarinic acetylcholine receptors in cortex of, 107, 108
stimulation of, may inhibit epileptic discharges, 139
Cerebral cortex, in Huntingdon's chorea, 5, 6
Cerebrospinal fluid (CSF),
composition of, as indiciator of brain metabolism, 257

folate in, 247, 248, 250
glutamate in, in epilepsy, 208
metabolites of transmitter amines in, 13, 15, 17, 19, 53, 75
need for study of, in human and experimental epilepsy, 256–257
Cervical ganglia (bovine superior),
dopamine stimulates formation of cAMP in, 35
receptors in, for GABA, but not for glycine, 221
Chloride, in neurotransmission, 220, 221, 222, 224
p-Chlorophenylalanine,
inhibits tryptophan hydroxylase, 149
in myoclonus, 19
and responses in photosensitive baboons, 149
p-Chlorophenyl-γ-aminobutyric acid, anticonvulsant, 258
Chlorpromazine, neuroleptic, 22, 58
accumulates in lipids, 115
accumulates in melanin-containing tissues, 74, and is detectable for up to 18 months after cessation of treatment, 50
antagonizes effects of L-dopa with MAO inhibitor, 114
catalepsy-producing potency of, and HVA in striatum, 60
catecholamines in brain and, 68, 98
corneal changes induced by, 52
HVA in CSF after, 75
5-HT in brain and, 68, 69
inhibits dopamine stimulation of adenylate cyclase, 38–39
long-term effects of (rat), 75–76
metabolism of, 73
and myoclonus (drug-induced), 99
Parkinsonism and, 48
tardive dyskinesia and, 48, 119
X-ray crystallography of structure of, 38
Chlorprothixene, neuroleptic, 60
Cholera enterotoxin (choleragen), activates adenylate cyclase, 42–43
Choline acetyltransferase, in different parts of brain in Huntingdon's chorea, 6, 18, 84, 104, 106, 107
Cholinergic drugs, and tardive dyskinesia, 10
Cholinergic receptors, muscarinic,
affinity labels for, 103–105
in Huntingdon's chorea, in different parts of brain, 6, 106, 107, 108
in monkey and human, 107, 108
relative potencies of drugs as antagonists of dopaminergic receptors and of, 41–42
Cholinergic systems
abnormal in Huntingdon's chorea? 98
two, affecting dopamine metabolism in opposite directions? 64
Chorathetosis, 50
Chorea, 4, 50
hormonal and other types of, 6
Huntingdon's, *see* Huntingdon's chorea
Cingulate gyrus of brain, muscarinic acetylcholine receptors in, 107, 108
"Circling", caused by dopamine agonists in rats with unilateral damage to dopamine tracts in substantia nigra, 16, 25–26, 27, 28, 97–98, 100
in reverse direction, induced by apomorphine, 27
Clonazepam, and responses of photosensitive baboons, 155, 156
Clothiapine, neuroleptic, 60
Clozapine, anticholinergic neuroleptic, 10, 40, 41–42, 60, 68
prolactin in blood and, 79
and tardive dyskinesia, 118, 119
Cobalt foci in frontal cortex, induce epilepsy, 138, 163, 261, 262
amino acids in, 197
compared with human epileptogenic foci, 165, 166
EEG after establishment of, 167
effects of, compared with human cortical epilepsy, 164
effects on ECoG of rats with, of taurine, 170–171, and of transaminase inhibitors, 168, 169
folate in region of, 249
glutamate activates, 208
glutamate released from, at onset of seizure, 205–206
secondary or "mirror" foci on opposite side of brain from, 138, 165, 171, 182–183, 262, 266
Computerized techniques, for study of brain, 129, 138
Contraceptives, oral,
dyskinesias induced by, 6, 50
Convulsant drugs, *see also individual drugs*
brain amino-acid levels after, 199
changes in metabolism of brain induced by, 235–236
and inhibitory processes, 224–229

Convulsant drugs (*continued*)
suggested modes of action of, 198, 223, 229
Copper, increased in brain and plasma after chlorpromazine, 74
Cornea, phenothiazine-induced changes in, 52
Corpus striatum,
acetylcholine metabolism in, 68
cAMP in, 32, 35, 36
in dyskinesias, 3
effects of injection of picrotoxin into, 94–95
firing rate in lesioned and normal cells of, 29–31
in Huntingdon's chorea, 5, 6, 18
Creatine phosphate, in brain preceding and during convulsions, 235, 238, 240
Cuneate nucleus,
key site in drug induction of seizures? 229
sensitivity of, to glutamate, 218
tests of convulsants on, 225, 227
Cyanide, as convulsant, 242

D

Decarboxylase inhibitors, extracerebral, 24–25
Dendritic arborization, loss of,
in alumina foci, 265
in epileptic brain, 158, 159, 265
Dentate nucleus of brain, in Huntingdon's chorea, 85
Deoxyglucose,
convulsions induced by, 239–240
interferes with energy metabolism, 256
Deoxyglucose 6-phosphate, inhibits hexose phosphate isomerase, 240
Deoxypyridoxine, convulsant,
induction of reflex epilepsy by, 147
inhibits GAD, 146, 238, 256
Depolarization, ganglionic, 227, 255, 264
effects of TETS and bicuculline on, 228, 229
Depolarization shifts, paroxysmal, 209, 210
in epileptogenic foci, 223–234
Depression,
recurrent endogenous, sodium in red blood cells in, 264
spreading, 208–209
Diazepam, anticonvulsant, 152
abolishes dystonia, 48, 122
depresses presynaptic inhibition, 229
prevents febrile convulsions, 140
and responses in photosensitive baboons, 155, 156
Dihydroergotoxine, and responses in photosensitive baboons, 150
Dihydroxyphenylacetic acid,
brain concentration of, increased by neuroleptics, 58
correlation of catalepsy with concentration of, after pimozide, 69
in striatum after different doses of pimozide, 61–62, 63
2-dimethylamino-ethanol (Deanol), tried for Huntingdon's chorea, 90
N-Dimethyldopamine, stimulates formation of cAMP, 36
Dimethyltryptamine, and responses in photosensitive baboons, 149, 150
Diphenylbutylpiperidines, block dopamine receptors, 22
Diphenylhydantoin, anticonvulsant, 122, 152
affects calcium metabolism, 242
antagonizes L-dopa, 122
chorea induced by excess of, 6
enhances Huntingdon's chorea, 122
in epilepsy, 150
in epileptogenic foci, 186
in experimental epilepsy, 138
may cause folate deficiency, 247
and post-tetanic potentiation in olfactory cortex, 190
and responses in photosensitive baboons, 153, 156
side effects of, 143
n-Dipropylacetate (Epilim), anticonvulsant, 90, 133, 148
effect of, on ECoG in cobalt-implanted rats, 168, 169
inhibits enzymic breakdown of GABA, 204, 205, 258
Donazepam, anticonvulsant, 152
L-Dopa,
dopaminefrom, 15, 16, 21
dyskinesia induced by, in Parkinsonism, 4, 7, 8–9, 14, 15, 16–17, 18, 50, 111
effects of, with decarboxylase inhibitor, 98, 114
effects of, with MAO inhibitor, 24, 114, 115, 148, 149
emesis induced by, 9, 24–25
exacerbates Huntingdon's chorea, 18
exacerbates tardive dyskinesia, 51, 53
and GAD activity in brain, 120

growth hormone in plasma after, 79, 87
metabolites of, 15, 112
in myoclonus, 20
as possible test for liability to Huntingdon's chorea, 91
prolactin in plasma after, 79
response to, of "circling" rats, 28
response to, of lesioned and normal sides of striatum, 31
structurally similar to glutamate, 111

Dopa decarboxylase,
chlorpromazine and response of rats to, 75–76
inhibitors of, 28, 70

Dopamine,
agonists of, and adenylate cyclase formation, 36–38
agonists of, in Parkinsonism, 20–22
agonists of, and tyrosine hydroxylase, 63
antagonists of, and adenylate cyclase formation, 38–40
antagonists of, in Parkinsonism, 22–23
in basal ganglia after phenazines, 76
content of, in different parts of brain, in Huntingdon's chorea, 18, 84, 86, 88, 89
differences between *in vitro* and *in vivo* effects of, 40
feedback control of synthesis of, 32
from L-dopa, 15, 16, 21
HVA in CSF as measure of brain turnover of, 13
neuroleptics and brain metabolism of, 11, 57–60, 63–64, 69
possible mode of action of, 33
protects against electroshock, 148
rate-limiting step in synthesis of, 16
relations of GABA and, 120
in striatum, 28–29, 32
in Huntingdon's chorea, 17–18
Parkinsonism, in, 4, 13, 15, 18, 83, 120
after pimozide, 62–62, 63
systems for studying, 35
turnover of, in drug-induced dyskinesias, 17
turnover of, in Parkinsonism, 15–16

Dopamine hydroxylase, in different parts of brain, in normals and Huntingdon's chorea, 84, 88, 89

Dopamine receptors,
cAMP in response of, 14, 23, 25, 32
drugs inhibiting, 9, 17
drugs inhibiting, induce dyskinesias and Parkinsonism, 22–23, and tardive dyskinesia, 10, 53
drugs inhibiting or stimulating have reciprocal effects on release of acetylcholine, 32
neuroleptic-induced dyskinesias due to stimulation of, 8, 9, 10, 67
relative potencies of drugs as antagonists of dopaminergic receptors and of, 41–42
sites of
hypothalamus, 87
mesolimbic region, 11
striatum, 111
stimulated by ergometrine, 26
supersensitive,
after drugs, 32–33
denervated, 23, 27, 28
in dyskinesias, 4, 9, 10, 17, 80
in Parkinsonism, 15–16
two types of, 9, 16, 21, 66
inhibitory and excitatory, with different responses to neuroleptics, 112, 113

Drug addicts, dyskinesias in, 51

Dyskinesias, 3 *see also individual dyskinesias*
at opposite end of spectrum from Parkinsonism, 4, 23
drug-induced, 3–4, 8–11, 14–17, 23
induction of, in animals, 6
as result of excessive dopaminergic function, 22, 120

Dystonia, 4, 6–7
frequency of, after neuroleptics, 53–54
in response to antipsychotic drugs, 47, 48

E

EDTA (ethylenediamine tetra-acetic acid), 256

Electrical impulses, factors impairing conduction of, 253–254

Electrocorticogram (ECoG), in photosensitive baboons, 168–172

Electro-encephalogram (EEG), 127
amine metabolism and, in photosensitive baboons, 148, 149
amino-acid release and, 215
before, during, and after operation for temporal lobe epilepsy, 134, 135, 136, 137
in diagnosis, 129
in epilepsy, not affected by anticonvulsants, 176

Electro-encephalogram (EEG), *continued*
five years after brain injury, 130
showing petit mal attack, 126
Electron acceptor, melanin as, 74
Electroshock,
protectants against effects of, 148
seizures induced by, 151
Emesis, induced by L-dopa, 9, 24–25
Endocrines, and epilepsy, 125
Energy metabolism of brain, 234, 235, 254
Epilepsy, 125–128
association of, with Huntingdon's chorea, 83
association of, with myoclonus, 7
basic lesions in, 179–182
classification of, 128–129
clinical aspects of, 129–132, 175–178
drug therapy of, 132–133, 186–192
glutamate release in, 210
incidence of, 128
inheritance of tendency to, 182
possible causative factors in, 253–254
prevention of, 140, 158–159
surgical treatment of, 133–137, 177–178, 180, 181
volitional control of, 177
Epilepsy, experimental, in animals, 137–140, 159–160
tests on,
of anticonvulsant drugs, 143–156
of drugs terminating prolonged seizures, 156–157
of prevention of brain damage by prolonged seizures, 157–158
of prevention of onset of epilepsy, 158–159
Epileptogenic foci in brain, 125, 127, 135, 139; *see also* alumina cream, cobalt, *and* tungstic acid gel foci
amino acids in, 196
cortical human, compared with sites of metal implants in animals, 165, 166
GABA and glutamate in excised, 203
not always distinguishable in epilepsy, 164
possible prevention or limitation of development of, after brain injury, 140
respiratory rates of, 202
secondary or "mirror", 138, 165, 171, 183, 262, 266
at sites of scarring, 181
Epinine, and cAMP formation, 36, 38
Ergocornine, dopamine agonist, 28
and responses in photosensitive baboons, 149, 150
Ergometrine, stimulates dopamine receptors, 26
Ergot alkaloids, and responses in photosensitive baboons, 149–150
Ethopromazine, anticholinergic, 39, 41
Ethosuximide, anticonvulsant, 151, 152, 156
and experimental epilepsy in rat, 138

F

Fatty acids, in CSF, 257
Febrile convulsions
in animals, 139
in childhood, epilepsy after, 127, 140, 159, 182
Fluoroacetate, fluorocitrate
intervene in energy metabolism, 256
α-Flupenthixol, weakly anticholinergic neuroleptic, 39–40, 41
Fluphenazine, weakly anticholinergic neuroleptic, 38, 39, 41, 60
condition like Parkinsonism induced in baboons by, 116
α- and β-Fluphenthixol, 114
Foetus, susceptible to phenothiazines, 49
Folate,
as methyl donor? 250
deficiency of, induced by some neuroleptics, 247–248
excitatory properties of, 248, 249
as glutamate carrier? 249, 261
Freeze lesion in brain, epileptogenic, 197, 202, 203
release of glutamate from, at onset of seizure, 206
uptake of potassium by, 264
Frog,
effects of dopa and 6-hydroxydopa on glutamate receptors of, 111
Frontal cortex of brain, *see also* metal implants in frontal cortex
GAD content of, in normals and Huntingdon's chorea, 85, 86
muscarinic acetylcholine receptors in, 107, 108

G

Galoperidol, and responses in photosensitive baboons, 149
Geriatric patients, oral dyskinesia in, 49
Gilles de la Tourette's syndrome, 8
Glial cells,
amino acids in, 215

taurine in cultures of, 214
Gliosis, in epileptogenic foci, 181, 262, 263
Globus pallidus of brain,
abnormal neuronal activity in, in dyskinesias, 111
"circling" produced by dopamine agonists after unilateral damage to, 97–98
effect of injection of picrotoxin into, 97
GABA cell bodies and nerve terminals in, 99
GAD activity in, in normals and Huntingdon's chorea, 85
muscarinic acetylcholine receptors in, 107, 108
removal of inhibition in striatum reflected in? 80
Glucose, *see also* hypoglycaemia
affinity of hexokinase for, decreased in epileptic brain, 237
after phenobarbitone, 242, 243, 244
brain receptors for? 240–241, 245
increased in brain during convulsions, 237–238, 240
after phenobarbitone, 242, 243, 244
rapidly abolishes hypoglycaemic convulsions, 238–239
utilization of, inhibited by anticonvulsants, 233–234
Glucose-6-phosphate
accumulates in brain before and during deoxyglucose convulsions, 240
after phenobarbitone, 242
Glutamate, excitatory transmitter, 191, 195, 204, 209–210, 213, 218, 254
in brain after phenobarbitone, 242
in brain excitation and inhibition 205–208, 209–210
cobalt foci activated by, 208
in CSF in epilepsy, 208
in epileptogenic foci, 196, 197, 222, 261, 262; excised, 202
folate as carrier of?, 249, 261
in freeze foci, 202, 203
involved in metabolism as well as neurotransmission, 205
promotes swelling of neural tissue, 208
release of, from cerebral cortex *in vivo*, 216, 217
release of, from cobalt and freeze foci at onset of seizure, 205–206, 207
sensitivity of neurons to, 112, 218, 220, 255, 256
in spinal cord, 214
in spreading depression, 208–209
whole brain after convulsants, in, 199, 200, 262; in preconvulsive state, 205
Glutamate antagonist (HA 966), 112, 218, 219
Glutamate decarboxylase (GAD)
allylglycine as inhibitor of, 95, 146, 198, 203, 294
in cobalt foci, 66
correlation between levels of GABA and of, 214
depleted in basal ganglia in Huntingdon's chorea, 6, 18, 85, 91, 93, 120
different parts of brain in, in normals and Huntingdon's chorea, 84–87
inhibition of, by convulsants, 145–146, 198, 293–204
inhibition of, enhances epileptic seizures, 147, 204
stable in postmorten brain, 84
in vitamin B_6-deficient rats given pyridoxine, 266
Glutamate dehydrogenase, in whole brain in preconvulsive state, 205
Glutamine,
converted to glutamate during extraction and analysis of brain tissue, 208
in CSF, 195
in epileptogenic foci, 196, 197, 222, 262
in epileptogenic foci after administration of GABA and glutamate, 202
release of, from cerebral cortex *in vivo*, 215, 216, 217
in whole brain after convulsants, 199, 200
Glutamine synthetase,
inhibited by methionine sulphoxime, 198
in metabolism of glutamate and of ammonia, 196
Glycine, main postsynaptic inhibitory transmitter, 221, 254
brain receptors for, 204
differential sensitivity of neurons to GABA and, 220
in epileptogenic foci, 222–223, 262
release of, from cerebral cortex *in vivo*, 215, 216, 217
strychnine blocks inhibition by, 145, 165, 198, 225, 226
Glycogen in brain, 238, 242
Glycolysis in brain,
decreased by barbiturates, 242–243
stimulated by amphetamines, 234
Grand mal, 127–128, 175, 185

Grand mal *continued*
  anticonvulsants in, 152
  prognosis of, 131
Griseum septum of brain,
  GAD activity in, in normals and Huntingdon's chorea, 85
Growth hormone,
  effects of single dose and continued dosage of L-dopa on blood level of, 79
  in Huntingdon's chorea, 87, 92

H

Haloperidol (butyrophenone), dopamine receptor antagonist, 17, 19, 22, 58
  catalepsy-producing potency of, and HVA in striatum, 60
  dystonia induced in baboon by, 115–116
  increased dopamine in brain perfusate after, 63
  increased HVA in CSF after, 75
  inhibits stimulation of adenylate cyclase by dopamine, 40; and by noradrenaline, 43
  inhibits locomotor activity after L-dopa with MAO inhibitor, 114
  myoclonus and, 99
  response of pigs to, 66, 67,
  suppresses tardive dyskinesia, 52, 119
Hamatomas, in temporal lobe epilepsy, 182
Helix aspersa, tests for dopamine agonists and antagonists on neurons of, 38
Hexokinase,
  from epileptic brain, shows lowered affinity for glucose, 237
  inhibited by glucose-6-phosphate, 240
  phosphorylates deoxyglucose, 238, 240
Hexose phosphate isomerase, inhibited by deoxyglucose-6-phosphate, 239, 240
Hippocampus,
  chronic induration of, in some epileptics, 181–182
  effect of picrotoxin injection into, 96
  GAD activity in, in normals and Huntingdon's chorea, 85, 86
  high levels of GABA and GAD in, and high density of inhibitory terminals, 214
  muscarinic acetylcholine receptors in, 107, 108
Histochemical methods, for locating transmitter amines, 13–14
Homeostasis in brain, 128, 235
Homocarnosine, in brain in Huntingdon's chorea, 18
Homocysteic acid,
  inhibits affinity binding of glutamate, 220
  stimulates excitatory amino acid receptors, 218, 256
Homovanillic acid (HVA),
  after anoxide episode, 19
  in CSF, as measure of brain dopamine turnover, 13
  after dopamine antagonists, 61–62, 63, 65, 66
  in epilepsy, 510
  folate and, 251
  in Huntingdon's chorea, 17
  after neuroleptics, 17, 58, 60–61, 63–65, 75
  in Parkinsonism, 16
  in Parkinsonism and tardive dyskinesia induced by drugs, 53
Huntingdon's chorea, 4–6, 13, 17–19, 83
  akinetic rigid (Westphal) form of, 5, 17, 84–86, 106
  apomorphine in treatment of, 112
  brain metabolism in, 83–84, 87–89
  growth hormone in, 92
  metabolism of glutamate in blood platelets in, 117
  muscarinic acetylcholine receptors of brain in, 103–108
  premature aging in? 119, 210
  question of test to discover liability to, 91–92
  treatment of, 89–90
Hydantoins, anticonvulsants, 151; *see also* diphenylhydantoin
Hydrazine derivatives,
  induction of reflex epilepsy by, 147
  inhibit GAD and in some cases GABA-T, 146, 200, 204
Hydrazinopropionic acid, anticonvulsant, inhibits breakdown of GABA, 148
1-Hydroxy-3-aminopyrrolid-2-one (HA 966), glutamate antagonists, 112, 218, 219
γ-Hydroxybutyric acid, affects dopamine metabolism, 68–69
6-Hydroxydopa, interacts with glutamate receptors in frog, 111
6-Hydroxydopamine, neurotoxin,
  damage to substantia nigra by, 16, 27, 29, 100
  increases sensitivity of adenylate cyclase

to dopamine, 79
increases supersensitivity to apomorphine, inhibits response to amphetamine, 32
trihydroxyphenylacetic acid (metabolite of dopa) not derived from, 112
5-Hydroxyindoleacetic acid (5-HIAA),
after anoxic episode, 19
in CSF, as measure of turnover of 5-HT in brain, 13
after diphenylhydantoin, 150
in epilepsy, 150
in Huntingdon's chorea 17, 91
after phenobarbitone, 150, 251
after reserpine, 58
relation between folate and, 250, 251
4-Hydroxy-3-methoxyphenylglycol, metabolite of noradrenaline, 13, 121
5-Hydroxytryptamine (serotonin: 5-HT),
anticonvulsants and synthesis of, 69, 251
aromatic amino acid decarboxylase in neurons containing, 15
deficient in Parkinsonism? 16
effect of dopamine in deficiency of, 17
histological location of brain sites of, 14
in Huntingdon's chorea and myoclonus, 19, 98
5-HIAA in CSF as measure of brain turnover of, 13
neuroleptics and brain metabolism of, 57–60, 68
precursors of, and tardive dyskinesia, 52
production of, sensitive to hypoxia, 242
protects against pentylenetetrazol, 148
5-Hydroxytryptophan,
increases 5-HT in neurons, 98
inhibits seizures in photosensitive baboons, 148, 149
in treatment of myoclonus, 7, 19, 20, 150
Hyoscine, chorea induced by, 50
Hyperpyrexia, in prolonged seizures, 157, 158
Hyperventilation, may activate epilepsy, 125
Hypoglycaemia,
convulsant effect of, 125, 156, 235, 238–241, 263, 266
may abolish drug-induced convulsions, 237, 266
in prolonged seizures, 157, 158
Hypoparathyroidism, chorea in, 6
Hypothalamus,
in Huntingdon's chorea, 85, 88, 89
possible hormonal effects of hyperresponsive dopamine receptors in, 87
Hypoxia, *see also* anoxia
abolishes metrazole-induced convulsions, 238
amine hydroxylases sensitive to, 242
mild systemic, in prolonged seizures, 157, 158

I

Imipramine, dyskinesia induced by, 50
Indoles, hallucinogenic,
and responses in photosensitive baboons, 149
Infections, may activate epilepsy, 176
Inheritance,
of Huntingdon's chorea, 5, 83
of tendency to epilepsy, 182
of torsion dystonia, 7
Insulin, convulsions induced by, 235, 240
Ion pumps,
conduction of electrical impulses and, 253–254
epilepsy, in, 264
Ischaemic cell change in brain, after prolonged epileptic seizures, 157
Isoniazid,
dyskinesia induced by, 50
reflex epilepsy induced by, 147

J

Jacksonian epilepsy, 177, 191

K

Kainic acid, stimulates excitatory amino acid receptors, 256
Kernicterus, torsion dystonia caused by, 6

L

Lactate, in brain,
after deoxyglucose, 240
after phenobarbitone, 242, 243
preceding and during convulsions induced by methionine sulphoxime, 238
Lacticacidaemia, in prolonged seizures, 157
Lead poisoning, epilepsy in, 125
Leptazol, blocks presynaptic inhibitory transmitters, 223, 226, 229
Levodopa, *see* dopa

Levomepromazine, neuroleptic, 60
Limbic cortex,
adenylate cyclase of, 43
neuroleptics act on dopamine turnover in? 69
sensitive to epileptogenic process, 182
Lipid storage diseases, epilepsy in, 125
Lipids of body, accumulation of chlorpromazine in, 115
Locus coeruleus of brain,
in Huntingdon's chorea, 88, 89
rats with indwelling electrodes in, 121
Loxapine, neuroleptic, 60
Lysergic acid diethylamide, dopamine agonist, 21
and responses in photosensitive baboons, 150

M

Magnesium, seizures associated with low plasma concentration of, 156
Manganese, increased in basal ganglia after phenothiazines, 76
Marihuana, dyskinesias induced by, 51
Melanin, as electron acceptor, 74
Membranes,
permeability of, 220, 221, 245
structural defects in, and conduction of electrical impulses, 253, 254
3-Mercaptopropionic acid, convulsant, activates GABA-T? 198
induces reflex epilepsy, 147
inhibits GAD, 146, 198, 203, 204, 256
Mesolimbic system, dopamine receptors in, 9, 11
Metal implants in frontal cortex, convulsant effect of, 257, 163–172; *see also* alumina, cobalt, tungsten
Metals, in CSF, 257
Methadone, induces stereotyped behaviour in rats, 67
Methergoline, and responses in photosensitive baboons, 150
Methionine sulphoxime, convulsant, 208
changes in brain constituents preceding and during convulsions induced by, 238
GABA and glutamate in whole brain after, 200
inhibits glutamine synthetase, 198, 238
Methiolepin, neuroleptic, blocks 5-HT receptors in brain, 68
Methotrexate, folate antagonist, as anticonvulsant, 249
3-Methoxyhydroxyphenylglycol, metabolite of noradrenaline, 13, 121
2-Methoxytyramine, neuroleptics enhance increase induced by MAO inhibitor in, 58
N-Methylaspartic acid, stimulates excitatory amino acid receptors, 256
Methyldithiocarbazinate, convulsant, inhibits GAD, 146
3-*O*-Methyldopa,
dyskinesia induced by, 50
in plasma during treatment with L-dopa, 15
α-Methyldopamine, stimulates formation of cAMP, 36
3-*O*-Methyldopamine, without effect on cAMP formation, 15
Methylergonovine, and responses in photosensitive baboons, 150
Methylphenidate, abolishes dystonia, 48
Methyl transferase in brain, folate as methyl donor for? 250
α-Methyltyrosine,
depletes brain catecholamines, 98
increases response to adrenergic drugs, 32
responses in photosensitive baboons and, 149
suppresses tardive dyskinesia, 52
Methysergide, blocks 5-HT receptors, 19, 98
and responses in photosensitive baboons, 150
Metoclopramide, 66, 122
effects of, on pigs, 66–67, 69, 122
Metrazole, convulsant, 139, 208
amino acids in whole brain after, 199, 200
hypoxia abolishes convulsions induced by, 238, 266
Monkeys, *see also* baboons
dopamine and manganese in basal ganglia of, after phenazines, 76
experimental epilepsy in, 127, 138, 139
muscarinic acetylcholine receptors in humans and, 107, 108
recovery from cobalt implant in, 167
tests of anticonvulsants on, 151–156
Monoamine oxidase (MAO), in cerebral capillary endothelium, 24
Monoamine oxidase inhibitors
effects of L-dopa combined with, 24, 114, 115, 148, 149
increase brain concentrations of monoamines and metabolites, 58, 67
raise seizure threshold, 148, 149
Monoamine transmitters *see also* catecholamines, *and individual amines*,

brain content of, and seizure threshold, 148
effects of anticonvulsants on brain content and turnover of, 150, 250–251
histological methods for locating, 13–14
metabolites of, in CSF, 13
production of, sensitive to hypoxia, 212
and responses in photosensitive baboons, 149
as transmitters for small fraction only of brain nerve endings, 213
Motor retardation, in depressive illness, 8
Myasthenia gravis, "brittle" form of, 23
Myoclonus, 4, 7, 19, 50
action, relieved by 5-hydroxytryptophan, 7, 19–20, 150
after striatal injection of picrotoxin, 94–95; abolished by injection of GABA, 95–96
catecholamines not involved in, 98

N

β-Naphthylamine analogues, and cAMP formation, 37
Neoplasms of brain,
epilepsy induced by, 128, 131, 180, 181
surgery of, 134
Neuroleptic drugs (tranquillizers), *see also inidividual drugs*,
antagonists of dopamine stimulation of adenylate cyclase, 38–40
catalepsy-producing potency of, parallel to increase in striatal HVA, 60–61
and dopamine receptors, 63, 67
and dopamine turnover, 69, 113
dyskinesias induced by, 4, 9–11, 111
dystonias induced by, 7, 53–54
effects of, on concentration of noradrenaline metabolite after MAO inhibitor, 67–68
mechanism and site of increase in HVA concentration after, 63–65
and monoamine metabolism in brain 57–60
Parkinsonism induced by, 4, 47
possible sites of action of, 59, 60
and prolactin in blood, 79
ratio of cholinergic to dopaminergic potencies of, 41–42
time course of effects of, in schizophrenia, 114–115
and tyrozine hydroxylase, 63
Neuromelanin, as electron acceptor, 74
Neurons,
alteration of environment of populations of, in epilepsy, 179
alteration of properties of populations of, in epilepsy, 179, 182–183
hypothetical network of, 254–255
Neurotransmission, inter-relations of different forms of, 16–17
Neurotransmitters, *see also individual transmitters*
defects in synthesis, release, or reuptake of, 254
Nickel, less epileptogenic than cobalt, 165
Nigrostriatal dopamine system, in dyskinesias and Parkinsonism, 4
Noradrenaline,
adenylate cyclase sensitive to, 32
aromatic amino acid decarboxylase in neurons containing, 15
deficient in Parkinsonism? 16
in different parts of brain, normals and Huntingdon's chorea, 88, 89
in locus coeruleus 121
3-methoxyhydroxyphenylglycol as metabolite of, 13, 121
neuroleptics and brain metabolism of, 57–60
neuroleptics and effects of MAO inhibitor on, 67–68
protects against electroshock, 148
stimulates cAMP synthesis, 36
Normetanephrine,
neuroleptics enhance increase of, in brain after MAO inhibitor, 58

O

Occipital cortex of brain,
GAD content of, in normals and Huntingdon's chorea, 86
Olfactory cortex (prepiriform) preparation (guinea pig), for locating site of antiepileptic activity, 186–191
effect of glucose concentration on, 240–241
Olfactory tubercle of brain,
adenylate cyclase in, 43
in Huntingdon's chorea, 85, 86–87, 88
Ouabain,
applied to brain, and epileptic seizures, 144, 158
inhibits (Na, K) ATPase, 236, 256
Oxazolidinediones, anticonvulsants, 151
side effects of, 143
Oxiperomide, neuroleptic, 113

Oxygen, *see also* hypoxia
hyperbaric, amino acids in whole brain after, 199, 200

P

P-wave amplitude in rat cuneate neuron, drugs and, 225, 226, 227, 229
Parkinsonism, at opposite end of spectrum from dyskinesias, 4, 23
brain in, 13, 15–16
"brittle", 23
condition resembling, in monkey, 116
depression in, 8
drug-induced, 4, 22, 23, 47–48, 118
dyskinesias induced by L-dopa in, *see under* L-dopa;
effects in, of dopamine agonists, 21–22 and antagonists, 22–23
striatal dopamine in, 4, 13, 15
testing of drugs for, on "circling" rats, 28
Penicillin,
epileptogenic foci produced by, 165, 197, 182
question of mode of action of, 223, 226
Pentobarbitone,
effects of, on olfactory cortex preparaquestion, 188, 190, 192
and presynaptic inhibition, 229
Pentylenetetrazole, convulsant, 198
activates reflex epilepsy, 125, 147
5-HT protects against, 148
hypoglycaemia may block convulsant effects of, 266
seizures induced by, in screening of anticonvulsants, 151
Perlapine, neuroleptic, 60
Perphenazine, neuroleptic, 60, 119
Petit mal, 125, 126, 175–176, 185
anticonvulsants in, 152, 205
may be provoked by hypoglycaemia, 266
Phenobarbitone, anticonvulsant, 148, 151, 152
effects of, on brain metabolites, 242
in epilepsy, 150
and epileptogenic foci, 186
and experimental epilepsy, 138
may cause folate deficiency, 247
and monoamine metabolism, 251
and responses in photosensitive baboons, 153, 156
side effects of, 143
Phenothiazines, neuroleptics, 5; *see also* chlorpromazine, *and other individual phenothiazines*
affect foetus, 49
block dopamine receptors, 17, 22
chelating action of, 74
dyskinesias induced by, 17, 19, 22
as electron donors, 74
induce pigmentation in skin and eye, 52–53, 74
may induce Parkinsonism, 4, 22, 48
metabolism of, 73, 75
pathological changes caused by, 74–75
potencies of, as neuroleptics, agree with potencies as inhibitors of dopamine stimulation of adenylate cyclase, 38–39, 40
slow clearance of, from body, 50, 73
and tardive dyskinesia, 49, 50, 52, 119
Phensuximide, anticonvulsant, 151
Phenytoin,
decreases presynaptic inhibition, 229
for epilepsy, 133
Phosphodiesterase, acting on cAMP, 23, 25
tests of inhibitors of, in Parkinsonism, 23, 25
Photic stimulation, may activate epilepsy, 125, 147
Physostigmine, anticholinesterase,
alleviates Huntingdon's chorea, 18
in dyskinesias, 52
Picrotoxin, convulsant, 208
amino acids in whole brain after, 199, 200
blocks GABA receptors, 198, 221, 256
blocks presynaptic inhibitory transmitter, 198, 223, 226
effects of injection of, into different parts of brain, 94–95, 96
enhances reflex epilepsy, 147
myoclonus induced by, 95–96, 99
seizures induced by, in screening of anticonvulsants, 151
Pigmentation,
induced in skin and eye by phenothiazines, 52–53, 74
of substantia nigra, increased in Huntingdon's chorea, 87
Pigs,
effects on, of apomorphine, 66
effects on, of metodopramide, 66–67, 69, 122
effects on, of neuroleptics, 67
Pimozide (diphenylbutyl piperidine), 5, 62–63
antagonizes noradrenaline, stimulates cAMP formation, 43

blocks dopamine receptors, 8, 22
catalepsy-producing potency of, and striatal HVA, 69
and dyskinesias, 5, 8, 113, 114
effects of striatal injection of, on dopamine and metabolites, 61–62
induces dystonia in baboon, 115–116
inhibits dopamine stimulation of adenylate cyclase, 40
ratio of anticholinergic to antidopaminergic potencies of, 41–42
and responses in photosensitive baboons, 149
and stereotyped behaviour induced by apomorphine, 65
and striatal HVA, 65, 66
and tardive dyskinesia, 119
Piperazine phenothiazines, tardive dyskinesia after, 118
Piribedil, dopamine agonist, 8
metabolite of, stimulates cAMP production, 37
in Parkinsonism, 21, 22, 23, 28
and responses in photosensitive baboons, 149
Postactivation depression in olfactory cortex, 189
pentobarbitone and, 190
Post-tetanic potentiation in olfactory cortex, phenobarbitone and, 189, 190
Potassium,
in CSF in seizures (monkey), 208
and effect of phenobarbitone on glucose transport, 243
energy requirements for transport of, 236
in epileptic brain tissue, 236, 264
extracellular content of, in brain, and epileptic seizures, 144, 224, 264, 265
in neurotransmission, 218, 222, 256
penicillin and, 223
in spreading depression, 209
Prepiriform cortex preparation, *see* olfactory cortex
Presynaptic inhibition, GABA as transmitter for, 221–222
Primidone, anticonvulsant, 152
may cause folate deficiency, 247
side effects of, 143
Probenecid, amine metabolites in CSF after treatment with, 19, 53
Prolactin, effect of L-dopa on blood level of, 79
Promazine, weak neuroleptic, 38, 39
Promethazine, weak neuroleptic, 38, 39
Propyl-2-aminoethylbenzilate, affinity label for muscarinic acetylcholine receptors, 104–105
Propylbenzilcholine mustard, affinity label for muscarinic acetylcholine receptors, 103
Protein synthesis in brain, electrical stimulation and, 203
Psilocybin, and responses in photosensitive baboons, 149, 150
Psychomotor epilepsy, 185
Psychotic patients,
clinical similarities between neurological patients and, 8
with epilepsy, 177
Putamen of brain,
dopamine receptors in, 63
in Huntingdon's chorea, 17, 85, 87–88, 106, 107
muscarinic acetylcholine receptors in, 106, 107, 108
Pyridoxal phosphate, pyridoxine,
amino-oxyacetic acid likely to inhibit all enzymes dependent on, 70
compounds inhibiting GAD do so by interference with, 145–146, 198
epileptic seizures induced by antagonists of, 147; and by deficiency of, 125, 156, 204
GABA and GAD in brain of rats after treatment of deficiency of, 266
without effect in Huntingdon's chorea, 117
Pyruvate dehydrogenase, thiamine and, 233

R

Radiographs of brain, in diagnosis of epilepsy, 129
Rats,
cholinergic systems of basal ganglia not fully developed in 11-day-old, 42
"circling", with unilateral damage to dopamine tracts in substantia nigra, 16, 25–26, 27, 28, 97–98, 100
experimental epilepsy in, 138
long-term effects of chlorpromazine on, 75–76
testing on, of drugs for Parkinsonism, 28
Red blood cells,
macrocytosis of, accompanying folate deficiency, 248
sodium content of, in epilepsy and depression, 264

Renal artery, responds by vasodilation to dopamine and related compounds, 35, 38
Reserpine and derivatives,
 cause supersensitivity to apomorphine 32
 deplete monoamines in brain, 58, 98, 148
 increase 5-HIAA in brain, 58, 68
 may induce Parkinsonism, 4, 22, 48
 rarely if ever cause dyskinesia, 50
 suppress tardive dyskinesia, 52
Respiratory rates,
 of brain after barbiturates, 242, 245
 of epileptogenic foci, 202
Retina, dopamine stimulates cAMP formation in, 35

S

Scarring, post-inflammatory, in brain, as cause of epilepsy, 180, 181, 182
Schizophrenia,
 movement disorders in, 51
 response to chlorpromazine in, 73
 time-course of effects of neuroleptics in, 114–115
Sclerosis,
 end-folium, 182
 mesial temporal,127–128, 181–182
 tuberose, 182
Scopolamine, anticholinergic, 98
Semicarbazide, convulsant, GAD inhibitor, 198, 199, 200
Serine, increased in epileptogenic foci, 196, 197
Serotonin, *see* 5-hydroxytryptamine
Sleep, may activate epilepsy, 125, 176
Snail, dopamine receptors in, 112–113; *see also Helix*
Sodium,
 in brain, increased by convulsants, decreased by anticonvulsants, 235
 energy requirements for transport of, 236
 in neurotransmission, 218, 222
 in red blood cells in recurrent endogenous depression, 264
 transport of, affected by barbiturates and diphenylhydantoin, 242
Spinal inhibitory mechanisms, 145
Spinal neurons,
 metabolism of, affects CSF, 13
Spiperone, neuroleptic, 60
 response of pigs to, 66, 67
Spiroperidol, weakly anticholinergic neuroleptic, 9–10, 40, 41
Spreading depression (not in primates), glutamate and, 208–209
Status epilepticus, 152, 157, 176–177
 barbiturates for, 188
 brain anoxia in, 182
 minor form of, 177
Stereotypies, 7–8, 61
Strychnine and related alkaloids, convulsants
 amino acids in whole brain after, 199, 200;
 as glycine antagonists, 220, 221, 223
 compete with glycine for receptor sites, 145, 165, 198, 226
 decrease sensitivity to glycine, 225
 seizures induced by, in screening of anticonvulsants, 151
Substantia nigra of brain
 "circling" after damage to, *see* circling
 dopamine system of, damaged by 6-hydroxydopamine, 16, 27, 29, 100
 in dyskinesia, 75, 111
 GABA terminals in, 99
 in Huntingdon's chorea, 5, 17, 85, 86, 87–89
 neuroleptics and dopamine system in, 58, 60
 in Parkinsonism, 4
 selective accumulation of chlorpromazine in, *in vitro*, 74
Succinic semialdehyde dehydrogenase,
 some anticonvulsants inhibit, 148; leading to increase in brain GABA level, and protecting against seizures, 204
Succinimides, anticonvulsants, 151
Sulthiame anticonvulsant, 151, 152
 and responses of photosensitive baboons, 154
Surgery, in epilepsy, 133–137, 177–178, 180, 181

T

Tardive dyskinesia, 118
 aetiology of, 52–53
 diagnosis of, 51
 drug-induced, 10–11, 17, 19, 47, 48–51, 118
 irreversible? 116–117
 neuroleptics controlling, 119
Taurine, inhibitory transmitter, anticonvulsant, 169–170, 171
 in cobalt foci, 169
 differential sensitivity of neurons to GABA and, 220, 221

effect of single or continued doses of, on epilepsy in cobalt-implanted rats, 170–171
in epileptogenic foci, 196, 197, 222
in glial cell cultures, 214
release of, from cerebral cortex *in vivo*, 215, 216, 217
in whole brain after convulsants, 199
Temporal lobe epilepsy, 127, 128, 159, 177
EEGs for, pre- and post-operative, 134, 137; and during operation, 135, 136
lesions found in, 182
in monkeys, 139
prognosis in, 131
surgery for, 136, 180
Tetanus toxin, decreases release of inhibitory transmitters, 223, 224, 226, 256
Tetrabenazine, amine depleter, 5, 17
in Huntingdon's chorea, 18
induces Parkinsonism, 22
suppresses tardive dyskinesia, 52, 119
Tetrahydroisoquinoline derivatives, and cAMP formation, 37
Tetramethylene disulphotetramine (TETS), convulsant
effects of, on neurotransmission, 223, 228, 229
Thalamus,
in dyskinesias, 111
effect of picrotoxin injections into, 96
in epilepsy, 127–128
in Huntingdon's chorea, 121
stereotactic lesioning of nuclei in, in rats with cobalt implant, 170–171, 172; and in treatment of human epilepsy, 171
Theophylline, weak inhibitor of phosphodiesterase, 25
Thiamine, brain acetyl CoA and acetylcholine in deficiency of, 233
Thiocarbohydrazide, convulsant and GAD inhibitor, 146
Thiopentone, in status epilepticus, 188
Thiopropazate, neuroleptic, 119
Thioridazine, anticholinergic neuroleptic 10, 41–42, 60
and Parkinsonism, 48
and prolactin in blood, 79
and tardive dyskinesia, 118, 119
Thiosemicarbazide, convulsant, 208
amino acids in whole brain after, 199
induction of reflex epilepsy by, 147
decreases release of inhibitory transmitters, 223
inhibits GAD, 146, 198, 256
Thioxanthenes, neuroleptics,
inhibit dopamine stimulation of adenylate cyclase, 39–40
Thyrotoxicosis, chorea in, 6, 18–19
Tics, 7–8, 51
Torsion dystonia, 4, 6–7
Torticollis, spasmodic; as torsion dystonia, 7
Tranquillizers, *see* neuroleptic drugs
Tranylcypromine, monoamine oxidase inhibitor, 24
and responses in photosensitive baboons, 149
Trauma of brain,
epilepsy after, 127, 128, 131, 180; especially after dural penetration, 158, 159
Tremor, in idiopathic and drug-induced Parkinsonism, 47, 49
Tridione, decreases presynaptic inhibition, 229
Trifluoperazine, weakly anticholinergic neuroleptic, 9–10, 38, 39, 41
and Parkinsonism, 48
and tardive dyskinesia, 119
Trifluperidol, and Parkinsonism, 48
Trihexiphenidyl, anticholinergic, exacerbates dyskinesia, 52
Trihydroxyphenylacetic acid, metabolite of L-dopa, 112
Trimethadione, anticonvulsant, 151, 152, 190
and epileptogenic foci, 186
N-Trimethyldopamine, stimulation of cAMP formation by, 36
Tryptamine, may act on nigrostriatal dopamine system, 4
Tryptophan,
chlorpromazine increases brain uptake of, 69
in serum of schizophrenics, chlorpromazine and, 115
not protective against seizures in photosensitive baboons unless combined with MAO inhibitor, 148, 149
Tryptophan hydroxylase,
after anoxic episode, 19
inhibitor of, and responses in photosensitive baboons, 149
sensitive to hypoxia, 242
Tubocurarine, GABA antagonist, 223
chorea induced by striatal injection of, 98–99
Tumours of brain, *see* neoplasms

Tungstic acid gel foci in frontal cortex, induce epilepsy, 163, 165
  effects of, compared with cortical epilepsy in humans, 164
Tyramine, without effect on cAMP formation, 36
Tyrosinase, copper-containing, 74
Tyrosine hydroxylase, 58, 60
  different parts of brain, in, in normals and Huntingdon's chorea, 84, 87–89
  in different parts of substantia nigra, 86, 87–89
  increased in locus coeruleus by stimulation, 121
  opposing effects of neuroleptics and dopamine agonists on, 63
  sensitive to hypoxia, 242

U

Uraemia, seizures in, 125, 136

V

Valium, anticonvulsant, 148
Vitamin B_6, *see* pyridoxal phosphate
Vitamin B_{12}, folate in metabolism of, 250

W

Writer's cramp, 7

Z

Zinc, in epilepsy, 237